Monographs in Theoretical Computer Science
An EATCS Series

Springer
Berlin
Heidelberg
New York
Barcelona
Budapest
Hong Kong
London
Milan
Paris
Singapore
Tokyo

Zoltán Fülöp
Heiko Vogler

Syntax-Directed Semantics

Formal Models Based on Tree Transducers

With 74 Figures

Springer

Authors

Dr. Zoltán Fülöp
Department of Computer Science
József Attila University
Árpad tér 2
H-6720 Szeged, Hungary
E-mail: fulop@inf.u-szeged.hu

Prof. Dr.-Ing. Heiko Vogler
Department of Computer Science
Dresden University of Technology
Mommsenstraße 13
D-01062 Dresden, Germany
E-mail: vogler@inf.tu-dresden.de

Series Editors

Prof. Dr. Wilfried Brauer
Institut für Informatik, Technische Universität München
Arcisstrasse 21, D-80333 München, Germany

Prof. Dr. Grzegorz Rozenberg
Department of Computer Science
University of Leiden, Niels Bohrweg 1, P.O. Box 9512
2300 RA Leiden, The Netherlands

Prof. Dr. Arto Salomaa
Data City
Turku Centre for Computer Science
FIN-20520 Turku, Finland

Library of Congress Cataloging-in-Publication Data

Fülöp, Zoltán.
Syntax-directed semantics : formal models based on tree transducers / Zoltán Fülöp, Heiko Vogler.
p. cm. -- (Monographs in theoretical computer science)
Includes bibliographical references and index.
ISBN 3-540-64607-8 (alk. paper)
1. Programming languages (Electronic computers)--Semantics.
2. Transducers. I. Vogler, Heiko, 1957- . II. Title.
III. Series.
QA76.7.F83 1998
005.13'1--dc21 98-38386
CIP

ISBN 3-540-64607-8 Springer-Verlag Berlin Heidelberg New York

Typesetting: Camera ready by authors
Cover Design: *design & production* GmbH, Heidelberg
SPIN: 10680470 45/3142-5 4 3 2 1 0 – Printed on acid-free paper

Dedicated to

Joost Engelfriet and Ferenc Gécseg

Preface

The subject of this book is the investigation of tree transducers. Tree transducers were introduced in theoretical computer science in order to study the general properties of formal models which give semantics to context-free languages in a syntax-directed way. Such formal models include attribute grammars with synthesized attributes only, denotational semantics, and attribute grammars (with synthesized and inherited attributes). However, these formal models share certain constituents which are irrelevant in the investigation of the general properties considered in this book. In particular, we can abstract (a) from derivation trees of the context-free grammar and take trees over some ranked alphabet, (b) from the semantic domain of the model and use the initial term algebra instead, and finally (c) from the machine-oriented computation paradigm, which maintains the incarnation information of recursive function calls, and take a term rewriting semantics instead. Applying these three abstraction steps to attribute grammars with synthesized attributes only, to denotational semantics, and to attribute grammars we obtain the concepts of *top-down tree transducer*, *macro tree transducer*, and *attributed tree transducer*, respectively. The *macro attributed tree transducer* combines the concepts of the macro tree transducer and the attributed tree transducer. This book explores the general properties of these four types of tree transducers.

The general properties which we examine are the transformational power of the tree transducers above, the restriction of their resources, and an intensive investigation into the composition and decomposition of the tree transformations induced by them. Our aim is to develop a theory of the these types of tree transducers and thereby contribute to theoretical computer science.

We intended this book for those who work in an academic environment, e.g., in universities or research institutes, and who deal with theoretical computer science. It is a small handbook of recent results on tree transducers that can be used both for carrying out further research and giving lectures in the topic. The subject can also be taught to graduate students in computer science. However, as prerequisites, some basic knowledge of universal algebra and of automata and formal language theory is necessary to use the book effectively.

The background to the book is as follows. We began to carry out research into the theory of tree transducers independently from each other many years ago. Since meeting for the first time in 1986 in Szeged, Hungary, we have continued a constant discussion about the state of the art. We have attempted to systematize the results from different research schools using a unified terminology. Some years ago we decided to write down our knowledge in the form of a book and the result is in your hands.

The authors are grateful to many colleagues who in one way or another contributed to this book. The first author was introduced to the subject by Ferenc Gécseg and benefited much from valuable discussions with Zoltán Ésik, Sándor Vágvölgyi, and Magnus Steinby. The second author had the luck to meet Joost and Louco Engelfriet and, also, to learn the theory of tree transducers from Joost. He had fruitful discussions about the subject with Armin Kühnemann and Thomas Noll. Both authors are very grateful to Armin Kühnemann for careful reading of the manuscript. They would also like to thank Sven Buchholz, Matthias Gette, and Christian Lescher for taking over the troublesome task of drawing the figures. Many thanks are also due to the computer system managers Károly Dévényi and Sebastian Maneth who maintained the intensive computer connection between our universities on which our joint progress was based.

Last but not least, the authors are indebted to Springer-Verlag, especially to Dr. Hans Wössner and his team, for the pleasant and highly professional cooperation during the preparation of the camera ready version of this monograph.

Our work was supported by different grants. We received much assistance from the German-Hungarian Research Project "Formal Languages, Automata and Petri-Nets" (1995-1997) supported by the *Te'T Foundation, Budapest, Hungary*, grant no. D/102 (formerly: No. OMFB-NPI-102) and by the *Forschungszentrum Karlsruhe, Germany*, grant no. 233.6, respectively. We were also supported by the Hungarian Scientific Foundation (OTKA), grant no. T 015843 and by the Ministry of Culture and Education of Hungary, grant no. FKFP 0095/1997.

May 1998

Zoltán Fülöp
Szeged, Hungary

Heiko Vogler
Dresden, Germany

Contents

1. Introduction

1.1 Syntax-Directed Semantics

In computer science one frequently encounters the task of specifying the semantics of a context-free language. In detail, this means that there is a language L generated by a context-free grammar and every string $w \in L$ has a meaning, called the semantics of w; the semantics is an element of the carrier set of some semantic domain. Let us call the whole machinery which computes the semantics of w the *semantics evaluator* (Fig. 1.1). A typical occurrence is where L is a programming language, i.e, the strings are programs, and the semantics of a program w is the assembly code which is obtained by translating w into some assembly language.

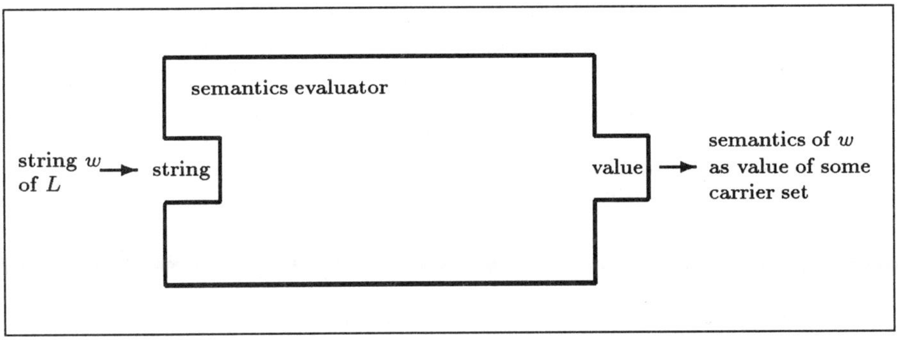

Fig. 1.1. The semantics evaluator

The abstract method

A well-known abstract method of specifying the semantics of a context-free language is the so-called syntax-directed semantics.

> *Syntax-directed semantics* is based on the idea that the semantics of a string of a context-free language is defined in terms of the semantics of the syntactic substructures of the string.

Thus, if the semantics is specified in a syntax-directed way, it is necessary that the string $w \in L$ is already parsed before the computation of its semantic value can begin. Hence the semantics evaluator is subdivided into a parser which constructs the derivation tree t_w of w, and a *syntax-directed semantics evaluator* which computes the value of t_w (Fig. 1.2; in figures we often abbreviate the term "syntax-directed semantics" to "sds").

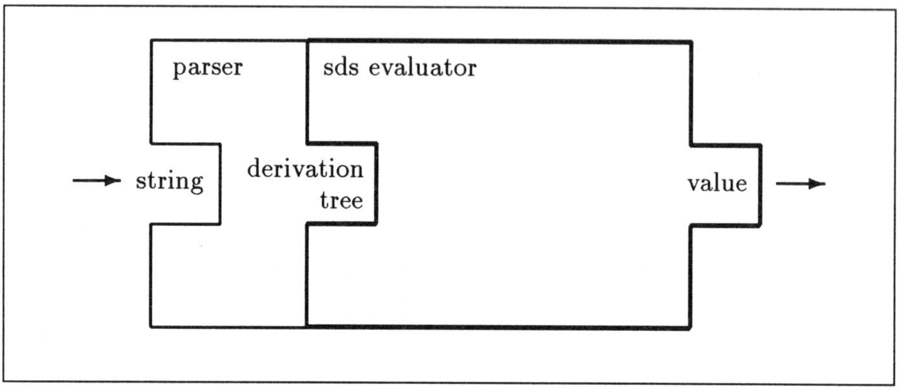

Fig. 1.2. The sds evaluator

In this book we are not interested in parsing. Instead we focus on formal models which obey the abstract specification method of syntax-directed semantics. In Fig. 1.2 we indicate this by drawing a box with bold lines around the syntax-directed semantics evaluator. Also in the other relevant figures in Sections 1.1 and 1.2 we indicate our focus by bold lines.

In order to compute the syntax-directed semantics of a derivation tree t_w we need two modules (Fig. 1.3):

- a specification $\mathcal{S}$ of the syntax-directed semantics in some specification language and
- a computation paradigm which defines how to compute the semantic value of a derivation tree t_w according to $\mathcal{S}$.

A combination of both, i.e., a specification language which is based on the concept of syntax-directed semantics, and an appropriate computation paradigm, is called a *formal model of syntax-directed semantics.* We will develop three different such formal models. More precisely, the first formal model is based on the abstract method as defined above, whereas the second and third formal models are based on a more flexible, extended abstract method which will be defined later.

In the following we will assume that the context-free language L is generated by the context-free grammar $G = (N, T, P, S)$.

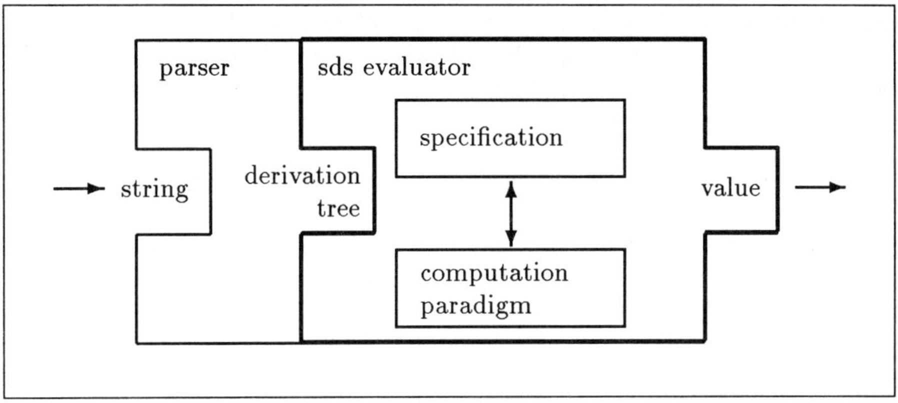

Fig. 1.3. The sds evaluator based on two modules

A first formal model: attribute grammars with synthesized attributes only

The first formal model of syntax-directed semantics which we are going to examine, is called *attribute grammars with synthesized attributes only*.

First we must realize that a string $w \in L$ may have several semantics, e.g., the length of w or the number of occurrences of a particular symbol in w. Hence, we need a finite set M the elements of which are called "meaning names" (e.g., *length* and *number-of-a* for some symbol a can be such meaning names). Roughly speaking, every meaning name represents a semantics of w. More precisely, for every meaning name $m \in M$, we specify a set V_m of values, called the carrier set. The m-meaning of w is then an element of V_m. Moreover, we distinguish a "principal" meaning name m_0 which represents the meaning of w in which we are interested. The other meaning names may be considered as auxiliary ones.

To every nonterminal $A \in N$ of the context-free grammar, some of the meaning names are associated by means of the function $\alpha : N \to \mathcal{P}(M)$, where $\mathcal{P}(M)$ denotes the power set of M. Moreover, to every production

$$p : A \to w_0 A_1 w_1 \dots w_{n-1} A_n w_n,$$

(with nonterminals $A, A_1, \dots, A_n$) and every meaning name $m \in \alpha(A)$, a semantic equation of the form

$$m(A) = \zeta$$

is associated which is called the *(m, A)-equation of p*. Its right-hand side ζ is an expression built up from basic functions over carrier sets and compound objects of the form $m_j(A_{i_j})$ with $m_j \in \alpha(A_{i_j})$ and $1 \leq i_j \leq n$. Intuitively, the (m, A)-equation of p expresses the m-meaning of a string w which has been derived from A by starting with production p. If $m_1(A_{i_1}), \ldots, m_r(A_{i_r})$ are the compound objects which occur in ζ, then for every $1 \leq j \leq r$, the m-meaning of w depends on the m_j-meaning of the substring of w derived from A_{i_j}. Hence, the computation of the m-meaning of A assumes that, for every $1 \leq j \leq r$, the m_j-meaning of A_{i_j} is known.

Let us denote the system of the productions of G together with the corresponding sets of associated semantic equations by $\mathcal{S}_{sem}$. This system clearly forms a specification of the semantics of $L(G)$, i.e., the language generated by G, in a syntax-directed way. Since $\mathcal{S}_{sem}$ itself is not executable, we first transcribe it into a system $\mathcal{S}_{fp}$ of recursive function procedures (Fig. 1.4) and then use the usual, well-known computation paradigm for recursive function procedures. More precisely, for every meaning name m, we construct a recur-

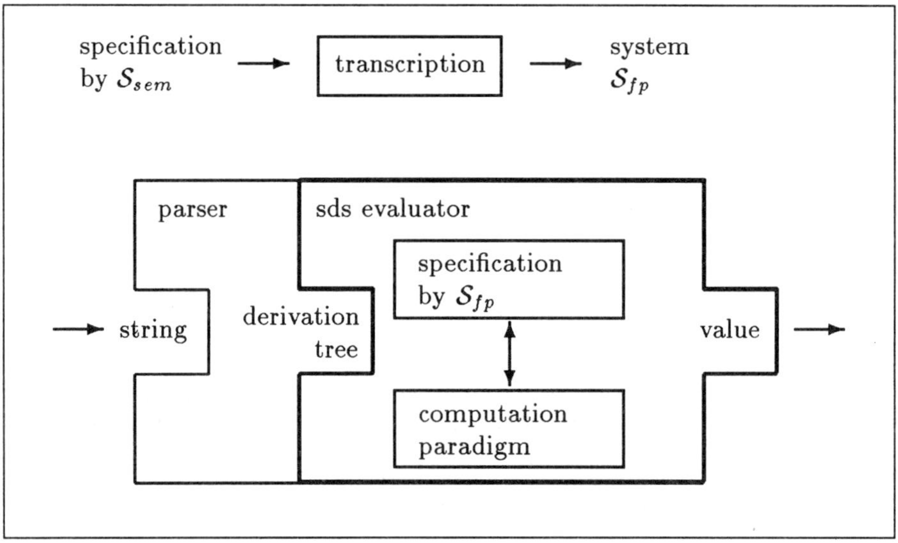

Fig. 1.4. Specification by $\mathcal{S}_{sem}$ and its transcription into $\mathcal{S}_{fp}$

sive function procedure *m-eval* which has one formal parameter d; the actual value of this parameter is a node of some derivation tree t. The body of *m-eval* just consists of one case statement. For every production p for which there is an (m, A)-equation of p, this case statement contains one clause. More precisely, if $m(A) = \zeta$ is the (m, A)-equation of p, then

$$p : m\text{-}eval := \zeta'$$

is a clause in the case statement and ζ' is obtained from ζ by replacing every compound object $m_j(A_{i_j})$ by the function call $m_j\text{-}eval(d\,i_j)$ where $d\,i_j$ denotes the i_j-th descendant of d. We formulate the function procedure in an imperative programming language style as shown in Fig. 1.5.

```
func m-eval(d: node): V_m ;
begin
   case production applied at d of
      ...
      p: m-eval := ζ';
      ...
   end
end m-eval
```

Fig. 1.5. The function procedure *m-eval*

As already mentioned, the semantics of w is obtained in the usual way by calling the function procedure m_0-*eval* at the root of the derivation tree t_w of w. In fact, the underlying computation paradigm is defined by a machine and a compiler; we will discuss this machine-oriented computation paradigm in a later subsection.

We note that in the area of attribute grammars the meaning names are called synthesized attributes.

An example of an attribute grammar with synthesized attributes only

Let us illustrate this formal model by discussing an example in detail. Let G_0 be the usual context-free grammar which generates arithmetic expressions built up from the variables a, b, and c, the operation symbols $+$, $-$, and $*$, and the two parentheses (and). Hence, G_0 uses the nonterminals

- E (start symbol) for generating expressions,
- T for generating tags of expressions,
- F for generating factors, and
- I for generating identifiers

and it contains the productions

$$
\begin{array}{lll}
p_1 : E \to E + T & p_2 : E \to E - T & p_3 : E \to T \\
p_4 : T \to T * F & p_5 : T \to F & \\
p_6 : F \to (E) & p_7 : F \to I & \\
p_8 : I \to a & p_9 : I \to b & p_{10} : I \to c.
\end{array}
$$

We note that G_0 can easily be extended such that it generates infinitely many identifiers. However, three identifiers are sufficient for illustrating this formal model.

It should be clear that a natural semantics can be associated with every expression e. Let us call this semantics the *expression semantics of* e and denote it by $\mathcal{E}[\![e]\!]$. It is a function such that, for every $n_1, n_2, n_3 \in \mathbb{N}$, where $\mathbb{N}$

is the set of natural numbers, $\mathcal{E}[\![e]\!](n_1, n_2, n_3)$ is the value of e if we replace a, b, and c by n_1, n_2, and n_3, respectively. Thus, e.g., $\mathcal{E}[\![a*(b+a)]\!](3,2,5) = 15$.

We now specify another semantics of an expression $e \in L(G_0)$ which we call the *code semantics of* e. It is an assembly program *ass-prog*(e) such that, if *ass-prog*(e) is executed on a machine $\mathcal{M}$, the machine computes the expression semantics $\mathcal{E}[\![e]\!]$ of e. In other words, we describe a small compiler for arithmetic expressions into assembly code which should be correct with respect to the expression semantics. Hence, the expression semantics is our reference semantics for the correctness of the compiler.

The machine $\mathcal{M}$ has an accumulator and a memory which consists of memory locations (Fig. 1.6). The accumulator and every memory location can store a value from $\mathbb{N}$. There are three memory locations referenced by the addresses a, b, and c, respectively. At the beginning of the computation, these locations are supposed to keep the values of the identifiers a, b, and c, respectively. Moreover, for every $i \in \mathbb{N}$, there is a memory location referenced by the address i. For every address $i \in \{a, b, c\} \cup \mathbb{N}$, let $c(i)$ denote the contents of the location with address i. The machine accepts the following assembly instructions:

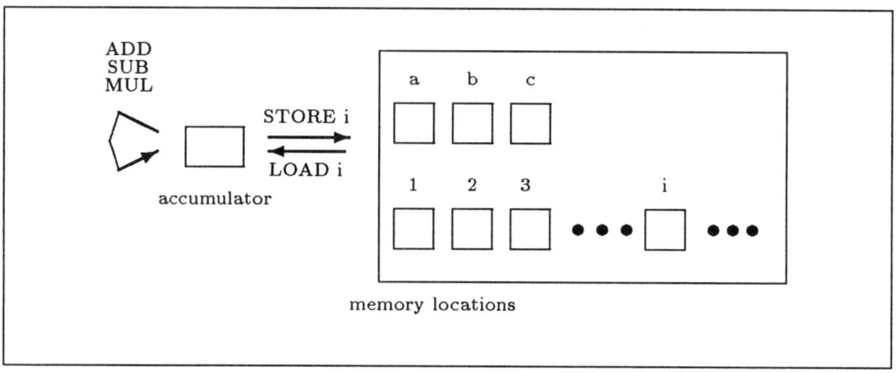

Fig. 1.6. The configuration of the machine $\mathcal{M}$

LOAD i	for loading $c(i)$ into the accumulator,
STORE i	for storing the value of the accumulator in location i,
ADD i	for adding $c(i)$ to the value of the accumulator,
SUB i	for subtracting $c(i)$ from the value of the accumulator,
MUL i	for multiplying $c(i)$ with the value of the accumulator.

If q is an assembly program and $n_1, n_2, n_3 \in \mathbb{N}$, then we denote by $\mathcal{M}[\![q]\!](n_1, n_2, n_3)$ the value of the accumulator after having executed q on $\mathcal{M}$ with n_1, n_2, n_3 as initial values of memory locations a, b, and c, respectively.

The code semantics of an arithmetic expression e is now specified in a syntax-directed way. For this purpose we will use three meaning names, viz., *code*, *temp*, and *name*, where *code* is the principal one, i.e., for every arithmetic expression e, we define *ass-prog*(e) to be the value of *code*(E) at the root of the derivation tree of e. The meaning names are associated with nonterminals as follows

$$\alpha(E) = \alpha(T) = \alpha(F) = \{code, temp\},$$
$$\alpha(I) = \{temp, name\}.$$

The carrier sets of *code*, *temp* and *name* are the sets

$$\begin{aligned} V_{code} &= \{q \mid q \text{ is a sequence of assembly instructions}\}, \\ V_{temp} &= \mathbb{N}, \\ V_{name} &= \{a, b, c\}. \end{aligned}$$

The semantic equations for *code*, *temp* and *name* are designed as follows. Let t be a derivation tree and d be a node of t labeled by E, F, or T. Moreover, let s be the subtree of t starting at node d and let v be the frontier of s (Fig. 1.7 where $X \in \{E, F, T\}$).

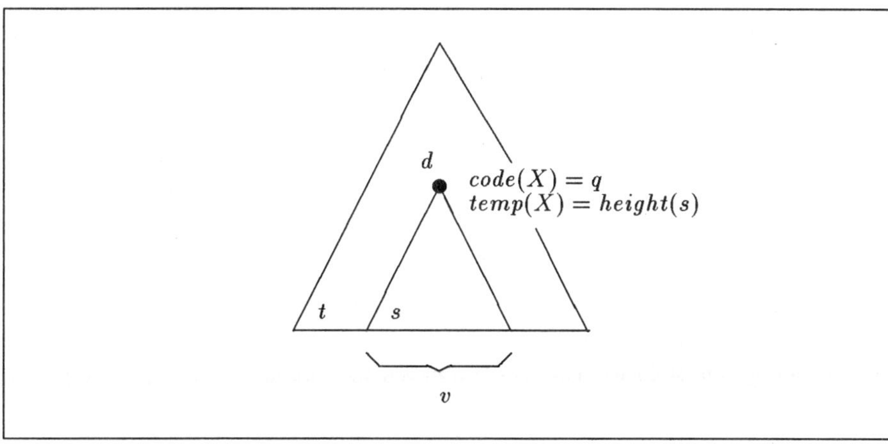

Fig. 1.7. A derivation tree t and the subtree s at node d

- The value of the meaning name *code* at node d (i.e., *code*(X)) is the assembly program q such that $\mathcal{M}[\![q]\!] = \mathcal{E}[\![v]\!]$ and
- the value of *temp* at d (i.e., *temp*(X)) is the height of s, i.e., the maximum of the lengths of all paths of s.

If d' is a node of t labeled by I, then

- the value of *name* at d' is the identifier which labels the immediate descendant of d' and

- the value of *temp* at d' is 1, because the height of the subtree rooted by d' is 1.

The meaning name *temp* is needed to compute the value of *code* at node d, in particular, if the production $E \rightarrow E + T$, $E \rightarrow E - T$, or $T \rightarrow T * F$ is applied at d. Since the machine has just one accumulator, the value of the first argument of the binary operation (+, −, or ∗) has to be stored temporarily in a safe location in the memory before computing the value of the second argument. A location is safe if it is not overwritten by the computation of the second argument. We specify the value of *temp* at a node d to be the height of the subtree starting at node d. One can easily check that this does provide a safe location.

First we consider the production p_1: $E_0 \rightarrow E_1 + T$ and associate semantic equations with it. We use the indices 0 and 1 in order to distinguish the two occurrences of E in the semantic equations. With p_1 the following two semantic equations are associated:

$$\begin{array}{llcl} p_1 : E_0 \rightarrow E_1 + T & code(E_0) & = & code(E_1) \text{ STORE } temp(T) \\ & & & code(T) \text{ ADD } temp(T) \\ & temp(E_0) & = & max(temp(E_1), temp(T)) + 1 \end{array}$$

Thus, for every expression e_0 of the form $e_1 + t$ (where e_1 and t are derived from E_1 and T, respectively), the assembly code $code(e_0)$ consists of the following pieces:

1. the value of $code(E_1)$, which is the code of the subexpression e_1 (the execution of $code(E_1)$ computes the expression semantics of e_1 in the accumulator),
2. the STORE $temp(T)$ instruction (its execution stores the value of the accumulator to the safe memory location with address $temp(T)$),
3. the value of $code(T)$, which is the code of the subexpression t (its execution computes the expression semantics of t in the accumulator), and
4. the ADD $temp(T)$ instruction (its execution retrieves the value from the memory location with address $temp(T)$ and adds it to the contents of the accumulator).

We note the semantic equation for $code(E_0)$ fits into the general form of semantic equations by letting ζ be the expression $f(code(E_1), temp(T), code(T))$ where f is the function

$$f : V_{code} \times V_{temp} \times V_{code} \rightarrow V_{code}$$

such that, for any codes $code_1, code_2 \in V_{code}$ and address $i \in V_{temp}$, we define

$$f(code_1, i, code_2) \quad = \quad code_1 \text{ STORE } i \; code_2 \text{ ADD } i$$

The semantic equations of all the productions are shown in Fig. 1.8.

$$
\begin{array}{llcl}
p_1 : E_0 \rightarrow E_1 + T & code(E_0) & = & code(E_1) \text{ STORE } temp(T) \; code(T) \\
 & & & \text{ADD } temp(T) \\
 & temp(E_0) & = & max(temp(E_1), temp(T)) + 1 \\
 & & & \\
p_2 : E_0 \rightarrow E_1 - T & code(E_0) & = & code(T) \text{ STORE } temp(E_1) \; code(E_1) \\
 & & & \text{SUB } temp(E_1) \\
 & temp(E_0) & = & max(temp(E_1), temp(T)) + 1 \\
 & & & \\
p_3 : E \rightarrow T & code(E) & = & code(T) \\
 & temp(E) & = & temp(T) + 1 \\
 & & & \\
p_4 : T_0 \rightarrow T_1 * F & code(T_0) & = & code(T_1) \text{ STORE } temp(F) \; code(F) \\
 & & & \text{MUL } temp(F) \\
 & temp(T_0) & = & max(temp(T_1), temp(F)) + 1
\end{array}
$$

$p_5 : T \rightarrow F$, $p_6 : F \rightarrow (E)$

The semantic equations are the same as the semantic equations for p_3 except that E and T are replaced appropriately.

$$
\begin{array}{llcl}
p_7 : F \rightarrow I & code(F) & = & \text{LOAD } name(I) \\
 & temp(F) & = & temp(I) + 1 \\
 & & & \\
p_8 : I \rightarrow a & name(I) & = & a \\
 & temp(I) & = & 1
\end{array}
$$

$p_9 : I \rightarrow b$, $p_{10} : I \rightarrow c$

The semantic equations are obtained from the equations of p_8 by replacing a by b and c, respectively.

Fig. 1.8. The semantic equations for *code*, *temp*, and *name* in the framework of attribute grammars with synthesized attributes only

As an example for the transcription of semantic equations to recursive function procedures, we show the recursive function procedure *code-eval* (Fig. 1.9). There are seven productions p such that there is a $(code, X)$-equation of p for some nonterminal X: p_1 through p_7. Hence, the case statement of *code-eval* consists of seven clauses. Note that, e.g., *temp-eval*$(d2)$ is the call of function procedure *temp-eval* to the second descendant of d.

We now give an example for computing the code semantics of a particular arithmetic expression generated by G_0. Let us consider $e = a * (b + a)$ and

```
func code-eval (d: node): V_code;
begin
  case production applied at d of
    p1: code-eval := code-eval(d1) STORE temp-eval(d2)
                     code-eval(d2) ADD temp-eval(d2);
    p2: code-eval := code-eval(d2) STORE temp-eval(d1)
                     code-eval(d1) SUB temp-eval(d1);
    p3: code-eval := code-eval(d1);
    p4: code-eval := code-eval(d1) STORE temp-eval(d2)
                     code-eval(d2) MUL temp-eval(d2);
    p5: code-eval := code-eval(d1);
    p6: code-eval := code-eval(d1);
    p7: code-eval := LOAD name-eval(d1)
  end
end code-eval
```

Fig. 1.9. The function procedure *code-eval*

assume that the derivation tree t_e of e is given (Fig. 1.10). We can then call the recursive function procedure *code-eval* at the root of t_e. In Fig. 1.10, we have attached to some nodes the values of the meaning names as they are computed by the corresponding function calls.

Thus, for the string $e = a * (b + a)$, we have obtained that $code(E)$ at the root of t_e is the following assembly program (clearly, this is not optimal); it is the code semantics $ass\text{-}prog(e)$ of e:

```
LOAD a
STORE 5
LOAD b
STORE 2
LOAD a
ADD 2
MUL 5
```

The extended abstract method

The foregoing specification of the code semantics of arithmetic expressions is a typical syntax-directed one. Often however, meaning names are not sufficient to specify the semantics of more complex context-free languages, e.g., high-level programming languages. This deficiency becomes obvious if we consider the while statement and its translation into assembly code. Let the syntax of the while statement be given by the context-free production

$$W \rightarrow \textbf{while}\ E > 0\ \textbf{do}\ L\ \textbf{od}$$

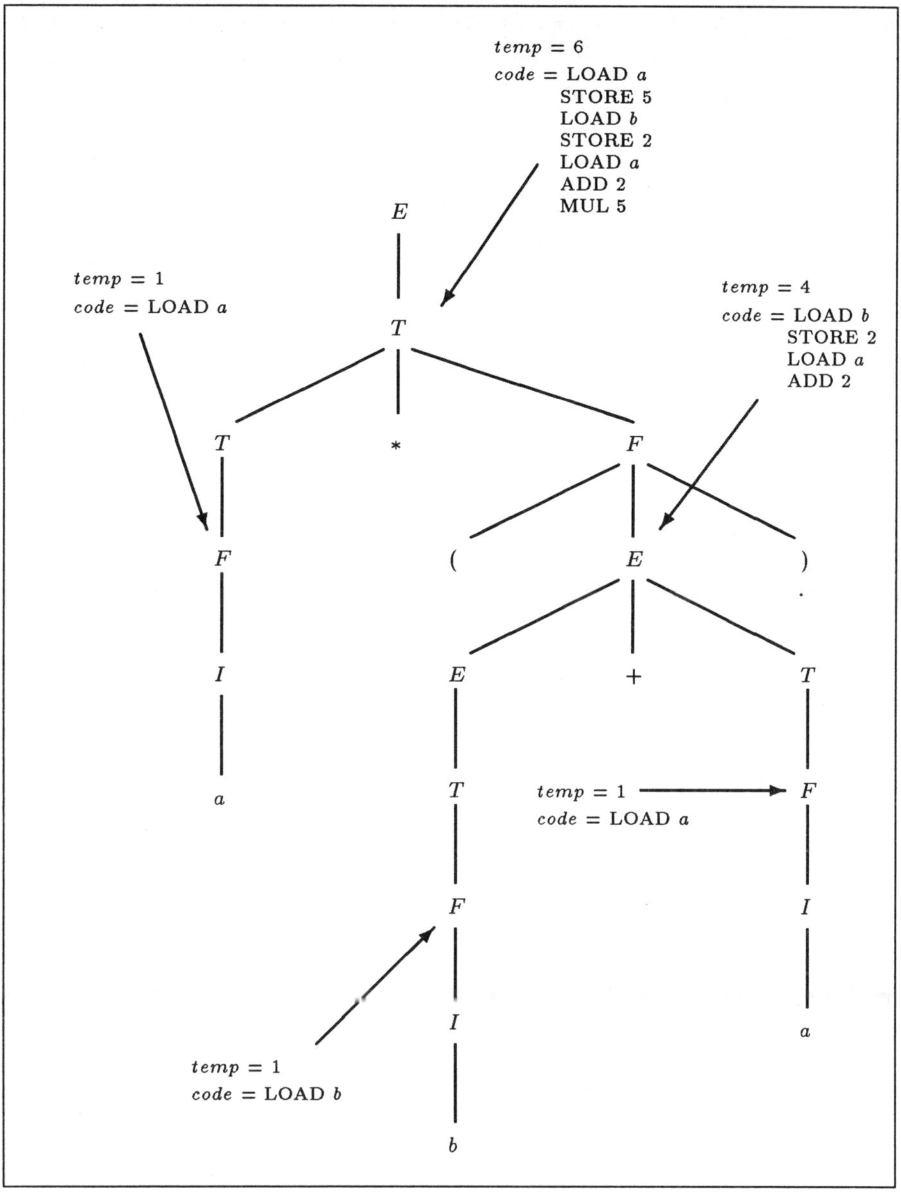

Fig. 1.10. The derivation tree t_e of the string $a * (b + a)$ with the computed values of *code* and *temp* at some nodes

where L generates any finite list of statements. Assume that the semantics of a string (i.e., program) of the language is an assembly program for the machine $\mathcal{M}$ as in the previous example. It should be clear that for such assembly programs a sequential control structure is not sufficient any more. Hence, we assume that $\mathcal{M}$ additionally accepts the following jump instructions:

JUMP i	for an unconditional jump to the i th statement of the assembly program and
JNP i	for a jump to the i-th statement of the assembly program provided that the value in the accumulator is not positive, otherwise the next assembly instruction is executed.

Then, as usual, for every while statement w of the form **while** $e > 0$ **do** l **od** (where e and l are an arithmetic expression and a list of statements, respectively), the assembly program *ass-prog*(w) of w is composed of the following pieces:

1. the code for computing the expression semantics of e in the accumulator,
2. the instruction JNP m, where m is the address of the next instruction behind *ass-prog*(w) (if the value of the accumulator is not positive, then its execution transfers the control of $\mathcal{M}$ to the next instruction behind *ass-prog*(w), otherwise the assembly instruction behind JNP is executed),
3. the code which results from the translation of l, and
4. the instruction JUMP n, where n is the address of the first instruction of *ass-prog*(w) (its execution transfers the control of $\mathcal{M}$ to the code for evaluating e again, because its value may have changed during the execution of l).

In particular in the fourth piece of *ass-prog*(w), we should know the address n of the first instruction of *ass-prog*(w). One can easily see that this address cannot be computed by using meaning names of E and L, because meaning names at a node can keep information only about the descendants of that node. So we should extend the original abstract method in such a way that a node d of the derivation tree t can also get information from its surrounding context, i.e., from that part of t which is outside the subtree rooted by d. Hence we refine our original abstract method as follows:

> *Syntax-directed semantics with context handling* is based on the idea that, for a word w of a context-free language, the semantics of a substring v of w is defined in terms of the semantics of the syntactic substructures of v and the context of v with respect to w.

Now there are two possible ways to formalize this extended abstract method:

- the implicit handling of context information and
- the explicit handling of context information.

The first way leads to formal models which are close to denotational semantics, and the second way leads to formal models which are close to attribute grammars. In the rest of this section we will develop instances of these two types of formal models. Since this development is based on the notions and notations of the original abstract method, we note that the first type of formal model will appear slightly different to the style in which denotational semantics usually appears in the literature. We will discuss this difference later.

A second formal model: denotational semantics

In this second formal model the pieces of information about the surrounding of some node d of a derivation tree t, i.e., the context information of d, do not have explicit names. Instead, meaning names are allowed to have parameters and the pieces of context information are passed via such parameters. In this sense context information is handled in an implicit way. Certainly, the context information coming via d from its context can also be passed down to the descendants of d as parameters of other meaning names.

As in our first formal model, the set of meaning names is denoted by M. However, for every meaning name $m \in M$, the carrier set V_m of m is no longer a set of elementary objects, but a set of functions over other carrier sets (we do not consider recursive domain equations here). Thus, m is a function of the type $V_1 \times \ldots \times V_k \rightarrow V$. As usual the mapping $\alpha : N \rightarrow \mathcal{P}(M)$ associates meaning names to nonterminals. To every production p of G a set of semantic equations is associated such that for every meaning name of the nonterminal on the left-hand side of p there is exactly one semantic equation which specifies the semantics. More precisely, let

$$p : A \rightarrow w_0 A_1 w_1 \ldots w_{n-1} A_n w_n$$

be a production of G and $m \in M$ be a meaning name. Then the m-meaning of every string which is derived from A by starting with production p is defined by a semantic equation of the form

$$m(A)(y_1, \ldots, y_k) = \zeta,$$

where $y_1, \ldots, y_k$ are the parameters of m, and ζ is an expression built up from the parameters $y_1, \ldots, y_k$, basic functions over carrier sets, and compound objects of the form $m_j(A_{i_j})$ where $m_j \in \alpha(A_{i_j})$ and $1 \leq i_j \leq n$. Clearly, since meaning names have parameters, they may also occur nested in ζ. The left-hand side of the equation should be parsed as follows: $m(A)$ is a function with k arguments, and it is applied to $y_1, \ldots, y_k$. We also refer to this equation here as the (m, A)-equation.

Intuitively, the (m, A)-equation of p defines the value of m at every node d at which the production p is applied. If the values of the parameters $y_1, \ldots, y_k$

are available from the surrounding of d, then the value of m is computed by evaluating ζ. This may naturally invoke the calculation of other meaning names on descendant nodes of d. To enable this computation, for every meaning name m, we again construct a recursive function procedure *m-eval*. It has one formal parameter of type node; additionally it has formal parameters y_1*-par*, ..., y_k*-par* of type $V_1, \ldots, V_k$, respectively, which represent the parameters of the meaning name m. Assume that V_m is the set of functions of the type $V_1 \times \ldots \times V_k \to V$. Then the function procedure *m-eval* is shown in Fig. 1.11 where ζ' is obtained from ζ by replacing

- every subexpression $m_j(A_{i_j})(\ldots)$ by m_j*-eval*$(d\,i_j, \ldots)$, where $d\,i_j$ is again the i_j-th descendant of d, and
- every y_j by y_j*-par*.

```
func m-eval(d: node, y1-par: V1, ..., yk-par: Vk): V ;
begin
   case production applied at d of
     ...
      p : m-eval := ζ';
     ...
   end
end m-eval
```

Fig. 1.11. The function procedure *m-eval* with formal parameters

We have called this formal model *denotational semantics*. However, we should note that denotational semantics allows for much more flexible and powerful constructions on the right-hand sides of semantic equations (as, e.g., λ-abstractions). In denotational semantics the meaning names are called semantic functions. As already mentioned, the usual notation slightly differs from the notation developed here. In particular, for a production

$$p : A \to w_0 A_1 w_1 \ldots w_{n-1} A_n w_n$$

and a semantic function m of A, the corresponding semantic equation looks like

$$m[\![w_0 u_1 w_1 \ldots w_{n-1} u_n w_n]\!](y_1, \ldots, y_k) = \zeta'$$

where it is assumed that, for every $1 \leq i \leq n$, u_i is a variable which is universally quantified over the set of terminal strings derived from A_i. The right-hand side ζ' is obtained from ζ by a similar notational transformation. Furthermore, rather than small letters (like m) capital calligraphic letters (like $\mathcal{C}$ or $\mathcal{E}$) are used to denote semantic functions. In the following example, however, we will retain our notational system.

An example of a denotational semantics

In this example we specify the code semantics of a small imperative programming language (including while statements) as assembly code by using the formal model based on denotational semantics. For this purpose we enlarge the context-free grammar G_0, which generates arithmetic expressions, to the context-free grammar G_1. The nonterminals of G_1 are the nonterminals E, T, F, and I of G_0, and additionally

P	(start symbol) for generating programs,
L	for generating lists of statements,
S	for generating statements,
W	for generating while statements,
A	for generating assignment statements.

The productions of G_1 are the productions p_1 through p_{10} of G_0 (see p. 5) and additionally

p_{11}: $P \rightarrow L$
p_{12}: $L \rightarrow L; S$ p_{13}: $L \rightarrow S$
p_{14}: $S \rightarrow W$ p_{15}: $S \rightarrow A$
p_{16}: $W \rightarrow$ **while** $E > 0$ **do** L **od**
p_{17}: $A \rightarrow I := E$.

Again it is quite obvious how to associate a natural semantics $\mathcal{P}[\![w]\!]$ with a program $w \in L(G_1)$: it is the partial function of type $\mathbb{N}^3 \rightarrow \mathbb{N}^3$ such that, for every $n_1, n_2, n_3 \in \mathbb{N}$, $\mathcal{P}[\![w]\!](n_1, n_2, n_3) = (n'_1, n'_2, n'_3)$, if

- the initial values of the program variables a, b, and c are n_1, n_2, and n_3, respectively,
- the program w is "executed",
- it terminates, and
- the final values of a, b, and c are n'_1, n'_2, and n'_3, respectively.

Clearly, the notion of "execution" is vague, but it can be substituted by "executed on a PC," or "executed on a sheet of paper," or "executed by hand waving," or "executed mentally." Henceforth this semantics is called the *program semantics of* w and it is used as a reference semantics for the correctness of the code semantics.

For every assembly program q (possibly with jump instructions), we define the partial function $\mathcal{M}[\![q]\!] : \mathbb{N}^3 \rightarrow \mathbb{N}^3$ in the following way: for every $n_1, n_2, n_3 \in \mathbb{N}$, $\mathcal{M}[\![q]\!](n_1, n_2, n_3) = (n'_1, n'_2, n'_3)$ if

- the initial values of the memory locations with addresses a, b, and c are n_1, n_2, and n_3, respectively,
- the assembly program q is executed on our extended machine,
- it terminates, and
- the final values of the memory locations a, b, and c are n'_1, n'_2, and n'_3, respectively.

Now, for a program $w \in L(G_1)$, we want its code semantics to be an assembly program *ass-prog*(w) such that $\mathcal{P}[\![w]\!] = \mathcal{M}[\![ass\text{-}prog(w)]\!]$, i.e., *ass-prog*$(w)$ is correct with respect to the program semantics of w.

We specify the code semantics by using denotational semantics as a formal model, i.e., the first formal model for the extended abstract method. The meaning names we need are *code*, $code_1$, *length*, *temp*, and *name*, where $code_1$ has one parameter, which keeps the context information, and the other meaning names do not have a parameter. The meaning names are associated to nonterminals by the mapping α as follows.

$$\begin{array}{ll} \alpha(P) & = \{code\} \\ \alpha(L) = \alpha(S) = \alpha(W) & = \{code_1, length\} \\ \alpha(A) & = \{code, length\} \\ \alpha(E) = \alpha(T) = \alpha(F) & = \{code, length, temp\} \\ \alpha(I) & = \{temp, name\} \end{array}$$

The carrier sets of the meaning names are defined as follows.

$$\begin{array}{lll} V_{code} & = & \{q \mid q \text{ is a sequence of assembly instructions}\} \\ V_{code_1} & = & \{f \mid f \text{ is a mapping from } \mathbb{N} \text{ to } V_{code}\} \\ V_{temp} & = & V_{length} \quad = \quad \mathbb{N} \\ V_{name} & = & \{a, b, c\} \end{array}$$

We now explain the intuition behind the meaning names. For this, let w be a program written in $L(G_1)$, let t_w be the derivation tree of w, and d be a node of t_w labeled by some nonterminal $X \in N$. Let v be the frontier of the subtree rooted by d, thus v is a substring of w derived from X.

- If $X \in \{P\}$ (i.e., d is the root of t_w and $v = w$), then the value of *code* at d will be an assembly program *ass-prog*(w) such that $\mathcal{M}[\![ass\text{-}prog(w)]\!] = \mathcal{P}[\![w]\!]$.
- If $X \in \{L, S, W\}$, then the value of $code_1$ at d, (i.e., $code_1(X)(n)$) will be an assembly program *ass-prog*(v) such that $\mathcal{M}[\![ass\text{-}prog(v)]\!] = \mathcal{P}[\![v]\!]$. Moreover, *ass-prog*$(v)$ is a substring of *ass-prog*(w) such that the first assembly instruction of *ass-prog*(v) is the n-th assembly instruction of *ass-prog*(w) (Fig. 1.12).
- If $X = A$, then the value of *code* at d will be an assembly program *ass-prog*(v) such that $\mathcal{M}[\![ass\text{-}prog(v)]\!] = \mathcal{P}[\![v]\!]$.
- If $X \in \{L, S, W, A\}$, then the value of *length* at d is the length of the value of $code_1$, i.e., the length of the assembly program *ass-prog*(v), where the length is measured in the number of assembly instructions. Note that the length of *ass-prog*(v) does not depend on the context information at d.
- If $X \in \{E, T, F\}$, then the value of *code* at d will be an assembly program *ass-prog*(v) such that $\mathcal{M}[\![ass\text{-}prog(v)]\!] = \mathcal{E}[\![v]\!]$.
- If $X \in \{E, T, F\}$, then the value of *length* at d is the length of the value of *code*, i.e., the length of the assembly program *ass-prog*(v).
- The meaning of *temp* is the same as in the previous example.

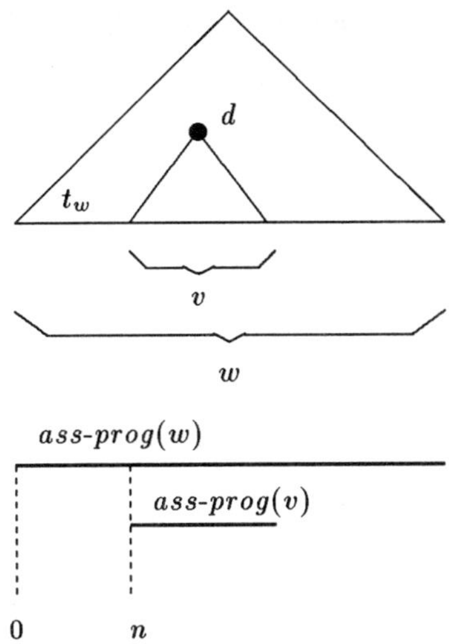

Fig. 1.12. The value of $code_1$ at d labeled by $X \in \{L, S, W, A\}$

- If $X = I$, then the value of *name* at d is the terminal which labels the immediate successor of d.

The semantic equations are presented in Fig. 1.13 and 1.14. We note that the assembly instruction STOP is used for technical reasons which will become clear later. Anyway, the effect of STOP is to stop assembly programs. We also note that the expression $code_1(L)(0)$ should be read as follows: $code_1(L)$ is a unary function which is applied to the expression 0.

Again, the semantic equations are transcribed into recursive function procedures. In Fig. 1.15 the recursive function procedure of the meaning name $code_1$ is shown.

Finally, we compute the code semantics of a simple program in the language $L(G_1)$. For this, let us consider the program

$$\begin{array}{lll} \textbf{while} & a - c > 0 & \textbf{do} \\ & & a := a - c\,; \\ & & b := b * b \\ & & \textbf{od} \end{array}$$

which we abbreviate by *exp*. The name *exp* is chosen because this program computes the function $F : \mathbb{N}^3 \to \mathbb{N}^3$ defined as

$$F(x, y, z) = \begin{cases} (x \bmod z, y^{x \operatorname{div} z}, z) \text{ if } x > z \\ (x, y, z) \qquad\qquad\qquad \text{otherwise,} \end{cases}$$

$$p_1 : E_0 \rightarrow E_1 + T \qquad \begin{aligned} code(E_0) &= code(E_1)\ \mathrm{STORE}\ temp(T) \\ &\ code(T)\ \mathrm{ADD}\ temp(T) \\ length(E_0) &= length(E_1) + length(T) + 2 \\ temp(E_0) &= max(temp(E_1) + temp(T)) + 1 \end{aligned}$$

$p_2 : E_0 \rightarrow E_1 - T$

The same equations as for p_1 except that we write SUB for ADD in the definition of $code(E_0)$ and exchange the roles of E_1 and T.

$$p_3 : E \rightarrow T \qquad \begin{aligned} code(E) &= code(T) \\ length(E) &= length(T) \\ temp(E) &= temp(T) + 1 \end{aligned}$$

$$p_4 : T_0 \rightarrow T_1 * F \qquad \begin{aligned} code(T_0) &= code(T_1)\ \mathrm{STORE}\ temp(F) \\ &\ code(F)\ \mathrm{MUL}\ temp(F) \\ length(T_0) &= length(T_1) + length(F) + 2 \\ temp(T_0) &= max(temp(T_1) + temp(F)) + 1 \end{aligned}$$

$p_5 : T \rightarrow F, p_6 : F \rightarrow (E)$

The equations for p_5 and p_6 can be obtained from that of p_3 by appropriate substitutions.

$$p_7 : F \rightarrow I \qquad \begin{aligned} code(F) &= \mathrm{LOAD}\ name(I) \\ length(F) &= 1 \\ temp(F) &= temp(I) + 1 \end{aligned}$$

$$p_8 : I \rightarrow a \qquad \begin{aligned} name(I) &= a \\ temp(I) &= 1 \end{aligned}$$

$p_9 : I \rightarrow b, p_{10} : I \rightarrow c$

The equations for p_9 and p_{10} can be obtained from that of p_8 by replacing a by b and c, respectively.

Fig. 1.13. The semantic equations for *code*, $code_1$, *length*, *name*, and *temp* in the approach based on denotational semantics

$$
\begin{array}{llcl}
p_{11}: P \rightarrow L & code(P) & = & code_1(L)(0)\ \mathrm{STOP} \\[1ex]
p_{12}: L_0 \rightarrow L_1; S & code_1(L_0)(y) & = & code_1(L_1)(y) \\
 & & & code_1(S)(y + length(L_1)) \\
 & length(L_0) & = & length(L_1) + length(S) \\[1ex]
p_{13}: L \rightarrow S & code_1(L)(y) & = & code_1(S)(y) \\
 & length(L) & = & length(S) \\[1ex]
p_{14}: S \rightarrow W & code_1(S)(y) & = & code_1(W)(y) \\
 & length(S) & = & length(W) \\[1ex]
p_{15}: S \rightarrow A & code_1(S)(y) & = & code(A) \\
 & length(S) & = & length(A) \\[1ex]
\multicolumn{4}{l}{p_{16}: W \rightarrow \textbf{while}\ E > 0\ \textbf{do}\ L\ \textbf{od}} \\
 & code_1(W)(y) & = & code(E) \\
 & & & \mathrm{JNP}\ y + length(E) + length(L) + 2 \\
 & & & code_1(L)(y + length(E) + 1) \\
 & & & \mathrm{JUMP}\ y \\
 & length(W) & = & length(E) + length(L) + 2 \\[1ex]
p_{17}: A \rightarrow I := E & code(A) & = & code(E)\ \mathrm{STORE}\ name(I) \\
 & length(A) & = & length(E) + 1
\end{array}
$$

Fig. 1.14. (cont.) The semantic equations for *code*, $code_1$, *length*, *name*, and *temp* in the approach based on denotational semantics

where $x \bmod z$ and $x \operatorname{div} z$ are the remainder and the result, respectively, of the integer-valued division of x by z. Thus, $\mathcal{P}[\![exp]\!] = F$.

The derivation tree of *exp*, denoted by t_{exp}, can be seen in Fig. 1.16, where the subtrees t_1, t_2, t_3, and t_4 of t_{exp} are defined in Figure 1.17. In both figures we have given names to some of the nodes so that we can refer to them. Note that the names d_4 through d_9 refer to the occurrence of t_1 in Fig. 1.16. We compute the code semantics of *exp* by applying the recursive function procedure *code-eval* to the root d_0 of t_{exp}. The protocol of the execution of *code-eval* looks as follows.

$$
\begin{array}{l}
\textbf{func}\ code_1\text{-}eval(d\text{: node; } y_1\text{-}par\text{: } V_{temp})\text{: } V_{code};\\
\textbf{begin}\\
\quad \textbf{case}\ \text{production applied at } d\ \textbf{of}\\
\qquad p_{12}\text{: } code_1\text{-}eval := code_1\text{-}eval(d\,1,\ y_1\text{-}par)\\
\qquad\qquad\qquad\qquad\quad code_1\text{-}eval(d\,2,\ y_1\text{-}par + length\text{-}eval(d\,1));\\
\qquad p_{13}\text{: } code_1\text{-}eval := code_1\text{-}eval(d\,1,\ y_1\text{-}par);\\
\qquad p_{14}\text{: } code_1\text{-}eval := code_1\text{-}eval(d\,1,\ y_1\text{-}par);\\
\qquad p_{16}\text{: } code_1\text{-}eval := code\text{-}eval(d\,1)\\
\qquad\qquad\qquad\qquad\quad \text{JNP } (y_1\text{-}par + length\text{-}eval(d\,1) +\\
\qquad\qquad\qquad\qquad\qquad length\text{-}eval(d\,2) + 2)\\
\qquad\qquad\qquad\qquad\quad code_1\text{-}eval(d\,2,\ y_1\text{-}par + length\text{-}eval(d\,1) + 1)\\
\qquad\qquad\qquad\qquad\quad \text{JUMP } y_1\text{-}par;\\
\quad \textbf{end}\\
\textbf{end}\ code_1\text{-}eval
\end{array}
$$

Fig. 1.15. The function procedure *code*$_1$*-eval*

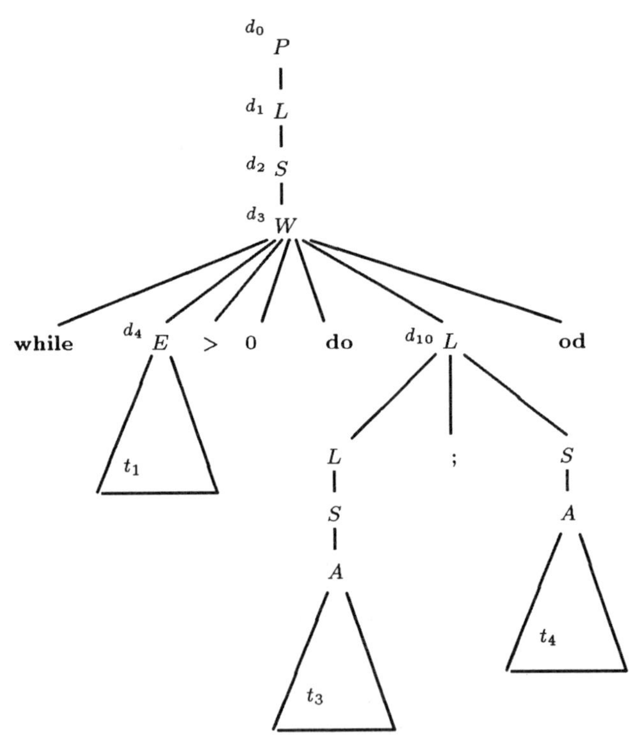

Fig. 1.16. The derivation tree t_{exp} of the program *exp*

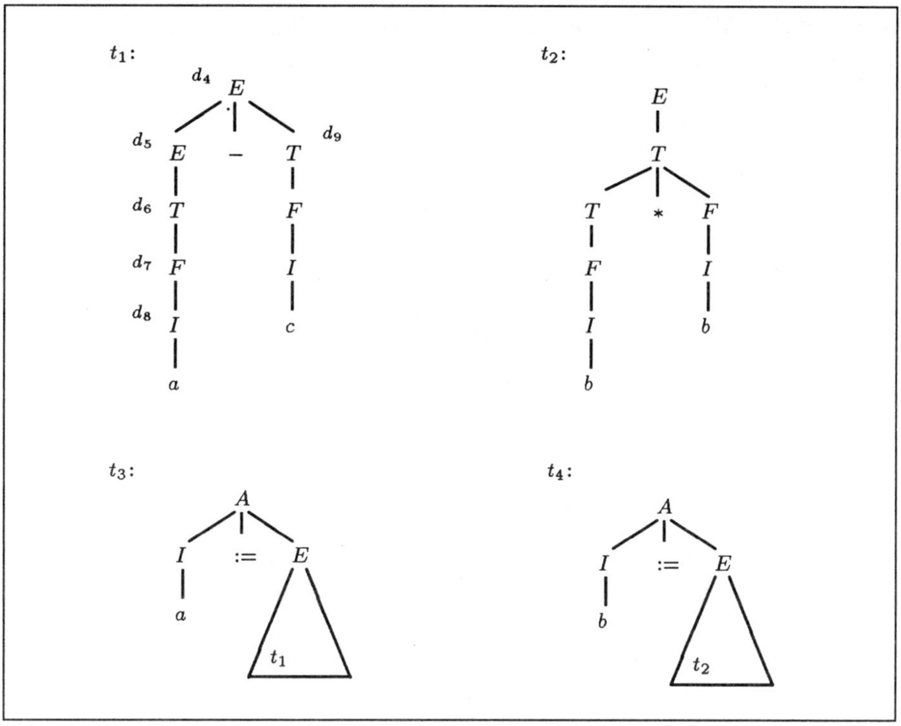

Fig. 1.17. The subtrees t_1, t_2, t_3, and t_4 of t_{exp}

$$
\begin{aligned}
\textit{code-eval}(d_0) \quad &= \quad \textit{code}_1\textit{-eval}(d_1)(0) \\
& \qquad \text{STOP} \\
&= \quad \textit{code}_1\textit{-eval}(d_2)(0) \\
& \qquad \text{STOP} \\
&= \quad \textit{code}_1\textit{-eval}(d_3)(0) \\
& \qquad \text{STOP} \\
&= \quad \textit{code-eval}(d_4) \\
& \qquad \text{JNP } \textit{length-eval}(d_4) + \textit{length-eval}(d_{10}) + 2 \\
& \qquad \textit{code}_1\textit{-eval}(d_{10})(\textit{length-eval}(d_4) + 1) \\
& \qquad \text{JUMP } 0 \\
& \qquad \text{STOP}
\end{aligned}
$$

Thus we have to execute the function procedure call $\textit{code-eval}(d_4)$:

$$
\begin{aligned}
\textit{code-eval}(d_4) \quad &= \quad \textit{code-eval}(d_9) \\
& \qquad \text{STORE } \textit{temp-eval}(d_5) \\
& \qquad \textit{code-eval}(d_5) \\
& \qquad \text{SUB } \textit{temp-eval}(d_5),
\end{aligned}
$$

where $code\text{-}eval(d_5) = code\text{-}eval(d_6) = code\text{-}eval(d_7) = code\text{-}eval(d_8) =$ LOAD a. Similarly to $code\text{-}eval(d_6)$, we can compute that $code\text{-}eval(d_9) =$ LOAD c. Next we compute $temp\text{-}eval(d_5) = 1 + temp\text{-}eval(d_6) = 2 + temp\text{-}eval(d_7) = 3 + temp\text{-}eval(d_8) = 4$. Hence we get

$$code\text{-}eval(d_4) = \begin{array}{l} \text{LOAD } c \\ \text{STORE } 4 \\ \text{LOAD } a \\ \text{SUB } 4 \end{array}$$

We now compute $length\text{-}eval(d_4)$ as follows: $length\text{-}eval(d_4) = length\text{-}eval(d_5) + length\text{-}eval(d_9) + 2$, where $length\text{-}eval(d_5) = length\text{-}eval(d_6) = length\text{-}eval(d_7) = length\text{-}eval(d_8) = 1$ and by similar computations we obtain $length\text{-}eval(d_9) = 1$. Hence $length\text{-}eval(d_4) = 4$. Similarly to $length\text{-}eval(d_4) = 4$, we can compute that $length\text{-}eval(d_{10}) = 10$. Since $length\text{-}eval(d_4) = 4$, we have $code_1\text{-}eval(d_{10})(length\text{-}eval(d_4)+1) = code_1\text{-}eval(d_{10})(5)$. Thus, we can compute

$$code_1\text{-}eval(d_{10})(5) = \begin{array}{l} \text{LOAD } c \\ \text{STORE } 4 \\ \text{LOAD } a \\ \text{SUB } 4 \\ \text{STORE } a \\ \text{LOAD } b \\ \text{STORE } 2 \\ \text{LOAD } b \\ \text{MUL } 2 \\ \text{STORE } b \end{array}$$

Finally, by substituting the computed values in $code\text{-}eval(d_0)$, we obtain

$code\text{-}eval(d_0) =$

0:	LOAD c	9:	STORE a
1:	STORE 4	10:	LOAD b
2:	LOAD a	11:	STORE 2
3:	SUB 4	12:	LOAD b
4:	JNP 16	13:	MUL 2
5:	LOAD c	14:	STORE b
6:	STORE 4	15:	JUMP 0
7:	LOAD a	16:	STOP
8:	SUB 4		

We note that the labels 0: through 16: are not part of the assembly program: they are just shown to emphasize the targets of the two jump instructions.

A third formal model: attribute grammars

In the literature our third formal model for syntax-directed semantics is known as *attribute grammars*. In this model the handling of context information is done explicitly. That is, we may use meaning names in their original sense, i.e., without parameters, and there is one principal meaning name m_0. Additionally, we introduce so-called *context names* which are used in order to transmit information in the derivation tree from a node to its descendants. (The other direction of information transmission is managed by the meaning names.) In this sense, the handling of context information is done in an explicit way.

The set C of context names is disjoint with M (M is again the set of meaning names). Also context names are associated with the nonterminals of the grammar. We use one mapping α, for both the association of meaning names and the association of context names; hence $\alpha : N \rightarrow \mathcal{P}(M \cup C)$. Context names also take their values from carrier sets: for every context name c, a carrier set V_c is defined. The values of context names are defined by semantic equations associated to the productions of the grammar. In fact, meaning names and context names may depend on each other, as can be seen if we show the form of the semantic equations. For every production

$$p : A \rightarrow w_0 A_1 w_1 \ldots w_{n-1} A_n w_n,$$

in P and meaning name $m \in \alpha(A) \cap M$, there is a semantic equation of the form

$$m(A) = \zeta,$$

where ζ is an expression built up from basic functions over carrier sets, compound objects of the form either $m_j(A_{i_j})$ or $z_j(A)$ where $m_j \in \alpha(A_{i_j}) \cap M$, $1 \leq i_j \leq n$, and $z_j \in \alpha(A) \cap C$. We call this the (m, A)-equation of p. Also, for every $1 \leq i \leq n$ and for every context name $z \in \alpha(A_i) \cap C$, there is a semantic equation of the form

$$z(A_i) = \zeta,$$

where ζ is an expression built up in the same way as in the (m, A)-equation. We call this the (z, A, i)-equation of p. Intuitively, it specifies the value of the context name z at the i-th descendant of every node at which p is applied.

For the computation of the semantics of a particular $w \in L(G)$, we again transcribe the system $\mathcal{S}_{sem}$ of semantic equations into a system $\mathcal{S}_{fp}$ of recursive function procedures. More precisely, for every meaning name there is one function procedure and for every context name there is also one function procedure. Both types of function procedures have one formal parameter which may take a node of the derivation tree of w as a value. There are no additional formal parameters. Let m and z be a meaning name and a context name, respectively. For a node d and its direct predecessor d' we define the *brother number of d* to be the position which d has in the sequence of all the direct descendants of d'. Then the form of the function procedures is shown in Fig. 1.18 where ζ' is obtained from the right-hand side ζ of the (m, A)-equation by replacing compound objects of the form $m_j(A_{i_j})$ and $z_j(A)$ by the function calls m_j-*eval*$(d\,i_j)$ and z_j-*eval*(d), respectively. Moreover, ζ_i' is obtained from the right-hand side of the (z, A, i)-equation in the same way as ζ' has been obtained from ζ and additionally d is substituted by d'.

```
func m-eval(d: node): V_m ;
begin
    case production applied at d of
        ...
        p: m-eval := ζ';
        ...
    end
end m-eval

func z-eval(d: node): V_z ;
begin
    let d' be the immediate predecessor of d
    case (production applied at d', brother number of d) of
        ...
        (p,i): z-eval := ζ'_i;
        ...
    end
end z-eval
```

Fig. 1.18. The function procedures *m-eval* and *z-eval*

The semantics of a string $w \in L(G)$ is then the value of the meaning name m_0 at the root of w's derivation tree t_w. This value is obtained by calling m_0-*eval* at the root of t_w, invoking a tree walk over t_w which may be very involved. It is even possible that the tree walk will loop, depending on the semantic equations. Fortunately, it is possible to decide whether, for a given system of semantic equations, there is a derivation tree t_w such that the induced tree walk is cyclic.

We note that in the theory of attribute grammars, meaning names and context names are called synthesized attributes and inherited attributes, respectively. However, we continue to use our terminology.

An example of an attribute grammar

As an example, we now specify the code semantics of programs in $L(G_1)$ in a syntax-directed way by means of an attribute grammar. It has the four meaning names *code*, *temp*, *length*, and *name* and one context name *first*. The carrier sets of *code*, *temp*, *length*, and *name* are the same domains as in the previous specification, and the carrier set of *first* is $\mathbb{N}$. Again, *code* is the principal meaning name.

The meaning names and context names are associated to the nonterminals by the mapping α as follows.

$$\begin{array}{lcl} \alpha(P) & = & \{code\} \\ \alpha(X) & = & \{code, length, first\}, \text{ for } X \in \{L, S, W\} \\ \alpha(A) & = & \{code, length\} \\ \alpha(I) & = & \{temp, name\} \\ \alpha(X) & = & \{code, length, temp\}, \text{ for } X \in \{E, F, T\} \end{array}$$

The intuitive idea of the meaning names should be clear by considering the previous two example specifications. The context name *first* corresponds to the parameter y in the specification based on denotational semantics. In fact, let $w \in L(G_1)$ be a program and t_w be the derivation tree of w. Let d be a node of t_w labeled by X and let v be the frontier of the subtree of t_w rooted by d (Fig. 1.12). Then, clearly, *ass-prog*(v) is a substring of *ass-prog*(w). It follows that the value of *first* at d will be the position of the first assembly instruction of *ass-prog*(v) with respect to the beginning of *ass-prog*(w).

Next we associate a system $\mathcal{S}_{sem}$ of semantic equations to the productions of G_1, see Fig. 1.19 and 1.20.

In order to illustrate the construction of the system $\mathcal{S}_{fp}$ of recursive function procedures, let us consider the context name *first*. The corresponding function procedure is shown in Fig. 1.21.

We finish this section by computing the code semantics *ass-prog*(exp) for the program exp which was shown in the previous example. We recall the schematic form of the derivation tree of exp, denoted by t_{exp}, in Fig. 1.22. The values of the meaning names and the context name at some nodes of t are indicated, although we could not show all values because of space limitations. The subtrees t_1 and t_2 of t_{exp} for the arithmetic expressions $a-c$ and $b*b$, respectively, are detailed in Fig. 1.23, where the values of *code* and *length* are computed at the roots of t_1 and t_2 in the same way as in the case of G_0. The codes *code*$(a-c)$ and *code*$(b*b)$ are then abbreviated by c_1 and c_2, respectively, and these abbreviations are used in Figs. 1.22 and 1.23, too. Thus the code of exp is the same assembly program as in the previous example where we specified the code semantics of $L(G_1)$ using a denotational semantics approach.

$p_1 : E_0 \to E_1 + T$, $p_2 : E_0 \to E_1 - T$

The same equations which were associated to p_1 and p_2 in G_0, in Figs. 1.13 and 1.14. Moreover, for both productions the equation $length(E_0) = length(E_1) + length(T) + 2$ is associated.

$p_3 : E \to T$

The same equations as for p_3 in G_0 and additionally the equation $length(E) = length(T)$.

$p_4 : T \to T * F$, $p_5 : T \to F$, $p_6 : F \to (E)$, $p_7 : F \to I$

The same semantic equations as associated with the productions of G_0 and additionally the equations for *length* of the nonterminal in the left-hand side. Note that $length(F) = 1$ for p_7.

$p_8 : I \to a$

The same semantic equations as associated with p_8 in G_0.

$p_9 : I \to b$, $p_{10} : I \to c$

The same semantic equations as associated with p_8 except that a is substituted by b and c, respectively.

$p_{11} : P \to L$

$$\begin{aligned} code(P) &= code(L)\ \mathrm{STOP} \\ first(L) &= 0 \end{aligned}$$

$p_{12} : L_0 \to L_1\ ;\ S$

$$\begin{aligned} code(L_0) &= code(L_1)\ code(S) \\ length(L_0) &= length(L_1) + length(S) \\ first(L_1) &= first(L_0) \\ first(S) &= first(L_0) + length(L_1) \end{aligned}$$

$p_{13} : L \to S$

$$\begin{aligned} code(L) &= code(S) \\ length(L) &= length(S) \\ first(S) &= first(L) \end{aligned}$$

Fig. 1.19. The semantic equations for *code*, *first*, and *length* in the framework of attribute grammars

$p_{14} : S \rightarrow W$

The same as for p_{13} except that L and S are substituted by S and W.

$p_{15} : S \rightarrow A$

Similarly as for p_{14} with the exception that there is no semantic equation for $first$.

$p_{16} : W \rightarrow$ **while** $E > 0$ **do** L **od**

$$\begin{aligned} code(W) &= code(E)\ \text{JNP}\ first(W) + length(E) + \\ &\quad\ length(L) + 2 \\ &\quad\ code(L)\ \text{JUMP}\ first(W) \\ length(W) &= length(E) + length(L) + 2 \\ first(L) &= first(W) + length(E) + 1 \end{aligned}$$

$p_{17} : A \rightarrow I := E$

$$\begin{aligned} code(A) &= code(E)\ \text{STORE}\ name(I) \\ length(A) &= length(E) + 1 \end{aligned}$$

Fig. 1.20. (cont.) The semantic equations for *code*, *first*, and *length* in the framework of attribute grammars

```
func first-eval(d: node): V_first;
begin let d' be the immediate predecessor of d
      case (production applied at d', brother number of d) of
         (p11, 1): first-eval := 0;
         (p12, 1): first-eval := first-eval(d');
         (p12, 2): first-eval := first-eval(d') +
                                 length-eval(d' 1);
         (p13, 1): first-eval := first-eval(d');
         (p14, 1): first-eval := first-eval(d');
         (p16, 2): first-eval := first-eval(d') +
                                 length-eval(d' 1) + 1
   end
end first-eval
```

Fig. 1.21. The function procedure *first-eval*

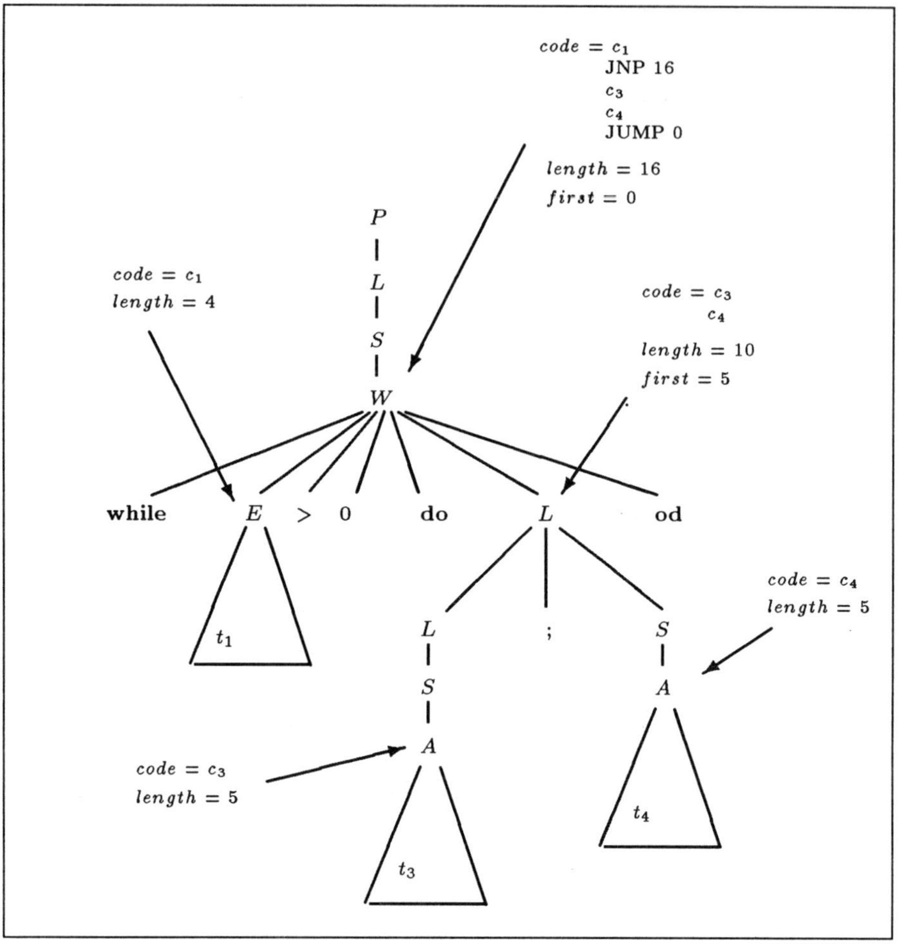

Fig. 1.22. The derivation tree t_{exp} with the computed values of the meaning names and the context name at some nodes

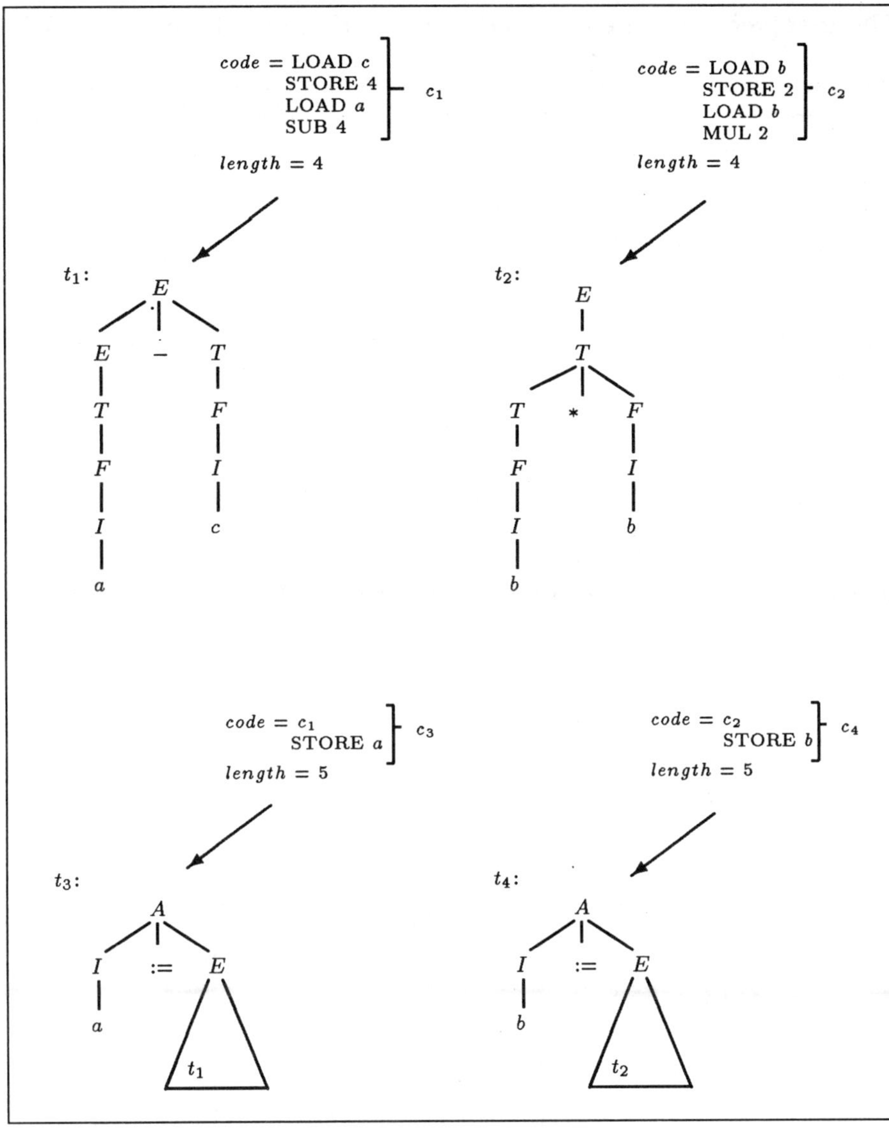

Fig. 1.23. The subtrees t_1 and t_2 of t_{exp} with the computed values of *code* and *length* at their roots

The involved machine-oriented computation paradigm

Up to now we have specified the intended semantics of a context-free language by means of three formal models. Each of these models consists of a system $\mathcal{S}_{sem}$ of semantic equations, or better: a system $\mathcal{S}_{fp}$ of recursive function procedures and a computation paradigm which can organize and maintain the sequence of calls of recursive function procedures.

Intuitively, it is clear how this calling process looks, and probably every first-year student of computer science will learn the effect of a call of a recursive function procedure. This is exactly the reason we have chosen this computation paradigm to give a semantics to the specification.

On the other hand, an abstract machinery which maintains recursive function procedure calls automatically is not so trivial to define, and consequently it is not likely that this is taught in detail to first-year students. The most common relevant machinery consists of a runtime stack machine $\mathcal{M}_{runtime}$ and a compiler *comp*. The machine $\mathcal{M}_{runtime}$ mainly consists of a stack which keeps track of the incarnations of the function procedures and the values of the parameters using static and dynamic links. The compiler *comp* translates the system $\mathcal{S}_{fp}$ into an assembly program $ass(\mathcal{S}_{fp})$ for $\mathcal{M}_{runtime}$. Thus the scenario of Fig. 1.4 has to be refined, and the more detailed situation is shown in Fig. 1.24. Later in our discussion we will return to the role of the computation paradigm.

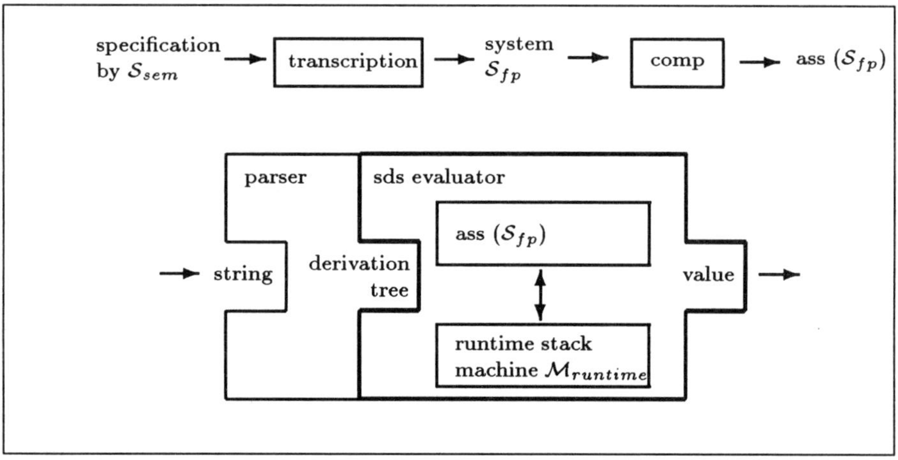

Fig. 1.24. The sds evaluator based on $\mathcal{M}_{runtime}$ and on *comp*

1.2 Tree Transducers

The aim of this book is to develop a theory of formal models of syntax-directed semantics. This theory contains abstract properties of such models. In particular, the properties do not depend on certain assumptions and details which are inherent in the three models of the previous subsections, and therefore we will abstract from them. More precisely,

- we abstract from derivation trees and use trees over some ranked alphabet instead,
- we abstract from the semantic domain and use the initial term algebra instead, and
- we abstract from the machine-oriented computation paradigm and use a rule-based derivation system instead.

This abstraction process produces formal models of syntax-directed semantics which we call *tree transducers*. In the following subsection we will explain the three abstraction steps in more detail.

Abstracting from derivation trees

Recall that, with every system $\mathcal{S}_{sem}$ of semantic equations, we have associated a system $\mathcal{S}_{fp}$ of recursive function procedures. Every function procedure has at least one formal parameter d which represents a node of the given derivation tree t_w (e.g., the function procedure *code*$_1$*-eval* in Fig. 1.15). The involved case statement represents a case analysis which depends on the production p applied at d. Thus, to be precise, we are not dealing with derivation trees but with abstract syntax trees. In such trees the nodes are labeled by productions of the underlying context-free grammar, and a node with label p has as many descendants as there are occurrences of nonterminals in the right-hand side of p. For instance, Fig. 1.25 shows the abstract syntax tree which corresponds to the derivation tree t_{exp} of *exp* in Fig. 1.16; t'_3 and t'_4 are the abstract syntax trees of t_3 and t_4 of Fig. 1.17.

Thus we should slightly adapt the scenario of Fig. 1.24 to catch the fact that the function procedures are working on abstract syntax trees (Fig. 1.26).

The function procedures also change slightly. Rather than testing the production applied at a node d, it tests the label of d. For instance, the adapted function procedure *code*$_1$*-eval* is shown in Fig. 1.27.

In Fig. 1.26 let us now consider the border between the parser and the syntax-directed semantics evaluator. The parser produces an abstract syntax tree t as output, and t is taken as input for the syntax-directed semantics evaluator. Clearly, the syntax-directed semantics evaluator does not have to check again whether the tree it receives from the parser is an abstract syntax tree or not, and hence this membership test is not an intrinsic part of the syntax-directed semantics evaluator. And since we want to study the abstract properties of formal models which can be used as syntax-directed semantics

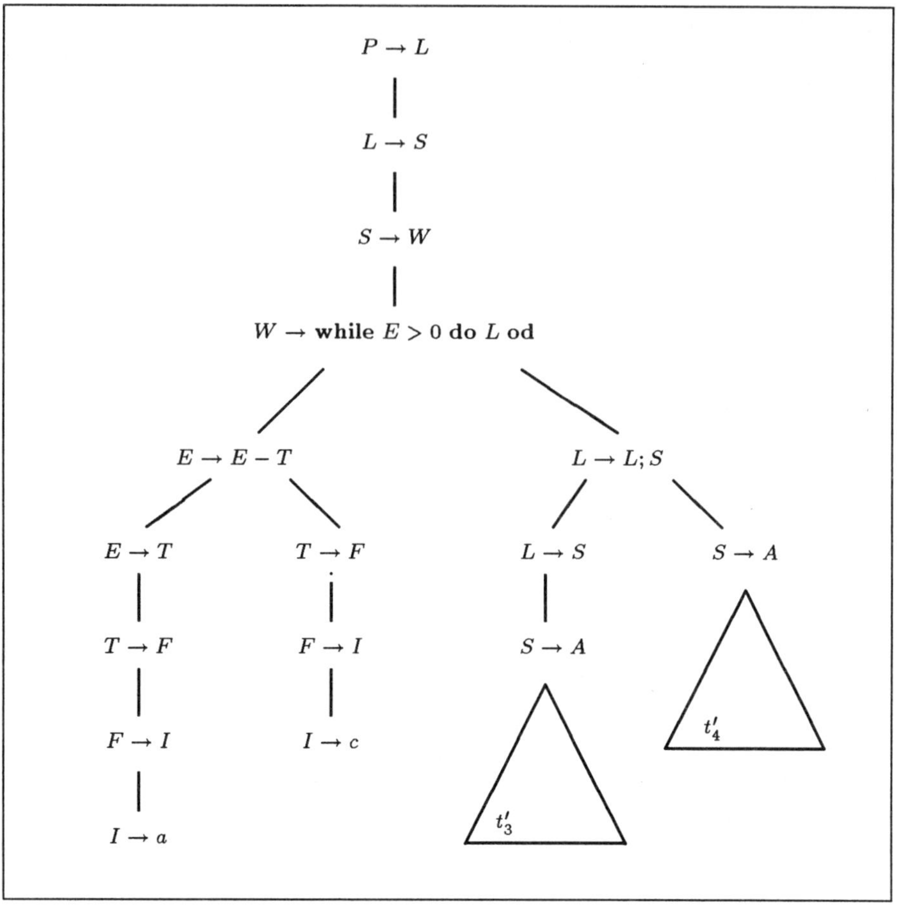

Fig. 1.25. The abstract syntax tree of *exp*

evaluators, we consider from now on only those formal models which accept arbitrary trees, i.e., trees over an arbitrary ranked alphabet as input. The syntax-directed semantics evaluator then becomes a *tree-to-value evaluator* which takes a tree as input and computes a value. The resulting scenario is shown in Fig. 1.28. In this sense we have abstracted from derivation trees.

In fact, the form of our function procedures does not even have to change. We just consider a production as a symbol which has a particular arity or rank. Note that there are no longer any sorts requiring that productions should fit in the usual way when they are glued together.

Abstracting from the semantic domain

Our investigations about formal models should also not depend on particular properties of the chosen semantic domain. For instance, we do not want

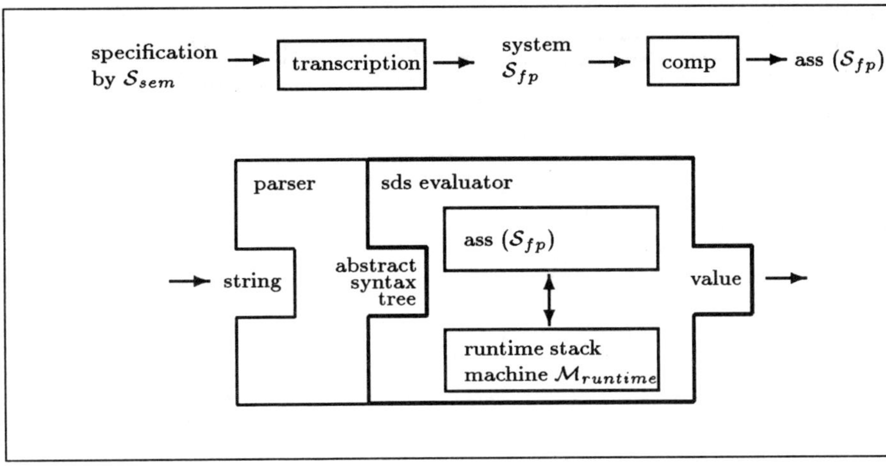

Fig. 1.26. The sds evaluator working on abstract syntax trees

func $code_1\text{-}eval(d\text{: node; } y_1\text{-}par\text{: } V_{temp})\text{: } V_{code}$;
begin
  case label of d **of**
    p_{12}: $code_1\text{-}eval := code_1\text{-}eval(d\,1, y_1\text{-}par)$
      $code_1\text{-}eval(d\,2, y_1\text{-}par + length\text{-}eval(d\,1))$;
    p_{13}: $code_1\text{-}eval := code_1\text{-}eval(d\,1, y_1\text{-}par)$;
    p_{14}: $code_1\text{-}eval := code_1\text{-}eval(d\,1, y_1\text{-}par)$;
    p_{16}: $code_1\text{-}eval := code\text{-}eval(d\,1)$
      JNP $(y_1\text{-}par + length\text{-}eval(d\,1) +$
        $length\text{-}eval(d\,2) + 2)$
      $code_1\text{-}eval(d\,2, y_1\text{-}par + length\text{-}eval(d\,1) + 1)$
      JUMP $y_1\text{-}par$;
  end
end $code_1\text{-}eval$

Fig. 1.27. The function procedure $code_1\text{-}eval$ working on abstract syntax trees

to deal with algebraic laws like associativity or commutativity, or with the question concerning whether the whole arithmetic on natural numbers can be performed in this model. Furthermore, we would like to compare different formal models, and the comparison should not depend on their semantic domain. Therefore we abstract from the semantic domain in the same way as in the theory of program schemes, and we choose the initial term algebra as the semantic domain. The tree-to-value evaluator then changes into a *tree-to-tree evaluator* which computes a tree. This is in fact not a restriction because, for every semantic domain D, there is a unique homomorphism which maps

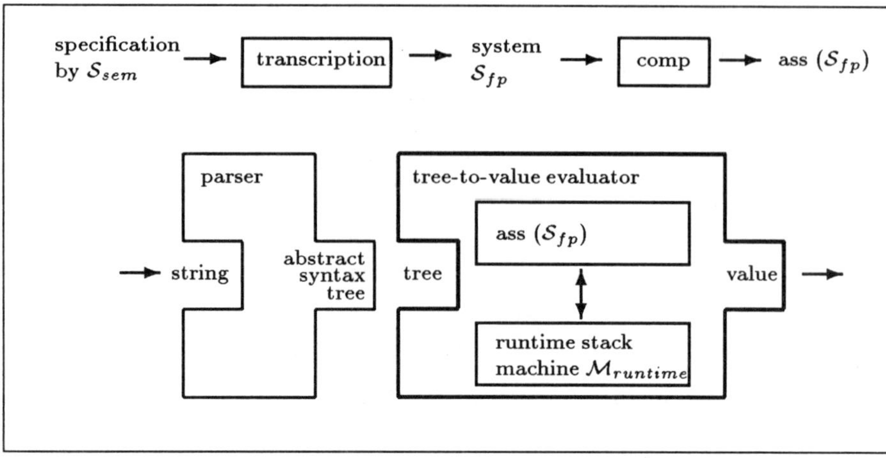

Fig. 1.28. The tree-to-value evaluator

trees over operation symbols of D to values of D. The homomorphism, so to speak, interprets the tree. The resulting scenario is shown in Fig. 1.29.

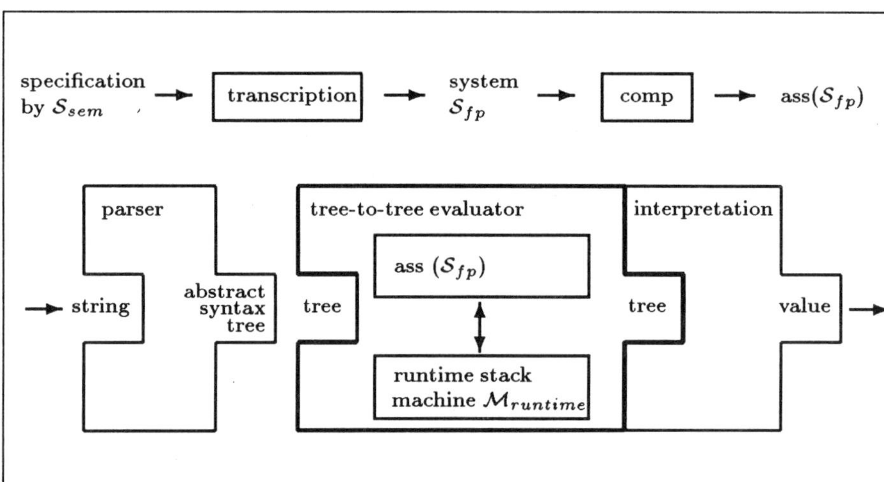

Fig. 1.29. The tree-to-tree evaluator

Now the function procedures have to change again slightly, because the right-hand sides of the case clauses are trees. We show an example in Fig. 1.30 which is derived from the function procedure in Fig. 1.27. We use the dot "." as a binary symbol which concatenates an assembly instruction with a list of assembly instructions. Note also + occurs in a prefix form.

```
func code₁-eval(d: node; y₁-par: V_temp): V_code;
begin
   case label of d of
      p12: code₁-eval := ·(code₁-eval(d 1, y₁-par),
              code₁-eval(d 2, +(y₁-par, lenght-eval(d 1))));
      p13: code₁-eval := code₁-eval(d 1, y₁-par);
      p14: code₁-eval := code₁-eval(d 1, y₁-par);
      p16: code₁-eval := ·(code-eval(d 1),
              ·(JNP +(y₁-par, +(length-eval(d 1), +
                      (length-eval(d 2), 2))),
              ·(code₁-eval(d 2, +(y₁-par, +(length-eval(d 1), 1))),
                      JUMP y₁-par)));
   end
end code₁-eval
```

Fig. 1.30. The function procedure $code_1$-*eval* working on arbitrary trees and producing trees

Abstracting from the machine-oriented computation paradigm

As discussed in the last subsection of Section 1.1, we have assumed a machine-oriented computation paradigm. We realize that, in order to let any abstract machinery calculate the semantic value of some string w, we first have to define an abstract runtime stack machine $\mathcal{M}_{runtime}$ which keeps track of the incarnations of the function procedures, and a compiler *comp* which translates a system $\mathcal{S}_{fp}$ of recursive function procedures into an assembly program $ass(\mathcal{S}_{fp})$ for the machine $\mathcal{M}_{runtime}$. But again, we are not interested in implementation issues and particular techniques for handling recursive procedures on a machine. Hence, we try to replace the machine-oriented computation paradigm by a more abstract computation paradigm.

Here we have chosen the framework of a derivation system (or: abstract reduction system). In general, a derivation system consists of a set A of intermediate results and a binary relation $\Rightarrow \subseteq A \times A$; this relation is called a derivation relation. It defines how one intermediate result is transformed to another intermediate result which is somehow "closer" to the final result. The relation $\Rightarrow$ does not have to be deterministic, confluent, or Noetherian. Thus, our scenario changes again and the tree-to-tree evaluator is replaced by a *tree-to-tree transducer* (Fig. 1.31). In this book, tree-to-tree transducers are referred to as just tree transducers.

In our context the derivation relation $\Rightarrow$ is based on a set R of rules and thus is denoted by $\Rightarrow_R$. The rules themselves are obtained by an easy transcription of the semantic equations. To illustrate this, reconsider the function procedure $code_1$-*eval* (Fig. 1.30). The body of $code_1$-*eval* is a case statement

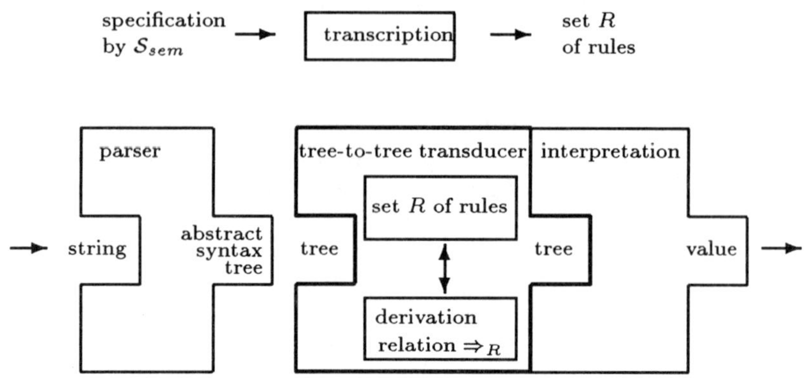

Fig. 1.31. The tree-to-tree transducer

in which the transcription of the $(code_1, X)$-equations of some production p is collected. Now we dissolve this case statement again and, for every clause (i.e., semantic equation), we construct a rule. In particular, for $code_1$-$eval$ we obtain the rules which are shown in Fig. 1.32.

$$
\begin{array}{rcl}
code_1(p_{12}(x_1, x_2), y_1) & \rightarrow & \cdot(code_1(x_1, y_1), \\
 & & code_1(x_2, +(y_1, length(x_1)))) \\
code_1(p_{13}(x_1), y_1) & \rightarrow & code_1(x_1, y_1) \\
code_1(p_{14}(x_1), y_1) & \rightarrow & code_1(x_1, y_1) \\
code_1(p_{16}(x_1, x_2), y_1) & \rightarrow & \cdot(code(x_1), \\
 & & \cdot(\mathrm{JNP}{+}(y_1, +(length(x_1), +(length(x_2), 2))), \\
 & & \cdot(code_1(x_2, +(y_1, +(length(x_1), 1))), \\
 & & \mathrm{JUMP}y_1)))
\end{array}
$$

Fig. 1.32. The rules for $code_1$

The result of the three abstraction steps

If we apply the abstraction steps of the three previous subsections to the formal models which were considered in Section 1.1, i.e., to

- attribute grammars with synthesized attributes only,
- denotational semantics (of a particular type, as mentioned), and
- attribute grammars,

then we obtain corresponding tree transducers:

- top-down tree transducers,
- macro tree transducers, and
- attributed tree transducers,

respectively. Figure 1.33 gives an intuitive idea about the relationship between these tree transducer types. That is, starting from top-down tree transducers and adding parameters (i.e., implicit handling of context), macro tree transducers are obtained, whereas adding inherited attributes (i.e., explicit handling of context) attributed tree transducers are obtained. Adding both, parameters and inherited attributes to top-down tree transducers, leads to macro attributed tree transducers. In fact, these four types of tree transducers are the subject of the investigations in this book; they are

formal models of syntax-directed semantics.

In the following let *TOP*, *MAC*, *ATT*, and *MAT* denote the classes of tree-to-tree transformations (for short: tree transformations) which are induced (or computed) by top-down tree transducers, macro tree transducers, attributed tree transducers, and macro attributed tree transducers, respectively.

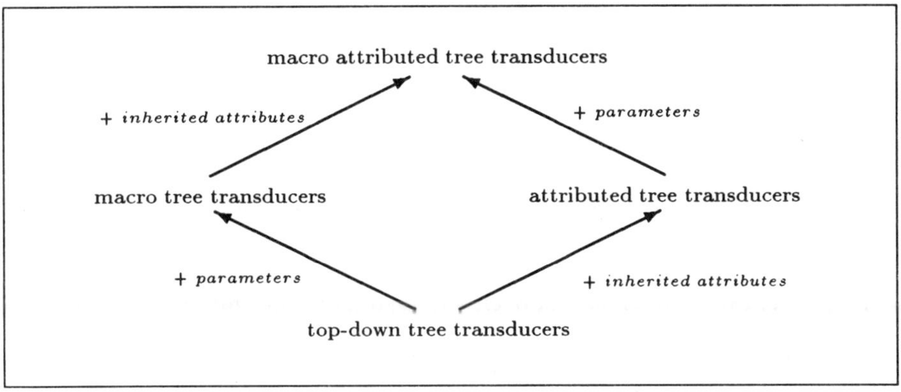

Fig. 1.33. The formal models considered and their syntactic relationship

1.3 Theory of Compositions of Classes of Tree Transformations

In the previous two sections we have concluded that tree transducers are formal models of syntax-directed semantics. In other words, every type of tree transducer can be considered as a specification language for syntax-directed semantics.

As usual in software engineering, it is desirable that a specification language is compositional in the sense that, if the specification is split up into several phases, the specification language should support the separate programming and testing of the phases. Moreover, it should be easy to integrate the phase specifications into one specification.

Formulating this in terms of tree transducers, we obtain the situation shown in the upper part of Fig. 1.34: the desired specification of syntax-directed semantics is composed of n subspecifications, and the i-th subspecification is written in the specification language U_i where U_i is a type of tree transducer. Then, the output of the i-th subspecification is the input of the $(i+1)$-th subspecification. In other words, the upper part of Fig. 1.34 shows the functional composition of n subspecifications.

In this setting, it is advantageous to know whether the composition of two consecutive subspecifications which are written as tree transducers of type U_i and U_{i+1}, respectively, can be specified by one tree transducer of some type U (see the lower part of Fig. 1.34). We support this observation with four reasons:

- Because of space restrictions, it might be desirable to avoid the computation and maintenance of the intermediate result after the i-th subspecification. In this respect, the knowledge about the composability of two consecutive subspecifications has been beneficially integrated, e.g., in the tool for compiler specification given in [GGV86] (also cf. [Gie88]).
- From the complexity of the subspecifications, one might wish to deduce the complexity of the whole specification. Here the complexity is measured by the membership of a specification in a particular class of tree transducers.
- Assume that there are program transformations available (like fold and unfold or the elimination of recursion) which are applicable to tree transducers of type U. Then one might wish to compose two consecutive subspecifications into one specification written in U, and to apply these transformations to the resulting specification and test the result for its efficiency. Such a program transformation has been successfully investigated for macro tree transducers in [MV95a, MV95b]. There, an automatic transformation strategy is defined which yields, for every given macro tree transducer M, a tree transducer M'. It has been proved that (1) M' is at least as efficient as M, (2) there is an algorithm which decides whether M' is more efficient than M, and (3) there are infinitely many macro tree transducers for which this transformation yields a more efficient tree transducer.
- There are very efficient implementations of tree transducers of type U. We refer the reader to [ERS80], [EV86], [Fil83] for machine characterizations of top-down tree transducers, macro tree transducers, and attributed tree transducers, respectively. In a certain abstract sense, these machines are the most efficient implementations, because their data structure cannot be simplified without losing the possibility of implementing every transducer

of the particular type. Now, if one knows that the composition of subspecifications of U_i and U_{i+1} yields a specification in U, then one can use the machine corresponding to the class U.

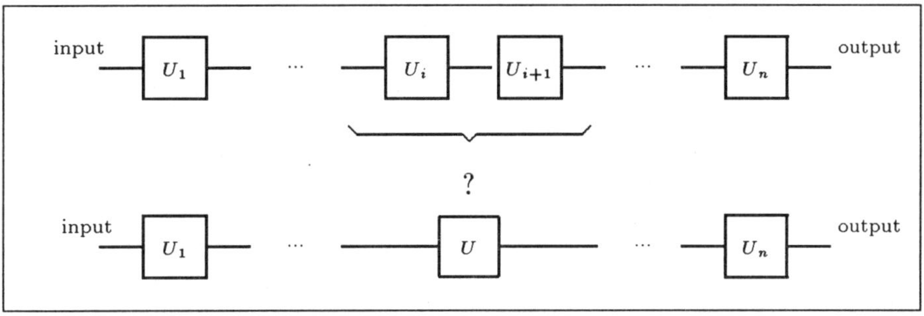

Fig. 1.34. Specification of syntax-directed semantics in n phases by tree transducers of type $U_1, \ldots, U_i, U_{i+1}, \ldots U_n$, respectively

These reasons motivate the study of the composition of tree transformations which are elements of the classes TOP, MAC, ATT and MAT. Typically our investigations concern the comparison of two classes

$$U_1 \circ \ldots \circ U_m \text{ and } V_1 \circ \ldots \circ V_n$$

of tree transformations where $m, n \geq 1$, $U_1, \ldots, U_m, V_1, \ldots, V_n$ are elements of the set $\{TOP, MAC, ATT, MAT\}$, and $\circ$ denotes the composition of tree transformations in the usual sense: if τ_1 and τ_2 are tree transformations, then $\tau_1 \circ \tau_2$ is a tree transformation such that for every tree t, $(\tau_1 \circ \tau_2)(t) = \tau_2(\tau_1(t))$; this composition of tree transformations is extended in an obvious way to classes of tree transformations. Thus, $U = U_1 \circ \ldots \circ U_m$ and $V = V_1 \circ \ldots \circ V_n$ are two classes of tree transformations.

With respect to the comparison of U and V, one of the following four cases is true:

- $U = V$,
- $U \subset V$,
- $U \supset V$, or
- U and V are incomparable.

If $U \subseteq V$, $m = 1$, $n > 1$, and the V_i's are "small" classes, then we quote this comparison as a *decomposition result*, because it shows how transducers of type U can be decomposed into transducers of simpler types. If $U \subseteq V$, $m > 1$, $n = 1$, and the U_i's are "large" classes, then we quote this comparison as a *composition result*. Clearly, in the case of equality we talk about a *characterization result*.

We establish certain techniques based on height properties and path languages to test whether $V - U \neq \emptyset$. This inequality together with $U \subseteq V$ yields $U \subset V$.

We consider two particular sets H_1 and H_2 of classes of tree transformations; more precisely, $H_1 = \{TOP, l\text{-}TOP, HOM\}$ where $l\text{-}TOP$ and HOM are the classes of tree transformations induced by *linear* top-down tree transducers and homomorphism tree transducers, respectively, and $H_2 = \{TOP, ATT, MAC\}$. For each of these sets we develop an algorithm which decides, for every two classes $U = U_1 \circ \ldots \circ U_m$ and $V = V_1 \circ \ldots \circ V_n$ of tree transformations such that $U_1, \ldots, U_m, V_1, \ldots, V_n \in H_i$ (for $i = 1$ or $i = 2$), which of the four cases mentioned above is true. In other words, for every $i \in \{1, 2\}$, we examine the semigroup generated by H_i with composition as the operation. We call this the *composition semigroup* based on H_i.

The shape of the developed algorithms is by no means restricted to the two particular base sets H_1 and H_2. It has been successfully applied to other such sets, cf. Sect. 3.8.

1.4 Other Formal Models

The previous sections indicate the formal models which we are concerned with in this book. Certainly, there are also other formal models which can be quoted as being syntax-directed. They can be classified according to the way in which they handle the input tree, i.e., the tree which directs the definition of the semantics. There are

- top-down models; here the input trees are processed by starting at the root and proceeding towards the leaves; e.g., generalized syntax-directed translation schemes [AU69a, AU69b, AU71], pushdown tree transducers [Gue83], graph translation schemes [Son87], high-level tree transducers [EV88], modular tree transducers [EV91], downwards accumulation [Gib91, Gib93], catamorphisms [Bir86, Mee86, Fok92], tree transducers with external functions [FHVV93], and top-down tree-to-graph transducers [EV94];
- bottom-up models; here the input trees are treated by starting at the leaves and proceeding towards the root; e.g., bottom-up tree transducers [Eng75], tree pushdown transducers [SG85], upwards accumulation [Gib91, Gib93], anamorphisms [Bir86, Mee86, Fok92], and bottom-up tree-to-graph transducers [EV96]; and
- tree-walking models; here the input trees are treated by starting at the root and by performing an arbitrary tree walk on them; e.g., pushdown assembler machines [LRS74], checking-tree pushdown transducers [ERS80], higher order attribute grammars [KSV89], and attributed nested stack tree transducers [FV92a].

With respect to this classification, top-down tree transducers and macro tree transducers are top-down models, and attributed tree transducers and macro attributed tree transducers are tree-walking models.

1.5 Outline of the Book

This book contains eight chapters, a list of figures, a bibliography, and an index. The last section of every chapter is devoted to bibliographic notes about the topics treated in that chapter.

After this introductory chapter, we collect in Chap. 2 all the basic notions and notations which are used in this book. Chapter 2 can be skipped on first reading.

In Chaps. 3, 4, 5, and 7 we develop the concepts of top-down tree transducers, macro tree transducers, attributed tree transducers, and macro attributed tree transducers, respectively. For every type of tree transducer the corresponding chapter has at least the following sections (not necessarily in this order):

- a section with *basic definitions* in which the tree transducer model is defined,
- a section about the *induced tree transformation* in which the derivation relation and the tree transformation computed by this type of tree transducer is defined,
- a section about an *inductive characterization* of the tree transformation,
- a section stating *height properties* in which the relationship between the heights of input trees and corresponding output trees is investigated, and
- a section with *composition and decomposition results* which involves the class of tree transformations computed by this type of tree transducer.

Additionally, Chapter 3 contains sections about different subclasses of TOP, an inclusion diagram of these subclasses, and a composition semigroup generated by the set of these subclasses.

Chapter 6 contains a detailed comparison of macro tree transducers and attributed tree transducers including the discussion about various restrictions. Moreover, it contains an inclusion diagram of the classes TOP, MAC, ATT, and the subclasses computed by the restrictions. At the end of Chap. 6 we investigate the composition semigroup generated by TOP, MAC, and ATT.

Finally, Chap. 8 gives two more complicated examples which are motivated by practical problems. The first example deals with the contextual analysis in block-structured programming languages, the second shows how insertion in 2–3 trees can be specified by means of a macro attributed tree transducer.

1.6 Bibliographic Notes

The concept of syntax-directed semantics goes back to [Iro61]. The development of Sects. 1.1 - 1.3 has been deeply influenced by [Eng81]. The algorithms for computing the values of meaning names and context names are taken from [Eng84]. The idea for our code semantics example is due to [AU73].

Attribute grammars were first introduced in [Knu68] and then studied intensively, for surveys cf., e.g., [DJL88, DJ90, AM91, Sch95, KV97, Paa95].

The concept of denotational semantics is due to [Str66, SS71]. It was further established in [Sto77]. Also cf. [Gor79, Ten91, Wat91, Win93, GM96]. For a survey paper see [Mos90].

The technique of keeping track of and maintaining calls of recursive function procedures is a subject treated in compiler theory, cf., e.g., [WG84, ASU86, Kas90, WM97].

For surveys and books about the theory of program schemes cf., e.g., [Pat72, Man73, CM73, Man74, Eng74, Gre75, Cou90, Nol95].

In [Hue80], several important, abstract properties of derivation systems (there called: abstract reduction systems) are proved.

For the origins of top-down tree transducers, macro tree transducers, attributed tree transducers, and macro attributed tree transducers we refer the reader to the bibliographic notes in Chaps. 3, 4, 5, and 7, respectively. Here we only mention the survey works [GS84, GS97].

For articles about program transformations, we refer the reader to [BD77, PP86, Par90].

For the theory of compositions of tree transformations, we also refer to [Küh97] in which several subclasses of MAC and ATT have been investigated from this point of view. In particular, subclasses based on the "syntactic single used requirement" of Ganzinger and Giegerich [Gan83, GG84, Gie88] are studied.

2. Basic Notions and Notations

2.1 Sets and Relations

We denote by $\mathbb{N}$ the set of nonnegative integers.

Let A and B be two sets. We denote by $card(A)$ the cardinality of A, by $\mathcal{P}(A)$ the power set of A, by $A \subseteq B$ that A is a subset of B, by $A \subset B$ that A is a proper subset of B, and by $A \not\subseteq B$ that A is not a subset of B. Moreover, $A \bowtie B$ stands for the fact that A and B are incomparable with respect to inclusion, i.e., that neither $A \subseteq B$ nor $B \subseteq A$ holds.

Given two sets A and B, any subset θ of the Cartesian product $A \times B$ is called a *relation* from A to B. We write $a\theta b$ to mean that $(a, b) \in \theta$. The *inverse* of θ is the relation $\theta^{-1} = \{(b, a) \mid (a, b) \in \theta\}$ from B to A.

Let $\theta \subseteq A \times B$ be a relation. If for every $a \in A$, there is exactly one $b \in B$ such that $a\theta b$, then θ is called a *function* or alternatively *a mapping from A to B*. The fact that θ is a function from A to B is denoted by $\theta : A \rightarrow B$. If θ is a function and $a\theta b$, then we also write $\theta(a) = b$. Moreover, for $B \subseteq A$ we put

$$\theta(B) = \{\, \theta(b) \mid b \in B \,\}.$$

The *range* of a function θ is denoted by $range(\theta)$ and it is defined by the equation $range(\theta) = \theta(A)$.

We now introduce the notion of *composition* for relations. Let θ be a relation from A to B and let τ be a relation from B to C. Then the *composition* of θ and τ is the relation $\theta \circ \tau$ from A to C defined by

$$\theta \circ \tau = \{\, (a, c) \mid a\theta b \text{ and } b\tau c \text{ for some } b \in B \,\}.$$

Note that if θ and τ are functions, then $(\theta \circ \tau)(a) = \tau(\theta(a))$.
We extend the concept of composition to classes of relations. If Y and Z are classes of relations, then

$$Y \circ Z = \{\, \theta \circ \tau \mid \theta \in Y \text{ and } \tau \in Z \,\}.$$

Moreover, we define the n-th power of Y as follows. Let $Y^1 = Y$ and for any $n > 1$, let $Y^n = Y \circ Y^{n-1}$.
A relation from A to A is also called a *relation over A*. Let θ be a relation over A. We say that θ is

- *reflexive* if, for every $a \in A$, it holds that $a\theta a$;
- *symmetric* if, for every $a, b \in A$, the condition $a\theta b$ implies $b\theta a$;
- *transitive* if, for every $a, b, c \in A$, the conditions $a\theta b$ and $b\theta c$ imply $a\theta c$.

If θ is reflexive, symmetric, and transitive, then it is an *equivalence relation* over A. For an element $a \in A$, the *equivalence class containing* a is the set $a/\theta = \{b \in A \mid a\theta b\}$. If θ is an equivalence relation over A, then a set $N \subseteq A$ is said to be a *set of representatives* for θ if, for every $a \in A$, there exists exactly one $b \in N$ such that $a\theta b$.

The *identical relation* $\{(a, a) \mid a \in A\}$ over A is denoted by $Id(A)$.

The *n-fold composition* of θ, denoted by θ^n, is defined by $\theta^0 = Id(A)$ and $\theta^n = \theta \circ \theta^{n-1}$, if $n > 0$. We introduce the notion $\theta^* = \bigcup_{n \geq 0} \theta^n$. The relation θ^* is called the reflexive, transitive closure of θ, meaning that it is the smallest reflexive and transitive relation that contains θ.

2.2 Partial Orders

A relation θ over a set A is *antisymmetric* if, for every $a, b \in A$, the conditions $a\theta b$ and $b\theta a$ imply $a = b$. A reflexive, antisymmetric, and transitive relation is called a *partial order*. A set A with a partial order θ over it is called a *partially ordered set*. Let A be a partially ordered set with the partial order θ. If an element $a \in A$ is such that, for every $b \in A$, the property $b\theta a$ implies $a = b$, then a is called a *minimal element of* A. Moreover, two elements a and b of A are called *incomparable* if neither $a\theta b$ nor $b\theta a$ holds for them.

Inclusion diagrams

Any finite partially ordered set has the nice property that it can be visualized by drawing its *Hasse diagram*. The Hasse diagram is in fact a graph presented as follows.

Let A be a finite set with the partial order $\leq$ on it. Let us represent every element a of A by a node which is labeled by a. We will call this node a-node. Draw the nodes representing the elements of A in such a way that whenever $a \leq b$ holds for two different elements a and b of A, the b-node is drawn above the a-node. Moreover, for every $a, b \in A$, connect the a-node and the b-node by an edge if a and b satisfy the conditions that $a \leq b$ and, for every $c \in A$, $a \leq c \leq b$ implies that $a = c$ or $b = c$. From this diagram we can reconstruct $\leq$ by noting that $a \leq b$ holds if and only if $a = b$ or the b-node is above the a-node and we can go from the b-node to the a-node via a sequence of descending edges.

In general, infinite partially ordered sets cannot be visualized so obviously. However those appearing in this book will be so special that one should be able to reconstruct them by having a finite piece of their Hasse diagram.

We will especially consider partially ordered sets of which the elements are classes and the partial order is the inclusion relation $\subseteq$. The Hasse diagram of such a partially ordered set will be called the *inclusion diagram.* So whenever we say inclusion diagram, it should be clear that we mean the Hasse diagram of a set H of classes, with the inclusion as a partial order over H.

To give an example, let A, B, C, and D be classes and let $\subseteq$ be the relation $\{(B, A), (B, C), (D, C)\}$. Then the inclusion diagram is shown is Fig. 2.1.

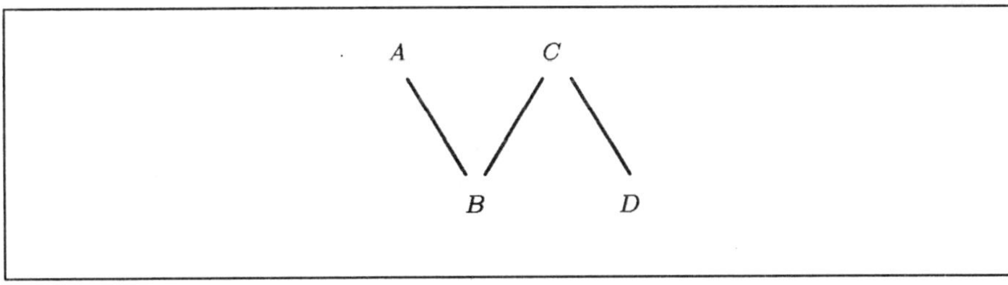

Fig. 2.1. An example of an inclusion diagram

We now present a method concerning inclusion diagrams which will be used later in the book. Assume that there is a finite set H of classes and that the elements of H are known to us. We would like to construct the inclusion diagram of H. However, sometimes we are not sure about all inclusion relations among the elements of H. In such a case, we can do the following. We make a conjecture about the inclusion diagram and we try to prove that the conjectured diagram is the inclusion diagram of H. By a conjectured diagram we mean a visualized graph whose nodes are labeled by elements of H and there may be edges between the nodes such that an edge can connect two nodes only if one of them is above the other. The proof of the claim that it is the inclusion diagram of H is in fact the proof of the claim that the inclusion relations are those and only those which are shown by the conjectured diagram. It should be clear that this implies the proof of the following three statements for every $A, B \in H$.

- If the A-node and the B-node are situated in the diagram such that the B-node is above the A-node and there is an edge from the B-node to the A-node, then the inclusion $A \subseteq B$ and the inequality $B - A \neq \emptyset$ hold.
- If the A-node and the B-node are the same, (i.e., A and B are conjectured to be equal), then $A \subseteq B$ and $B \subseteq A$ hold.
- If there is no descending path between the A-node and the B-node (i.e., A and B are guessed to be incomparable), then the inequalities $A - B \neq \emptyset$ and $B - A \neq \emptyset$ hold.

Hence, given a conjecture for the inclusion diagram of H, we can construct a finite number of inclusions and inequalities whose validity is necessary and sufficient to verify that our conjecture is the inclusion diagram of H. However, this finite number may be large even if the size of the conjectured diagram is relatively small. Therefore it would be nice to reduce this number. Unfortunately nothing can be done about the number of inclusions because we should verify all inclusions. However, the number of inequalities can sometimes be reduced using the observation presented below.

With the following procedure, we can construct the set $INCL$ of formal inclusions and the set $INEQ$ of formal inequalities from the conjectured diagram where "formal" means that the inclusions and inequalities are considered as formulas which are either true or false in H.

For every $A, B \in H$,

- If the A-node and the B-node are situated in the diagram such that the B-node is above the A-node and there is an edge from the B-node to the A-node, then let $A \subseteq B \in INCL$ and $B - A \neq \emptyset \in INEQ$.
- If the A-node is the same as the B-node, then let $A \subseteq B \in INCL$ and $B \subseteq A \in INCL$.
- If there is no descending path between the A-node and the B-node in the diagram, then let $A - B \neq \emptyset \in INEQ$ and $B - A \neq \emptyset \in INEQ$.

As we mentioned above, the conjectured diagram is the inclusion diagram of H if and only if all formulas in $INCL$ and $INEQ$ are valid.

Now assume that we could verify all formulas in $INCL$. We observe that the number of formal inequalities to be verified may be reduced. If we can verify $A - B \neq \emptyset \in INEQ$ and there is another formula $A' - B' \neq \emptyset \in INEQ$ such that $A \subseteq A'$ and $B' \subseteq B$ hold due to previously verified inclusion relations in $INCL$, then $A' - B' \neq \emptyset$ is also verified.

Hence we can define the relation $\leq$ over the set $INEQ$ such that

$$(A - B \neq \emptyset) \leq (A' - B' \neq \emptyset) \quad (*)$$

if $A \subseteq A'$ and $B' \subseteq B$ hold. We can easily see that $\leq$ is a partial order over $INEQ$. Moreover, by the above reasoning, for any formal inequalities $g, g' \in INEQ$, if $g \leq g'$ and g is valid, then g' is also valid. Thus it is sufficient to prove the formal inequalities which are minimal with respect to $\leq$.

Hence, we can construct the inclusion diagram of a finite set H of classes with the partial order $\subseteq$ applying the following method.

Algorithm 2.1. The construction of the inclusion diagram of a finite set H of classes.

1. Conjecture the inclusion diagram of H.
2. Construct the sets of formal inclusions $INCL$ and formal inequalities $INEQ$ from the conjectured diagram applying the three steps described above.
3. Verify that all formulas in $INCL$ hold.

4. Construct the set $MINEQ$ of minimal elements of $INEQ$ with respect to $\leq$. (The construction of $MINEQ$ can be made by simple manual calculation or, when there are too many formulas in $INEQ$, by invoking a standard algorithm which computes the minimal elements of a finite partially ordered set.)
5. Verify that all formulas in $MINEQ$ hold. □

Let us apply this algorithm to an example and assume that Fig. 2.1 shows a conjectured diagram. Then $INCL$ contains the following formal inclusions:

$$B \subseteq A, \; B \subseteq C, \; D \subseteq C$$

and the set $INEQ$ contains the following nine formal inequalities:

$$\begin{array}{lll} A - B \neq \emptyset & C - B \neq \emptyset & C - D \neq \emptyset \\ B - D \neq \emptyset & D - B \neq \emptyset & A - D \neq \emptyset \\ D - A \neq \emptyset & A - C \neq \emptyset & C - A \neq \emptyset \end{array}$$

Then, for example, $D - A \neq \emptyset \leq C - A \neq \emptyset$ and $C - A \neq \emptyset \leq C - B \neq \emptyset$. Next, we can compute the set $MINEQ$ which consists only of the following five formal inequalities:

$$\begin{array}{lll} A - B \neq \emptyset & C - D \neq \emptyset & B - D \neq \emptyset \\ D - A \neq \emptyset & A - C \neq \emptyset & \end{array}$$

Topological sorting

A partial order $\leq$ on a set A is called *linear order* if in addition, for every $a, b \in A$, $a \leq b$ or $b \leq a$ holds. A set with a linear order on it is called *linearly ordered set*.

Clearly, if A is a finite set, then every linear order $\leq_l$ on A can be represented as a sequence $\mu = a_1 a_2 \ldots a_n$ such that $A = \{a_1, a_2, \ldots, a_n\}$ and, for every i with $1 \leq i \leq n-1$, $a_i \leq_l a_{i+1}$.

Every partial order $\leq$ on A can be embedded into a linear order $\leq_l$ such that $\leq$ is a subset of $\leq_l$. The algorithm which constructs such a $\leq_l$ is called *topological sorting*.

Algorithm 2.2. Topological Sorting
Input: a finite set A and a partial order $\leq$ on A
Output: a linear order $\leq_l$ on A such that $\leq$ is a subset of $\leq_l$
Method: The algorithm needs the following variables:

T: subset of $A \times A$
T_l: a sequence of elements of A

The algorithm works as follows:

$T := \leq$;
$T_l :=$ the empty sequence;

<u>while</u> $T \neq \emptyset$ <u>do</u>
choose an element $b \in A$ such that there is no $a \in A$ with $a \neq b$ and $(a, b) \in T$;
$T_l := T_l \cdot b$;
$T := T - \{(b, c) \mid (b, c) \in T\}$
<u>end</u>;

$\leq_l := T_l$.

□

Due to the choice in the body of its while-statement, the topological sorting is a nondeterministic algorithm. This nondeterminism reflects the fact that, for a partial order $\leq$, there may be more than one linear order $\leq_l$ such that $\leq$ is a subset of $\leq_l$; the algorithm constructs one such linear order.

In this book, a frequent instance of the linearly ordered set is the *hierarchy*. A hierarchy is a family of classes $\{ C_k \mid k \geq 0 \}$ such that $C_k \subseteq C_{k+1}$, for every $k \geq 0$. A hierarchy is said to be *proper* if every inclusion is proper, i.e., $C_k \subset C_{k+1}$, for every $k \geq 0$. It should be clear that the inclusion diagram of a hierarchy is a vertical chain.

2.3 Directed Graphs

In the book we will also work with directed graphs. A *directed graph* is a pair $D = (V, E)$, where V is a set, called the set of nodes, and $E \subseteq V \times V$ is the set of edges. If $(a, b) \in E$, then we say that there is an edge from a to b in the graph. (Note that edges are unlabeled and there are no multiple edges between nodes.) A path in the graph is a sequence $a_1, \ldots, a_n$ of nodes such that, for every $1 \leq i < n$, there is an edge from a_i to a_{i+1}, i.e., $(a_i, a_{i+1}) \in E$. Moreover, a cycle in the graph is a path with $a_1 = a_n$.

2.4 Derivation Systems

A *derivation system* (or *reduction system*, see for example [Hue80], [Boo83], and [Klo92]) is a system $(A, \Rightarrow)$, where $\Rightarrow$ is a binary relation over A called a *derivation relation* (or *reduction relation*).

For an arbitrary $a \in A$, a *derivation starting from* a is a sequence $a_1, \ldots, a_n \in A$ for some $n \geq 1$ such that $a = a_1$ and, for every $1 \leq i \leq n - 1$, $a_i \Rightarrow a_{i+1}$. The *length* of this derivation is $n - 1$. Clearly, in general there may be more than one derivation starting from a particular element a. A *derivation* is a derivation starting from some $a \in A$.

An element $a \in A$ is *irreducible (with respect to* $\Rightarrow$*)* if there is no $b \in A$ such that $a \Rightarrow b$. If for some $a \in A$ there is exactly one b such that $a \Rightarrow^* b$ and b is irreducible, then b is called the *normal form of* a, and it is denoted by $nf(\Rightarrow, a)$. We will show that a has this property shortly by saying that $nf(\Rightarrow, a)$ exists.

We put $\Leftrightarrow = (\Rightarrow \cup \Rightarrow^{-1})$. We note that $\Leftrightarrow^*$ is an equivalence relation over A.

We say that

(a) $\Rightarrow$ *is terminating on* A if there is no infinite sequence of the form $a_1 \Rightarrow a_2 \Rightarrow \ldots$ with $a_1, a_2, \ldots \in A$;

(b) $\Rightarrow$ *is locally confluent on* A if, for every $a, b, c \in A$, the derivations $a \Rightarrow b$ and $a \Rightarrow c$ imply that $b \Rightarrow^* d$ and $c \Rightarrow^* d$ for some $d \in A$;

(c) $\Rightarrow$ *is confluent on* A if, for every $a, b, c \in A$, the derivations $a \Rightarrow^* b$ and $a \Rightarrow^* c$ imply that $b \Rightarrow^* d$ and $c \Rightarrow^* d$ for some $d \in A$;

(d) $\Rightarrow$ is *Church-Rosser on* A if, for every $a, b \in A$, the equivalence $a \Leftrightarrow^* b$ implies that $a \Rightarrow^* c$ and $b \Rightarrow^* c$ for some $c \in A$.

Now let $B \subseteq A$. We say that B *is closed under* $\Rightarrow$ if, for every $b \in B$ and $a \in A$, the derivation $b \Rightarrow a$ implies $a \in B$. Note that if $B \subseteq A$ is closed under $\Rightarrow$, then $(B, \Rightarrow)$ is also a derivation system. (Hence, if this is the case, an expression like "$\Rightarrow$ is terminating (locally confluent, confluent, Church-Rosser) on B" should be understood.)

We recall some fundamental results concerning an arbitrary derivation system $(A, \Rightarrow)$ which will be used in the book.

Lemma 2.3. The derivation relation $\Rightarrow$ is confluent on A if and only if it is Church-Rosser on A.

Proof. First suppose that $\Rightarrow$ is Church-Rosser. Let a, b and c be elements of A such that $a \Rightarrow^* b$ and $a \Rightarrow^* c$. Then certainly $b \Leftrightarrow^* c$ and, since $\Rightarrow$ is Church-Rosser, there exists a $d \in A$ with $b \Rightarrow^* d$ and $c \Rightarrow^* d$, which proves that $\Rightarrow$ is confluent.

Next assume that $\Rightarrow$ is confluent and let $a, b \in A$ be such that $a \Leftrightarrow^* b$. That means $a \Leftrightarrow^m b$, for some $m \geq 0$. We prove by an induction on m, that there exists a $c \in A$ with $a \Rightarrow^* c$ and $b \Rightarrow^* c$.

For $m = 0$, we have $a = b$, hence the proof is complete by letting $c = a (= b)$.

Now let $m > 0$. Then there is a b', for which $a \Leftrightarrow^{m-1} b' \Leftrightarrow b$ and thus, by the induction hypothesis, there exists a $c' \in A$ with $a \Rightarrow^* c'$ and $b' \Rightarrow^* c'$. On the other hand, by the definition of $\Leftrightarrow$ we have either $b' \Rightarrow b$ or $b \Rightarrow b'$.

If $b \Rightarrow b'$, then, for $c = c'$, we get $a \Rightarrow^* c$ and $b \Rightarrow b' \Rightarrow^* c$.

If the converse, i.e., $b' \Rightarrow b$ holds, then, since also $b' \Rightarrow^* c'$ and since $\Rightarrow$ is confluent, there exists a $c \in A$ such that $c' \Rightarrow^* c$ and $b \Rightarrow^* c$. Then we get $a \Rightarrow^* c' \Rightarrow^* c$, too, implying that $\Rightarrow$ is Church-Rosser. $\square$

The following lemma can be proved by the so-called *Noetherian induction* of which the principle is stated first.

Principle 2.4. Let $(A, \Rightarrow)$ be a derivation system such that $\Rightarrow$ is terminating on A and let P be a predicate over A. The *principle of proof by Noetherian induction* is the following formula.

If
1. for every irreducible element a the formula $P(a)$ is true and
2. for every $a \in A$, if, for every b with $a \Rightarrow^+ b$, the formula $P(b)$ is true, then the formula $P(a)$ is also true,

then, for every $a \in A$, the formula $P(a)$ is true.

□

The next lemma is usually referred to as *Newman's lemma.*

Lemma 2.5. Let $\Rightarrow$ be terminating on A. Then $\Rightarrow$ is confluent on A if and only if it is locally confluent on A.

Proof. Certainly, if $\Rightarrow$ is confluent, then it is locally confluent, whether $\Rightarrow$ is terminating or not.

Next suppose that $\Rightarrow$ is locally confluent and terminating. By Noetherian induction we prove that, for every $a \in A$, the predicate $P(a)$:

If there exist $b, c \in A$ such that $a \Rightarrow^* b$ and $a \Rightarrow^* c$, then there exists a $d \in A$, for which $b \Rightarrow^* d$ and $c \Rightarrow^* d$.

is true.

First we prove that $P(a)$ is true for every irreducible $a \in A$. In fact, if a is irreducible, then we have $a = b$ and $a = c$ and hence, $P(a)$ is true with $d = a (= b = c)$.

Now let $a \in A$ be an arbitrary element and suppose that for every $b \in A$, with $a \Rightarrow^+ b$, the predicate $P(b)$ is true.

Assume that b and c are such that $a \Rightarrow^* b$ and $a \Rightarrow^* c$, i.e., that $a \Rightarrow^m b$ and $a \Rightarrow^n c$, for some $m, n \geq 0$.

Consider the case $m = 0$. Then by definition $b = a$, and for $d = c$, we have $b = a \Rightarrow^* c = d$ and $c \Rightarrow^* c = d$, proving that $P(a)$ is true.

The case $n = 0$ can be handled similarly.

Next suppose that $m > 0$ and $n > 0$ (Fig. 2.2). Then there exist b' and c' in A such that $a \Rightarrow b' \Rightarrow^{m-1} b$ and $a \Rightarrow c' \Rightarrow^{n-1} c$. Then, using that $\Rightarrow$ is locally confluent, we have a d' with $b' \Rightarrow^* d'$ and $c' \Rightarrow^* d'$. Next we use that, by the Noetherian induction hypothesis, $P(b')$ is true, i.e., there exists a d'' such that $b \Rightarrow^* d''$ and $d' \Rightarrow^* d''$. Note that for d'' we also have $c' \Rightarrow^* d' \Rightarrow^* d''$. Finally, again by the Noetherian induction hypothesis, $P(c')$ is true, hence since $c' \Rightarrow^* d''$ and $c' \Rightarrow^* c$, there exists a $d \in A$, with $d'' \Rightarrow^* d$ and $c \Rightarrow^* d$. Then we get that $P(a)$ is also true, since $b \Rightarrow^* d'' \Rightarrow^* d$ holds. By the principle of Noetherian induction, this proves that $\Rightarrow$ is confluent. □

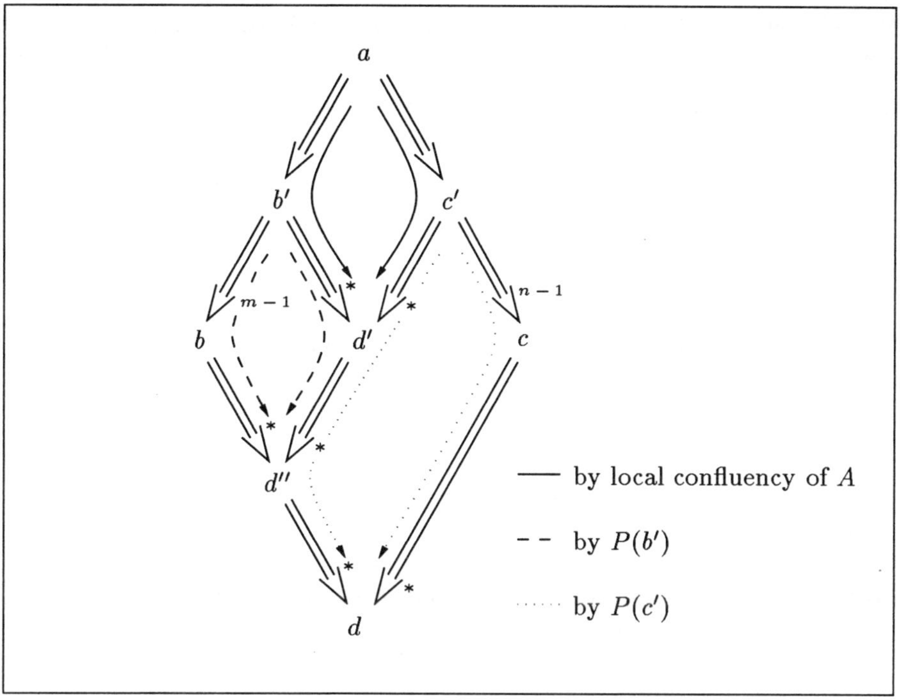

Fig. 2.2. The commuting diamond

Lemma 2.6. Let $\Rightarrow$ be terminating on A. Then $\Rightarrow$ is confluent on A if and only if every class of $\Leftrightarrow^*$ contains exactly one irreducible element.

Proof. First assume $\Rightarrow$ to be confluent. Take an arbitrary $\Leftrightarrow^*$-class and let a be an element of that class. Then either a is irreducible or, by the terminating property, there exists an irreducible b, such that $a \Rightarrow^* b$. Hence $a \Leftrightarrow^* b$ and thus b is an irreducible element in the class of a.

Now suppose that b and c are two irreducible elements in the $\Leftrightarrow^*$-class of a. Then $b \Leftrightarrow^* c$. Since, by Lemma 2.3, $\Rightarrow$ is Church-Rosser on A, we have $b \Rightarrow^* d$ and $c \Rightarrow^* d$, for some $d \in A$. However b and c are irreducible, hence $b = d$ and $c = d$ and thus $b = c$.

Next assume that each $\Leftrightarrow^*$-class contains exactly one irreducible element. Moreover, let $a, b, c \in A$ be such that $a \Rightarrow^* b$ and $a \Rightarrow^* c$. Since $\Rightarrow$ is terminating, there are irreducible elements b' and c' such that $b \Rightarrow^* b'$ and $c \Rightarrow^* c'$. Then also $b' \Leftrightarrow^* c'$, and hence $b' = c'$ implying that $\Rightarrow$ is confluent. □

Corollary 2.7. Let $\Rightarrow$ be terminating and confluent on A. Then, for every $a \in A$, the normal form $nf(\Rightarrow, a)$ exists.

Proof. Let $a \in A$. Then $nf(\Rightarrow, a) = b$, where b is the unique irreducible element in the $\Leftrightarrow^*$-class of a. □

Corollary 2.8. Let $\Rightarrow$ be terminating and confluent on A. Moreover, let $a, b \in A$, such that $a \Rightarrow^* b$. Then, $nf(\Rightarrow, a) = nf(\Rightarrow, b)$.

Proof. Since $a \Rightarrow^* b$, we also have $a \Leftrightarrow^* b$. Hence, by Lemma 2.6, both $nf(\Rightarrow, a)$ and $nf(\Rightarrow, b)$ are equal to the unique irreducible element in the $\Leftrightarrow^*$-class of a (and of b). □

2.5 Semigroups, Strings and Languages

We recall some basic notions and notations concerning semigroups.

A *semigroup* is a system $(M, \cdot)$, where M is an arbitrary set and $\cdot : M \times M \rightarrow M$ is an *associative* binary operation over M meaning that, for all elements $a, b, c \in M$, the equation $(a \cdot b) \cdot c = a \cdot (b \cdot c)$ holds. When the operation of a semigroup $(M, \cdot)$ is clear, then we drop it and write simply M.

Let $(M, \cdot)$ be an arbitrary semigroup and let θ be an equivalence relation over M. Then we say that θ is a *congruence relation* (or *congruence* for short), if for all elements $a, a', b, b' \in M$, the property $a\theta a'$ and $b\theta b'$ imply that $(a \cdot b)\theta(a' \cdot b')$. If θ is a congruence on M, then $M/\theta = \{a/\theta \mid a \in M\}$ is also a semigroup, called the *factor semigroup of M determined by θ*, where we define $a/\theta \cdot b/\theta = (a \cdot b)/\theta$.

Now let $(M, \cdot)$ and $(N, \circ)$ be two semigroups and let $h : M \rightarrow N$ be a mapping. Then h is a *homomorphism from $(M, \cdot)$ to $(N, \circ)$* if for all $a, b \in M$, the equation $h(a \cdot b) = h(a) \circ h(b)$ holds. If h is surjective, i.e., $h(M) = N$, then *N is the homomorphic image of M*. If N is the homomorphic image of M and h is a bijection, then we say that *M and N are isomorphic* and we write $(M, \cdot) \cong (N, \circ)$ or rather $M \cong N$.

For every homomorphism $h : M \rightarrow N$, the equivalence relation θ_h over M defined by $a\theta_h b$ if and only if $h(a) = h(b)$, for all $a, b \in M$, is a congruence as well. It is called the *kernel of h*. Moreover, if h is surjective, then $M/\theta_h \cong N$.

Let $(M, \cdot)$ be a semigroup. Then every subset $H \subseteq M$ *generates* a subset $[H] = \{a_1 \cdot \ldots \cdot a_m \mid m \geq 1, a_i \in H, \text{ for } 1 \leq i \leq m\}$ of M. If, especially, $[H] = M$, then we say that *H generates M*, or alternatively, that *H is a set of generators of M*.

Now we introduce a special semigroup called *the semigroup of strings* or *words*. Let A be a nonempty, finite set of symbols, called an *alphabet*. A *string* or *word* over A is a sequence of symbols of the form $a_1 \ldots a_m$, where $m \geq 1$ and $a_1, \ldots, a_m \in A$. Let A^+ denote the set of all strings over A. Then A^+ is a semigroup with the operation *concatenation* of strings, which for every pair of strings $w, z \in A^+$, yields the string $wz \in A^+$. Obviously, the concatenation is associative. It is also clear that A generates A^+.

It is a characteristic of the semigroup A^+ that for every other semigroup M and mapping $f : A \rightarrow M$, there is a unique homomorphism $h : A^+ \rightarrow M$

which extends f, i.e., for which $f(a) = h(a)$, for every $a \in A$. Hence, A^+ is called the free semigroup generated by A.

Finally, we introduce the concept of a monoid. An element e of an arbitrary semigroup $(M, \cdot)$ is called a *unit element* if, for every $a \in A$, the equalities $e \cdot a = a \cdot e = a$ hold. It can easily be seen that, if a unit element exists, then it is unique. A *monoid* is a semigroup with a unit element.

Let A be an arbitrary alphabet. Let ε be the empty string over A, i.e., the string which contains no element of A, and let $A^* = A^+ \cup \{\varepsilon\}$. Then every subset L of A^* is called a *language over* A. The concept of concatenation is extended for languages as follows. For every pair of languages $L_1, L_2 \subseteq A^*$, we put

$$L_1 L_2 = \{wz \mid w \in L_1 \text{ and } z \in L_2\}.$$

It is obvious that ε is a unit element with respect to concatenation because, for every string $w \in A^*$, we have $\varepsilon w = w\varepsilon = w$. Hence A^* is a monoid which is called the *free monoid generated by* A.

We define the reverse w^{-1} of a string $w \in A^*$.

(i) For $w = \varepsilon$ we put $w^{-1} = w$.
(ii) For a string $w = av$, where $a \in A$ and $v \in A^*$, we define $w^{-1} = v^{-1}a$.

The length $length(w)$ of a string $w \in A^*$ is defined as follows.

(i) For the empty string ε we define $length(\varepsilon) = 0$.
(ii) For a string $w = av$, where $a \in A$ and $v \in A^*$, we define $length(w) = 1 + length(v)$.

For $w \in A^*$ and $1 \le j \le length(w)$, we denote by $w(j)$ the j-th letter of w.

We frequently abbreviate consecutive concatenations of a string $w \in A^*$ as follows. For an arbitrary integer $n \ge 0$, by w^n, we mean ε, if $n = 0$, or ww^{n-1}, if $n \ge 1$.

Let $u, v \in A^*$. We say that *u is a prefix of v* if there is a string x such that $v = ux$.

For two strings u and w, we say that *u is a substring of w* (or *u occurs in w*), if there are strings x and y such that $w = xuy$. In particular u occurs in w, if u is a prefix of w. In general, the position of u in w is not unique. When we refer to a particular occurrence of u in w we always mean that the surroundings x and y are also given such that $w = xuy$.

Let now u, v, and w be strings such that u and v occur in w. Take a particular occurrence of both u and v, which are defined by $w = xuy$ and $w = x'vy'$. We say that this particular occurrence of u and v do not overlap in w if either

- $x'v$ is a prefix of x, or
- xu is a prefix of x'.

We note that "do not overlap" is a symmetric relation.

We now define the concept of the *string substitution*. Therefore, let w be a string, I be a finite index set, and let $U = \{u_i \mid i \in I\}$ be a set of nonempty strings. Let U be such that

- for every $i, j \in I$, if $u_i = u_j$, then $i = j$ (i.e., the u_i's are pairwise disjoint), and
- for every $i, j \in I$, if both u_i and u_j occur in w, then any two occurrences of u_i and u_j do not overlap in w.

Moreover, let $V = \{v_i \mid i \in I\}$ be a set of strings. Then we denote by

$$w[u_i \leftarrow v_i; i \in I]$$

the string which is obtained from w by replacing every occurrence of u_i by v_i for every $i \in I$. (We note that the denotation $w[u_i \leftarrow v_i; i \in I]$ does not necessarily imply that, for every $i \in I$, u_i is a substring of w.)

If $U = \{u_1, \ldots, u_n\}$ and $V = \{v_1, \ldots, v_n\}$, then $w[u_i \leftarrow v_i; i \in I]$ is also denoted by $w[u_1 \leftarrow v_1, \ldots, u_n \leftarrow v_n]$ or by $w[u_i \leftarrow v_i; u_i \in U]$.

2.6 String Rewrite Systems

Let A be an arbitrary alphabet. A *string rewrite system over* A is a finite set $S \subseteq A^* \times A^*$. Elements of S are called *(rewrite) rules* and are written in the form $u \rightarrow v$, rather than (u, v).

Any string rewrite system S determines a derivation system $(A^*, \Rightarrow_S)$, where the derivation relation $\Rightarrow_S$ is defined in the following way. For $w, z \in A^*$, we have $w \Rightarrow_S z$ if and only if

- there is a rule $u \rightarrow v$ in S,
- there are x and y in A^*,

such that $w = xuy$ and $z = xvy$. If this is the case, then we say that the rule $u \rightarrow v$ is applicable to w.

We denote the set of irreducible elements of A^+ with respect to $\Rightarrow_S$ by $IRR(A, S)$. Moreover, we say that S is terminating (locally confluent, confluent, Church-Rosser), if $\Rightarrow_S$ is so.

Let S be a string rewrite system over A. Then $\Leftrightarrow_S^*$ is a congruence over the free semigroup A^+ and hence $A^+/\Leftrightarrow_S^*$ is a semigroup too, called *the semigroup presented by* S. Alternatively, S presents the semigroup $A^+/\Leftrightarrow_S^*$.

A semigroup M is said to be *finitely presentable* if there is an alphabet A and a string rewrite system S over A such that $M \cong A^+/\Leftrightarrow_S^*$.

2.7 Grammars

A grammar is a 4-tuple $G = (N, \Sigma, P, S)$, where

(1) N is an alphabet, called the ***nonterminal alphabet,***
(2) Σ is another alphabet, disjoint from N, called the *terminal alphabet,*
(3) P, called the set of *rules,* is a finite subset of the set

$$(N \cup \Sigma)^* N (N \cup \Sigma)^* \times (N \cup \Sigma)^*,$$

i.e., it is a finite set of elements of the form $\alpha \to \beta$, where $\alpha, \beta \in (N \cup \Sigma)^*$, such that α contains at least one nonterminal.
(4) S is a designated element of N called the *start symbol.*

The grammar G induces a derivation system $((N \cup \Sigma)^*, \Rightarrow_G)$, where the derivation relation $\Rightarrow_G$, or shortly $\Rightarrow$ is defined in the following manner. For any $\gamma, \delta \in (N \cup \Sigma)^*$, we define $\gamma \Rightarrow \delta$ if and only if

- there is a rule $\alpha \to \beta$ in P, and
- there are γ_1 and γ_2 in $(N \cup \Sigma)^*$ such that $\gamma = \gamma_1 \alpha \gamma_2$ and $\delta = \gamma_1 \beta \gamma_2$.

Then G generates a language which is defined by

$$L(G) = \{w \in \Sigma^* \mid S \Rightarrow_G^* w\}.$$

In this book we are interested in language classes generated by some restricted grammars. First, we define context-free grammars. A grammar is *context-free* if every rule in P has the form $A \to \alpha$, where $A \in N$. A language $L \subseteq \Sigma^*$ is called context-free if it can be defined by a context-free grammar, i.e., there is a context-free grammar G such that $L = L(G)$. The class of all context-free languages is denoted by $\mathcal{L}_{cf}$.

Further, a context-free grammar is called *linear* if, for every rule $A \to \alpha$, the string α contains at most one symbol from N. In other words $\alpha \in (\Sigma^* N \Sigma^*) \cup \Sigma^*$ holds. The class of linear languages is defined in an obvious way and it is denoted by $\mathcal{L}_{lin}$.

The third language class we will consider is defined as follows. A grammar G as above is called *regular* if every rule in P has either the form $A \to wB$ or the form $A \to w$, where $A, B \in N$ and $w \in \Sigma^*$. The class of languages generated by regular grammars is called the class of regular languages and it is denoted by $\mathcal{L}_{reg}$.

2.8 Notions on Trees

In this section we collect definitions, notions, and notations concerning trees, the principles of definition and proof by structural induction, properties of trees, and tree transformations.

Trees

A *ranked set* is a tuple $(\Omega, rank)$ where Ω is a (possibly infinite) set and $rank : \Omega \to \mathbb{N}$ is a mapping called the rank function. If $\omega \in \Omega$ is such that $rank(\omega) = k$, then we say that the rank of ω is k or that ω has rank k. For every $k \geq 0$, we define $\Omega^{(k)} = \{\omega \in \Omega \mid rank(\omega) = k\}$. Thus, the fact that ω has rank k can also be expressed by writing $\omega \in \Omega^{(k)}$. Hence we can drop *rank* from the specification $(\Omega, rank)$ by just saying that Ω is a ranked set. Moreover, we also write $\omega^{(k)}$ to denote the fact that ω has rank k, i.e., that $\omega \in \Omega^{(k)}$. Sometimes we also denote the rank function by $rank_\Omega$.

In this book we will need an infinite supply of ranked symbols.

> Hence we now take and fix an infinite ranked set Ω having the property that, for every $k \geq 0$, $\Omega^{(k)}$ is also infinite.

By a *ranked alphabet* we mean a finite subset Σ of Ω, where ranks are taken over from Ω, i.e., we put $\Sigma^{(k)} = \Sigma \cap \Omega^{(k)}$ for every $k \geq 0$. With this definition of a ranked alphabet we state that the rank of a symbol σ is the same in all ranked alphabets to which it belongs, i.e., the one taken over from Ω.

A ranked alphabet Σ is called *unary* if $\Sigma = \Sigma^{(1)}$, i.e., every $\sigma \in \Sigma$ has rank 1. Moreover, Σ is *monadic* if $\Sigma^{(0)}$ contains exactly one element and, for every $n \geq 2$, the set $\Sigma^{(n)}$ is empty.

Let Σ be a ranked alphabet and let A be a set disjoint with Σ. The set of *(finite, labeled, and ordered) trees over Σ indexed by A*, denoted by $T_\Sigma(A)$, is the smallest subset T of $(\Sigma \cup A \cup \{(,)\} \cup \{,\})^*$, such that

(i) $A \subseteq T$ and
(ii) if $\sigma \in \Sigma^{(k)}$ with $k \geq 0$, and $t_1, \ldots, t_k \in T$, then $\sigma(t_1, \ldots, t_k) \in T$.

In the case $k = 0$, we denote $\sigma(\,)$ by σ. Moreover, $T_\Sigma(\emptyset)$ is denoted by T_Σ. It should be clear that $T_\Sigma = \emptyset$ if and only if $\Sigma^{(0)} = \emptyset$. Since we are not interested in this particular case, we assume that $\Sigma^{(0)} \neq \emptyset$ for every ranked alphabet appearing in this book.

A set $L \subseteq T_\Sigma$ is called a *tree language*.

For trees over a monadic ranked alphabet, we introduce a simpler denotation as follows. Let Σ be a monadic ranked alphabet and $t \in T_\Sigma$. Then either $t \in \Sigma^{(0)}$ or $t = \sigma_1(\sigma_2(\ldots\sigma_n(\sigma)\ldots))$, where $n \geq 1, \sigma_1, \ldots, \sigma_n \in \Sigma^{(1)}$ and $\sigma \in \Sigma^{(0)}$. In the latter case we also write t as $\sigma_1\sigma_2\ldots\sigma_n(\sigma)$, or as $\sigma_1\sigma_2\ldots\sigma_n\sigma$ for the sake of simplicity.

We will need sets of *variable symbols* to specify different kinds of tree transducers. Such a set of variable symbols is $X = \{x_1, x_2, \ldots\}$ and later we will specify some of its isomorphic copies. The elements of X will be called variables. For every $k \geq 0$, the set of the first k variables in X will be denoted by X_k, hence $X_k = \{x_1, \ldots, x_k\}$. Note that $X_0 = \emptyset$. These notations for variables will be kept fixed throughout the book. The variables will occur in trees, so we will frequently consider trees from sets of the form $T_\Sigma(X)$, $T_\Sigma(X_k)$, etc., where Σ is a ranked alphabet.

If Q is a unary ranked alphabet and A is a set of symbols, then $Q(A)$ denotes the set $\{q(a) \mid q \in Q, a \in A\}$.

Since trees are also strings, the concept of string substitution defined in Sect. 2.5 applies to trees as well. This operation will frequently be used in the special case when variables in a tree will be substituted by other trees. For example, let $t \in T_\Sigma(X_k)$, for some ranked alphabet Σ and $k \geq 0$. Moreover, let $t_1, \ldots, t_k$ also be trees over a (possibly different) ranked alphabet indexed by a set. Then, according to the concept of string substitution defined in Sect. 2.5, the notation $t[x_1 \leftarrow t_1, \ldots, x_k \leftarrow t_k]$ means the tree which is obtained from t by substituting, for every $1 \leq i \leq k$, the tree t_i for each occurrence of x_i in t.

Now let Σ be a ranked alphabet. Moreover, let $Z = \{z_1, z_2, \ldots\}$ be another set of variables, disjoint with X, and let $k \geq 0$. The *set of (Σ, k)-contexts*, denoted by $C_{\Sigma,k}$, is the set of trees t over Σ indexed by Z_k such that for every $1 \leq i \leq k$, the symbol z_i occurs exactly once in t. Note that a (Σ, k)-context t can also be considered as a string over the alphabet $\Sigma \cup \{z_1, \ldots, z_k\} \cup \{,\} \cup \{(,)\}$.

For contexts, we use a simpler notation for the substitution. Namely, let t be a (Σ, k)-context and let $t_1, \ldots, t_k$ be trees. Then we abbreviate $t[z_1 \leftarrow t_1, \ldots, z_k \leftarrow t_k]$ to $t[t_1, \ldots, t_k]$.

Structural Induction

Next we recall the *principle of structural induction* which is a straightforward generalization of the (usual) principle of induction on the set $\mathbb{N}$ of natural numbers. Let us briefly recall the principles of definition by induction and proof by induction.

Principle 2.9. Let C be a set. By *induction*, a mapping $f : \mathbb{N} \to C$ can be defined. The *principle of definition by induction* is:

> If
> IB: $f(1)$ is defined and
> IS: for every $n \geq 1$, if $f(n)$ is defined, then $f(n+1)$ is defined,
> then for every $n \geq 1$, $f(n)$ is defined.

□

The abbreviations IB and IS stand for induction base and induction step, respectively.

Principle 2.10. By *induction*, a predicate P over $\mathbb{N}$ can be proved. The *principle of proof by induction* is:

> If
> IB: the formula $P(1)$ is true and

IS: for every $n \geq 1$, if the formula $P(n)$ is true, then the formula $P(n+1)$ is true,

then for every $n \geq 1$, the formula $P(n)$ is true.

□

Now the generalization means replacing the initial algebra which is generated by the particular ranked alphabet $\{suc^{(1)}, zero^{(0)}\}$, by the initial algebra which is generated by an arbitrary ranked alphabet Σ.

Principle 2.11. Let Σ be a ranked alphabet and let A, C be sets. By *structural induction*, a function

$$f : T_\Sigma(A) \to C$$

can be defined.

The *principle of definition by structural induction* is:

If

IB: for every $\alpha \in A \cup \Sigma^{(0)}$, the value $f(\alpha)$ is defined and

IS: for every $k \geq 1$, $\sigma \in \Sigma^{(k)}$, and $s_1, \ldots, s_k \in T_\Sigma(A)$, if the values $f(s_1), \ldots, f(s_k)$ are defined, then the value $f(\sigma(s_1, \ldots, s_k))$ is defined,

then for every $s \in T_\Sigma(A)$, the value $f(s)$ is defined.

□

Principle 2.12. Let Σ be a ranked alphabet and A be a set. By *structural induction*, a predicate P over $T_\Sigma(A)$ can be proved.

The *principle of proof by structural induction* is:

If

IB: for every $\alpha \in A \cup \Sigma^{(0)}$, the formula $P(\alpha)$ is true and

IS: for every $k \geq 1, \sigma \in \Sigma^{(k)}$, and $s_1, \ldots, s_k \in T_\Sigma(A)$, if the formulas $P(s_1), \ldots, P(s_k)$ are true, then the formula $P(\sigma(s_1, \ldots, s_k))$ is true,

then for every $s \in T_\Sigma(A)$, the formula $P(s)$ is true.

□

We note that the principle of proof by structural induction is a special case of principle of proof by Noetherian induction (Principle 2.4), where the basic set is $T_\Sigma(A)$ and the derivation relation is the following: for a pair of trees $t, t' \in T_\Sigma(A)$ we define $t \Rightarrow t'$ if and only if t' is a direct subtree of t.

Properties of Trees

Let Σ be a ranked alphabet and A be an arbitrary set. By structural induction, we define four characteristics of a tree $t \in T_\Sigma(A)$: the *height of* t, the *size of* t, the *set of occurrences of* t, and the *set of subtrees* of t. Formally these can be described with the functions $height : T_\Sigma(A) \to \mathbb{N}$, $size : T_\Sigma(A) \to \mathbb{N}$, $occ : T_\Sigma(A) \to \mathcal{P}(\mathbb{N}^*)$, and $sub : T_\Sigma(A) \to \mathcal{P}(T_\Sigma(A))$, respectively, as follows.

IB: If $t \in \Sigma^{(0)} \cup A$, then
- $height(t) = 1$,
- $size(t) = 1$,
- $occ(t) = \{\varepsilon\}$,
- $sub(t) = \{t\}$.

IS: If $t = \sigma(s_1, \ldots, s_k)$ where $\sigma \in \Sigma^{(k)}$, $k \geq 1$ and $s_1, \ldots, s_k \in T_\Sigma(A)$, then
- $height(t) = 1 + max\{height(s_i) \mid 1 \leq i \leq k\}$,
- $size(t) = 1 + \sum_{1 \leq i \leq k} size(s_i)$,
- $occ(t) = \{\varepsilon\} \cup \{w \mid w = iv, 1 \leq i \leq k, v \in occ(s_i)\}$,
- $sub(t) = \{t\} \cup \bigcup_{1 \leq i \leq k} sub(s_i)$.

We note that this definition of *height* is different from that used in Chapter 1.

Also we define the *label at an occurrence* of a tree $t \in T_\Sigma(A)$. More precisely, the mapping

$$label : \{(t, w) \mid t \in T_\Sigma(A), w \in occ(t)\} \to \Sigma \cup A$$

is defined by structural induction for every $t \in T_\Sigma(A)$ and $w \in occ(t)$.

IB: If $t \in \Sigma^{(0)} \cup A$ (and thus $w = \varepsilon$), then $label(t, w) = t$.

IS: If $t = \sigma(s_1, \ldots, s_k)$ for some $\sigma \in \Sigma^{(k)}$ with $k \geq 1$ and $s_1, \ldots, s_k \in T_\Sigma(A)$, and
- if $w = \varepsilon$, then $label(t, w) = \sigma$, otherwise
- if $w = iv$ for some $1 \leq i \leq k$, then $label(t, w) = label(t_i, v)$.

We will also need the concept of the *set of variables occurring in a tree* $t \in T_\Sigma(X)$. This is defined with the function $Var_X : T_\Sigma(X) \to \mathcal{P}(X)$ in the following way.

IB: If $t \in \Sigma^{(0)} \cup X$, then if $t \in \Sigma^{(0)}$, $Var_X(t) = \emptyset$, otherwise $Var_X(t) = \{t\}$.

IS: If $t = \sigma(t_1, \ldots, t_k)$, for some $\sigma \in \Sigma^{(k)}$ with $k \geq 1$ and $t_1, \ldots, t_k \in T_\Sigma(X)$, then $Var_X(t) = \bigcup_{1 \leq i \leq k} Var_X(t_i)$.

The following lemma states a basic relation between the height and the size of a tree over a ranked alphabet.

Lemma 2.13. Let Σ be a ranked alphabet. Then there is an integer $c \geq 1$ such that, for every $s \in T_\Sigma$, $size(s) \leq c^{height(s)}$.

Proof. In fact, we prove somewhat more than we have stated. Namely, we show that $size(s) \leq c^{height(s)}$ if $\Sigma = \Sigma^{(0)}$, and that $size(s) < c^{height(s)}$ if $\Sigma \neq \Sigma^{(0)}$.

Let us define c by $c = max\{\, k \mid \Sigma^{(k)} \neq \emptyset \,\} + 1$.

If $\Sigma = \Sigma^{(0)}$, then $c = 1$ and, for every $s \in T_\Sigma$, $height(s) = size(s) = 1$. Then our statement obviously holds for every s.

Assume now that $\Sigma \neq \Sigma^{(0)}$. (Note that in this case $c > 1$.) We prove by structural induction on s.

Let $s \in \Sigma^{(0)}$. Then, since $size(s) = height(s) = 1$ and $c > 1$, we have $size(s) < c^{height(s)}$.

Let $s = \sigma(s_1, \ldots, s_k)$, for some $k \geq 1$, $\sigma \in \Sigma^{(k)}$, and $s_1, \ldots, s_k \in T_\Sigma$. Moreover, let i_0 with $1 \leq i_0 \leq k$ be such that $height(s_{i_0}) \geq height(s_i)$, for every $1 \leq i \leq k$. Then we can compute as follows.

$$\begin{array}{lll} & size(s) & \\ = & 1 + \sum_{i=1}^{k} size(s_i) & \\ \leq & \sum_{i=1}^{k} c^{height(s_i)} & \text{(by the induction hypothesis and by } k \geq 1\text{)} \\ \leq & k \cdot c^{height(s_{i_0})} & \text{(by the definition of } i_0\text{)} \\ < & c \cdot c^{height(s_{i_0})} & \text{(by } c > k\text{)} \\ = & c^{height(s_{i_0})+1} & \\ = & c^{height(s)} & \text{(by the definition of } height(s)\text{)} \end{array}$$

This proves our lemma. □

Tree Transformations

The concept of the tree transformation is central to this book. A *tree transformation* is a mapping $\tau : T_\Sigma \longrightarrow T_\Delta$ for some ranked alphabets Σ and Δ. The class of all *identity tree transformations*, denoted by ID, is the class of all relations $Id(T_\Sigma)$ for some ranked alphabet Σ.

The following lemma expresses the obvious property of tree transformations that every tree transformation can be written as the composition of itself and a suitable identical tree transformation. Since this fact will be used in the book, we state it explicitly.

Lemma 2.14. Let C, D, and E be classes of tree transformations. If $C \subseteq D$ and $ID \subseteq E$, then $C \subseteq D \circ E$ and $C \subseteq E \circ D$.

Proof. Assume that the conditions of our lemma hold. Take a tree transformation $\tau : T_\Sigma \longrightarrow T_\Delta$ from C. Then also $\tau \in D$. Moreover, the identical tree transformations $\iota = Id(T_\Sigma)$ and $\iota' = Id(T_\Delta)$ are in E and both $\tau = \tau \circ \iota'$ and $\tau = \iota \circ \tau$ hold. Therefore, both $\tau \in D \circ E$ and $\tau \in E \circ D$. □

The above lemma will be frequently used in the special case where $C = D$. Then it expresses that both $C \subseteq C \circ E$ and $C \subseteq E \circ C$ providing $ID \subseteq E$. Moreover, if $ID \subseteq C$, then $C \subseteq C \circ C$ also.

3. Top-Down Tree Transducers

In this chapter we introduce the concept of a top-down tree transducer. It is a formal model for syntax-directed semantics and it is created by abstraction from the model called attribute grammar with synthesized attributes only.

The structure of the chapter is as follows.

1. Basic definitions
2. Induced tree transformation
3. Characterization of top-down tree transformations
4. Height property
5. Subclasses of TOP
6. Composition and decomposition results
7. Composition semigroup generated by TOP, l-TOP, and HOM
8. Bibliographic notes

3.1 Basic Definitions

We start by defining a set, called the set of right-hand sides, of which the elements may be right-hand sides of rules of a top-down tree transducer.

Definition 3.1. Let Q be a unary ranked alphabet, Δ be a ranked alphabet disjoint with Q, and $k \geq 0$ be an integer. The set $RHS(Q, \Delta, k)$ of *right hand sides over* Q, Δ, *and* k is the smallest subset RHS of $T_{Q \cup \Delta}(X_k)$ satisfying the following two conditions:

(i) For every $p \in Q$ and $1 \leq i \leq k$, the term $p(x_i)$ is in RHS.
(ii) For every $\delta \in \Delta^{(l)}$ with $l \geq 0$ and $\xi_1, \ldots, \xi_l \in RHS$, the term $\delta(\xi_1, \ldots, \xi_l)$ is in RHS. □

We note that, for every $k \geq 0$, $T_\Delta \subseteq RHS(Q, \Delta, k)$ and that in particular $T_\Delta = RHS(Q, \Delta, 0)$.

We now define the concept of a top-down tree transducer.

Definition 3.2. A *top-down tree transducer* is a tuple $T = (Q, \Sigma, \Delta, q_0, R)$, where

- Q is a unary ranked alphabet, called the set of *states*,
- Σ and Δ are ranked alphabets with $Q \cap (\Sigma \cup \Delta) = \emptyset$, called the *input ranked alphabet* and *output ranked alphabet*, respectively,
- q_0 is a designated element of Q, called the *initial state*,
- R is a finite set of *rules* such that, for every $q \in Q$ and $\sigma \in \Sigma^{(k)}$ with $k \geq 0$, there is exactly one rule of the form

$$q(\sigma(x_1, \ldots, x_k)) \rightarrow \xi$$

in R with $\xi \in RHS(Q, \Delta, k)$. □

A rule as above is called the (q, σ)*-rule* or the q rule for σ. Moreover, ξ is called the *right-hand side of the rule* and is also specified as $rhs(q, \sigma)$. Thus the expression $\xi = rhs(q, \sigma)$ reads as ξ is the right-hand side of the (q, σ)-rule. Obviously, not every element of $RHS(Q, \Delta, k)$ is the right-hand side of some rule in R.

A top-down tree transducer can transform input trees into output trees and hence it realizes or induces a tree transformation. (The exact definition of the tree transformation induced by T will be defined later, after the necessary groundwork.) Trees over Σ, i.e., the elements of T_Σ are called *input trees to* T. An input tree will be transformed into an *output tree*, which is an element of T_Δ. Thus output trees are those trees over Δ which can be obtained by the tree transformation induced by T from an input tree.

We mention that in the theory of top-down tree transducers, a more general terminology is used. Among others there is no restriction on the number of (q, σ)-rules in R. Moreover, if a top-down tree transducer is such that, for every $q \in Q$ and $\sigma \in \Sigma$, there is at least one (q, σ)-rule in R, then it is called *total*. On the other hand, if, for every $q \in Q$ and $\sigma \in \Sigma$, there is at most one (q, σ)-rule in R, then it is called *deterministic*. Thus, the top-down tree transducer we have defined is both total and deterministic. In fact, this is sufficient as was shown by the discussion in Chap. 1.

Next we give an example for a top-down tree transducer.

Example 3.3. Let $T' = (Q, \Sigma, \Delta, q_0, R)$ be such that

- $Q = \{q_0, q_1, q_2\}$,
- $\Sigma = \{\alpha^{(0)}, \gamma^{(1)}\}$ and $\Delta = \{\alpha^{(0)}, \beta_1^{(0)}, \beta_2^{(0)}, \gamma^{(1)}, \delta^{(2)}\}$,
- R is the set of the following rules
  1) $q_0(\alpha) \rightarrow \alpha$,
  2) $q_1(\alpha) \rightarrow \beta_1$,
  3) $q_2(\alpha) \rightarrow \beta_2$,
  4) $q_0(\gamma(x_1)) \rightarrow \delta(q_1(x_1), q_2(x_1))$
  5) $q_1(\gamma(x_1)) \rightarrow \gamma(q_1(x_1))$
  6) $q_2(\gamma(x_1)) \rightarrow \gamma(q_2(x_1))$. □

3.2 Induced Tree Transformation

The discussion in this section concerns an arbitrary, but fixed top-down tree transducer

$$T = (Q, \Sigma, \Delta, q_0, R).$$

We define the tree transformation induced by the top-down tree transducer T according to the following schedule. First we define the derivation relation induced by T. Then we define the set of sentential forms of T and show that the derivation relation is terminating and confluent on sentential forms. Finally, the tree transformation induced by T will be defined as a mapping that associates to every input tree $s \in T_\Sigma$ the normal form of the tree $q_0(s)$ with respect to the derivation relation.

Hence we start by defining the derivation relation $\Rightarrow_T$ induced by T. Before inspecting the definition, the reader is advised to consult Sect. 2.8 for the definition of a context. Figure 3.1 illustrates the definition of the derivation relation.

Definition 3.4. The *derivation relation induced by* T is the binary relation $\Rightarrow_T$ over the set $T_{Q\cup\Sigma\cup\Delta}$ such that, for every $\varphi, \psi \in T_{Q\cup\Sigma\cup\Delta}$ we define $\varphi \Rightarrow_T \psi$, if

- there is a context $\beta \in C_{Q\cup\Sigma\cup\Delta,1}$,
- there is a state $q \in Q$,
- there is a $k \geq 0$ and a $\sigma \in \Sigma^{(k)}$,
- there are trees $s_1, \ldots, s_k \in T_\Sigma$,

such that $\varphi = \beta[q(\sigma(s_1, \ldots, s_k))]$ and $\psi = \beta[\xi[x_1 \leftarrow s_1, \ldots, x_k \leftarrow s_k]]$, where $\xi = rhs(q, \sigma)$. □

In the above case we say that the rule $q(\sigma(x_1, \ldots, x_k)) \to \xi$ is applicable to φ. It can be seen that $(T_{Q\cup\Sigma\cup\Delta}, \Rightarrow_T)$ is a derivation system in the sense of Section 2.4.

The necessary action to obtain ψ from φ, i.e., finding the context β and the subtree $q(\sigma(s_1, \ldots, s_k))$ to be substituted as well as substituting $\xi[x_i \leftarrow s_i; 1 \leq i \leq k]$ for $q(\sigma(s_1, \ldots, s_k))$, is called a *derivation step*. We now show an example for a single derivation step and a sequence of such steps.

Example 3.5. Let us consider again the top-down tree transducer of Example 3.3. Let $\varphi = \delta(q_1(\gamma(\alpha)), \gamma(q_2(\alpha)))$. Then, for the context $\beta = \delta(z_1, \gamma(q_2(\alpha)))$ and the tree $s_1 = \alpha$, we have $\varphi = \beta[q_1(\gamma(s_1))]$. Moreover, $rhs(q_1, \gamma) = \gamma(q_1(x_1))$. Then the relation $\varphi \Rightarrow_{T'} \psi$ holds , where

$$\begin{aligned}
\psi &= \beta[rhs(q_1, \gamma)[x_1 \leftarrow s_1]] \\
&= \delta(z_1, \gamma(q_2(\alpha)))[\gamma(q_1(x_1))[x_1 \leftarrow \alpha]] \\
&= \delta(z_1, \gamma(q_2(\alpha)))[\gamma(q_1(\alpha))] \\
&= \delta(\gamma(q_1(\alpha)), \gamma(q_2(\alpha))).
\end{aligned}$$

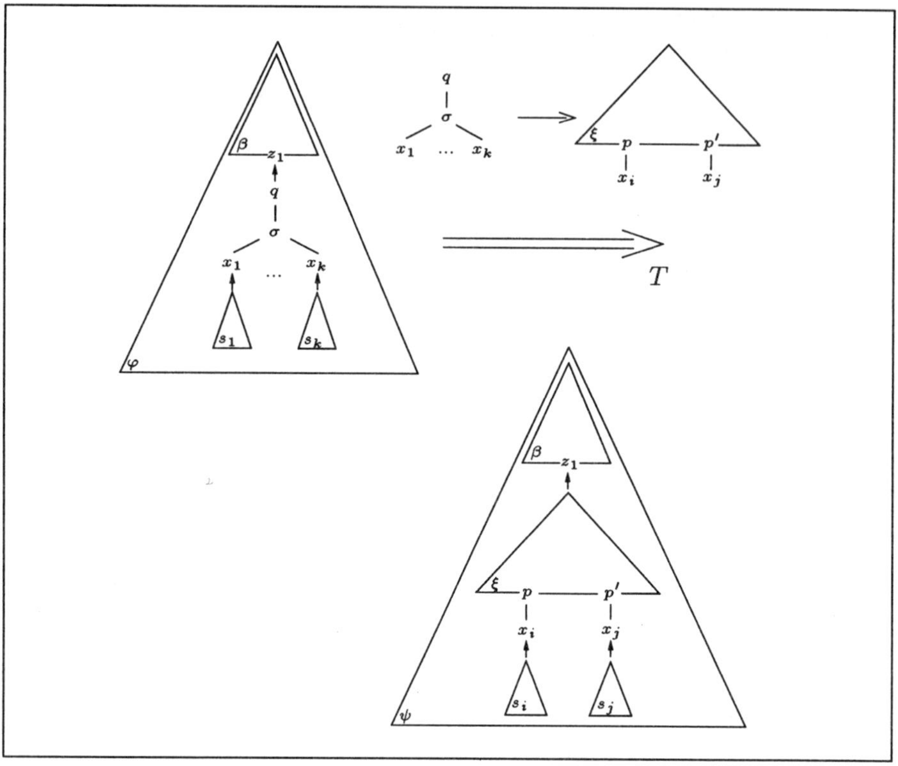

Fig. 3.1. A derivation step induced by T

Next we show a sequence of derivation steps for the top-down tree transducer appearing in Example 3.3 (Fig. 3.2). We can write the following.

$$
\begin{array}{llll}
q_0(\gamma^2(\alpha)) & \Rightarrow_{T'} & \delta(q_1(\gamma(\alpha)), q_2(\gamma(\alpha))) & \text{(by rule 4)} \\
 & & & \text{with } \beta = z_1 \\
 & \Rightarrow_{T'} & \delta(\gamma(q_1(\alpha)), q_2(\gamma(\alpha))) & \text{(by rule 5)} \\
 & & & \text{with } \beta = \delta(z_1, q_2(\gamma(\alpha))) \\
 & \Rightarrow_{T'} & \delta(\gamma(q_1(\alpha)), \gamma(q_2(\alpha))) & \text{(by rule 6)} \\
 & & & \text{with } \beta = \delta(\gamma(q_1(\alpha)), z_1) \\
 & \Rightarrow_{T'} & \delta(\gamma(\beta_1), \gamma(q_2(\alpha))) & \text{(by rule 2)} \\
 & & & \text{with } \beta = \delta(\gamma(z_1)), \gamma(q_2(\alpha))) \\
 & \Rightarrow_{T'} & \delta(\gamma(\beta_1), \gamma(\beta_2)) & \text{(by rule 3)} \\
 & & & \text{with } \beta = \delta(\gamma(\beta_1), \gamma(z_1))
\end{array}
$$

This example shows a particular property of a top-down tree transducer: it can make copies of a subtree of the input tree (viz. the two copies of $\gamma(\alpha)$) and then process these copies with different states (viz. the states q_1 and q_2) and produce different subtrees (viz. $\gamma(\beta_1)$ and $\gamma(\beta_2)$) of the output tree. □

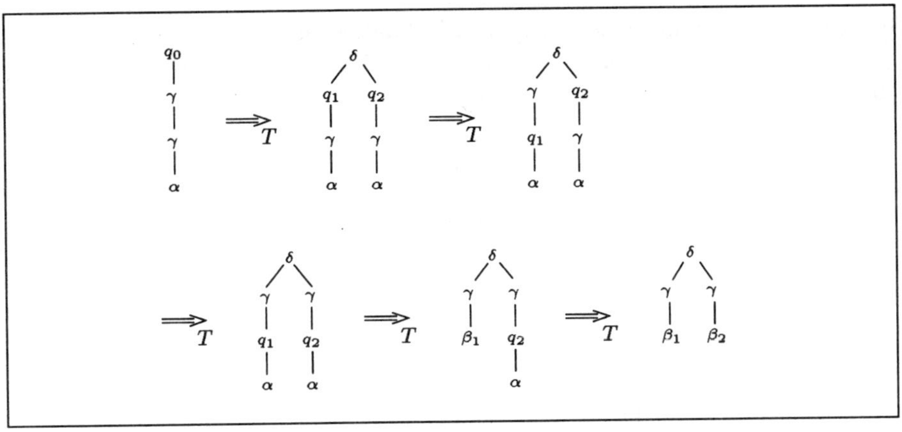

Fig. 3.2. A sequence of derivation steps

In order to define the tree transformation induced by T we do not need to consider $\Rightarrow_T$ over the whole set $T_{Q\cup\Sigma\cup\Delta}$. A smaller set which is called the set of sentential forms of T will be sufficient. The set of sentential forms can be thought of as the set of trees that can appear in derivations concerning the top-down tree transducer T.

Definition 3.6. The set of *sentential forms of* T, denoted by $SF(T)$, is the smallest subset SF of $T_{Q\cup\Sigma\cup\Delta}$ for which the following two conditions hold.

(i) For every $q \in Q$ and $s \in T_\Sigma$, the tree $q(s)$ is in SF.
(ii) For every $\delta \in \Delta^{(l)}$ with $l \geq 0$ and $\varphi_1, \ldots, \varphi_l \in SF$, the tree $\delta(\varphi_1, \ldots, \varphi_l)$ is in SF.

Moreover, a tree of the form $q_0(s)$, where $s \in T_\Sigma$, is called *initial sentential form*. We note that $SF(T) \subseteq T_{Q\cup\Sigma\cup\Delta}$. □

The following lemma is needed for technical reasons. It expresses the close connection between the set of possible right-hand sides and the set of sentential forms. Namely, each sentential form can be obtained by substituting input trees into a suitable right-hand side and vice versa; each tree obtained by the substitution described above is a sentential form.

Lemma 3.7. The following two statements hold for the top-down tree transducer T.

(a) For every $\xi \in RHS(Q, \Delta, k)$ with $k \geq 0$ and $s_1, \ldots, s_k \in T_\Sigma$ it holds that $\xi[x_1 \leftarrow s_1, \ldots, x_k \leftarrow s_k] \in SF(T)$.
(b) For every $\varphi \in SF(T)$, there exist $k \geq 0$, $\xi \in RHS(Q, \Delta, k)$, and $s_1, \ldots, s_k \in T_\Sigma$ such that $\varphi = \xi[x_1 \leftarrow s_1, \ldots, x_k \leftarrow s_k]$.

Proof. (a) The proof is by structural induction on ξ.

(i) If $\xi = q(x_i)$ for some $1 \leq i \leq k$, then $\xi[x_1 \leftarrow s_1, \ldots, x_k \leftarrow s_k] = q(s_i)$ and hence $\xi[x_1 \leftarrow s_1, \ldots, x_k \leftarrow s_k] \in SF(T)$.

(ii) Now let $\xi = \delta(\xi_1, \ldots, \xi_l)$ for some $l \geq 0, \delta \in \Delta^{(l)}$, and $\xi_1, \ldots, \xi_l \in RHS(Q, \Delta, k)$. Then we have $\xi[x_1 \leftarrow s_1, \ldots, x_k \leftarrow s_k] = \delta(\xi_1[x_1 \leftarrow s_1, \ldots, x_k \leftarrow s_k], \ldots, \xi_l[x_1 \leftarrow s_1, \ldots, x_k \leftarrow s_k])$. Since, by induction hypothesis, for every $1 \leq i \leq l$, the term $\xi_i[x_1 \leftarrow s_1, \ldots, x_k \leftarrow s_k]$ is in $SF(T)$ we find that $\xi[x_1 \leftarrow s_1, \ldots, x_k \leftarrow s_k]$ is also in $SF(T)$.

(b) Structural induction on φ.

(i) Let $\varphi = q(s)$ for some $q \in Q$ and $s \in T_\Sigma$. Then letting $k = 1, \xi = q(x_1)$, and $s_1 = s$ we have $\varphi = \xi[x_1 \leftarrow s_1]$.

(ii) Suppose now that $\varphi = \delta(\varphi_1, \ldots, \varphi_l)$ for some $\delta \in \Delta^{(l)}$ with $l \geq 0$ and $\varphi_1, \ldots, \varphi_l \in SF(T)$. By induction hypothesis, for every $1 \leq i \leq l$, there exist $k_i \geq 0$, $\xi_i \in RHS(Q, \Delta, k_i)$, and $s_{i1}, \ldots, s_{ik_i} \in T_\Sigma$ such that $\varphi_i = \xi_i[x_1 \leftarrow s_{i1}, \ldots, x_{k_i} \leftarrow s_{ik_i}]$. Let, for every $1 \leq i \leq l$,

$$\xi_i' = \xi_i[x_1 \leftarrow x_{k_1+\ldots+k_{i-1}+1}, \ldots, x_{k_i} \leftarrow x_{k_1+\ldots+k_i}],$$

$k = k_1 + \ldots + k_l$, and $s_1 = s_{11}, \ldots, s_k = s_{lk_l}$. Then, for every $1 \leq i \leq l$, it holds that

$$\xi_i[x_1 \leftarrow s_{i1}, \ldots, x_{k_i} \leftarrow s_{ik_i}] = \xi_i'[x_1 \leftarrow s_1, \ldots, x_k \leftarrow s_k].$$

Now define $\xi = \delta(\xi_1', \ldots, \xi_l')$, and, let k and $s_1, \ldots, s_k$ be as above. Then we can compute as follows

$$\begin{aligned} & \xi[x_1 \leftarrow s_1, \ldots, x_k \leftarrow s_k] \\ = \; & \delta(\xi_1'[x_1 \leftarrow s_1, \ldots, x_k \leftarrow s_k], \ldots, \xi_l'[x_1 \leftarrow s_1, \ldots, x_k \leftarrow s_k]) \\ = \; & \delta(\xi_1[x_1 \leftarrow s_{11}, \ldots, x_{k_1} \leftarrow s_{1k_1}], \ldots, \xi_l[x_1 \leftarrow s_{l1}, \ldots, x_{k_l} \leftarrow s_{lk_l}]) \\ = \; & \delta(\varphi_1, \ldots, \varphi_l) \\ = \; & \varphi. \end{aligned}$$

□

In the following lemma we prove that the set of sentential forms is closed with respect to the derivation relation, i.e., that $(SF(T), \Rightarrow_T)$ is also a derivation system. This will also prove that, starting from a sentential form, we can always obtain sentential forms by applying derivation steps. In particular, this will be true for initial sentential forms.

Lemma 3.8. *$SF(T)$ is closed under $\Rightarrow_T$.*

Proof. We show that, for every sentential form $\varphi \in SF(T)$ and term $\psi \in T_{Q \cup \Sigma \cup \Delta}$, if $\varphi \Rightarrow_T \psi$, then $\psi \in SF(T)$. The proof is performed by structural induction on φ.

(i) Let $\varphi = q(s)$, where $q \in Q$ and $s = \sigma(s_1, \ldots, s_k) \in T_\Sigma$, for some $\sigma \in \Sigma^{(k)}$ and $s_1, \ldots, s_k \in T_\Sigma$. Then, letting $\varphi = c[q(\sigma(s_1, \ldots, s_k))]$, where c

is a context in $C_{Q\cup\Sigma\cup\Delta,1}$, we must have $c = z_1$. Hence, by the definition of $\Rightarrow_T$,

$$\psi = c[\xi[x_1 \leftarrow s_1, \ldots, x_k \leftarrow s_k]] = \xi[x_1 \leftarrow s_1, \ldots, x_k \leftarrow s_k],$$

where $\xi = rhs(q, \sigma)$ and thus, by Lemma 3.7, $\psi \in SF(T)$.

(ii) Now let $\varphi = \delta(\varphi_1, \ldots, \varphi_l)$, for some $\delta \in \Delta^{(l)}$ with $l \geq 0$ and $\varphi_1, \ldots, \varphi_l \in SF(T)$. Then, by the definition of $\Rightarrow_T$, $\varphi \Rightarrow_T \psi$ means that $\varphi_i \Rightarrow_T \psi_i$, for some $1 \leq i \leq l$ and $\psi_i \in T_{Q\cup\Sigma\cup\Delta}$. By induction hypothesis, $\psi_i \in SF(T)$. On the other hand,

$$\psi = \delta(\varphi_1, \ldots, \varphi_{i-1}, \psi_i, \varphi_{i+1}, \ldots, \varphi_l),$$

and hence $\psi \in SF(T)$. □

In the following we are going to prove that the derivation relation is terminating and confluent on $SF(T)$. We start by proving local confluency; then confluency follows from the termination and Lemma 2.5.

Lemma 3.9. The derivation relation $\Rightarrow_T$ is locally confluent on $SF(T)$.

Proof. Let φ,φ_1, and φ_2 be sentential forms of T such that $\varphi \Rightarrow_T \varphi_1$ and $\varphi \Rightarrow_T \varphi_2$. We show that there exists a sentential form $\psi \in SF(T)$ such that $\varphi_1 \Rightarrow_T^* \psi$ and $\varphi_2 \Rightarrow_T^* \psi$.

By Def. 3.4, for $i = \{1, 2\}$,

- there is a context $c_i = C_{Q\cup\Sigma\cup\Delta,1}$,
- there is a state $q_i \in Q$
- there is a $k_i \geq 0$ and a $\sigma_i \in \Sigma^{(k_i)}$,
- there are trees $s_{i1}, \ldots, s_{ik_i} \in T_\Sigma$,

such that

$$\begin{aligned} \varphi &= c_1[q_1(\sigma_1(s_{11}, \ldots, s_{1k_1}))] = c_2[q_2(\sigma_2(s_{21}, \ldots, s_{2k_2}))], \\ \varphi_1 &= c_1[\xi_1[x_1 \leftarrow s_{11}, \ldots, x_{k_1} \leftarrow s_{1k_1}]], \\ \varphi_2 &= c_2[\xi_2[x_1 \leftarrow s_{21}, \ldots, x_{k_2} \leftarrow s_{2k_2}]], \end{aligned}$$

where $\xi_i = rhs(q_i, \sigma_i)$, for $i = \{1, 2\}$.

Consider the occurrences w_i leading from the root of c_i to the only z_1 in c_i. We make two observations about w_1 and w_2.

First we observe that if $w_1 = w_2$, then $c_1 = c_2, q_1 = q_2, \sigma_1 = \sigma_2, k_1 = k_2$, and $s_{11} = s_{21}, \ldots, s_{1k_1} = s_{2k_2}$. Consequently $\xi_1 = \xi_2$, hence $\varphi_1 = \varphi_2$ and we have $\psi = \varphi_1 (= \varphi_2)$.

Next we show that neither w_2 is a proper prefix of w_1, nor w_1 is a proper prefix of w_2; then we say that w_1 and w_2 are disjoint occurrences. Suppose on the contrary that w_2 is a proper prefix of w_1. Then there would exist a context $c \in C_{Q\cup\Sigma\cup\Delta,1}$ such that $c \neq z_1$ and $c_1 = c_2[c]$. Moreover, from

$$\begin{aligned}\varphi &= c_1[q_1(\sigma_1(s_{11},\ldots,s_{1k_1}))]\\ &= c_2[c[q_1(\sigma_1(s_{11},\ldots,s_{1k_1}))]]\\ &= c_2[q_2(\sigma_2(s_{21},\ldots,s_{2k_2}))]\end{aligned}$$

we would find that

$$c[q_1(\sigma_1(s_{11},\ldots,s_{1k_1}))] = q_2(\sigma_2(s_{21},\ldots,s_{2k_2})).$$

This would yield that the root of c is also q_2 which contradicts the supposition that $\varphi \in SF(T)$. In a similar way, we can show that w_1 cannot be a proper prefix of w_2 either.

Hence we may assume that w_1 and w_2 are disjoint occurrences. Then, since both w_1 and w_2 are also occurrences of φ, there is a context $c \in C_{Q\cup\Sigma\cup\Delta,2}$ such that

$$\begin{aligned}\varphi &= c[z_1 \leftarrow q_1(\sigma_1(s_{11},\ldots,s_{1k_1})), z_2 \leftarrow q_2(\sigma_2(s_{21},\ldots,s_{2k_2}))],\\ \varphi_1 &= c[z_1 \leftarrow \xi_1[x_1 \leftarrow s_{11},\ldots,x_{k_1} \leftarrow s_{1k_1}], z_2 \leftarrow q_2(\sigma_2(s_{21},\ldots,s_{2k_2}))],\\ \varphi_2 &= c[z_1 \leftarrow q_1(\sigma_1(s_{11},\ldots,s_{1k_1})), z_2 \leftarrow \xi_2[x_1 \leftarrow s_{21},\ldots,x_{k_2} \leftarrow s_{2k_2}]].\end{aligned}$$

Then it is easy to see that, for

$$\psi = c[z_1 \leftarrow \xi_1[x_1 \leftarrow s_{11},\ldots,x_{k_1} \leftarrow s_{1k_1}], z_2 \leftarrow \xi_2[x_1 \leftarrow s_{21},\ldots,x_{k_2} \leftarrow s_{2k_2}]],$$

we have $\varphi_1 \Rightarrow_T \psi$ and $\varphi_2 \Rightarrow_T \psi$. □

In order to prove statements concerning trees, like the terminating property of $\Rightarrow_T$ on $SF(T)$, we need another kind of proof technique. This is called proof by simultaneous induction, the principle of which will be presented now. As a preparation, we define two families of predicates, which will be required to form the statements precisely.

Definition 3.10. Let Σ be a ranked alphabet. Let A be a set and, for every $a \in A$, let $K_a : T_\Sigma \to \{true, false\}$ be a predicate over T_Σ. Moreover, for every $k \geq 0$, let B_k be a set, and, for every $b \in B_k$, let $L_{b,k} : (T_\Sigma)^k \to \{true, false\}$ be a predicate. Let

$$K = \{K_a \mid a \in A\}$$

and let

$$L = \{L_{b,k} \mid b \in B_k \text{ for some } k \geq 0\}.$$

We say that K holds if, for every $a \in A$ and $s \in T_\Sigma$, $K_a(s) = true$. Similarly, we say that L holds if, for every $k \geq 0, b \in B_k$, and $s_1,\ldots,s_k \in T_\Sigma$, $L_{b,k}((s_1,\ldots,s_k)) = true$. □

Thus simultaneous induction is a technique which can prove simultaneously that families of predicates like K and L hold. This technique generalizes the proof by structural induction (see Principle 2.12). The principle of simultaneous induction is as follows.

Principle 3.11. Let K and L be two families of predicates as in Def. 3.10. The *principle of proof by simultaneous induction* is:

If
IB: for every $b \in B_0$, the formula $L_{b,0}(()) = true$ and
IS1: for every $a \in A$, $k \geq 0$, and $s_1, \ldots, s_k \in T_\Sigma$, if for every $b \in B_k$, $L_{b,k}((s_1, \ldots, s_k)) = true$, then, for every $\sigma \in \Sigma^{(k)}$, $K_a(\sigma(s_1, \ldots, s_k)) = true$ and
IS2: for every $k \geq 1$ and $s_1, \ldots, s_k \in T_\Sigma$, if for every $a_1, \ldots, a_k \in A$, $K_{a_1}(s_1) = true, \ldots, K_{a_k}(s_k) = true$, then, for every $b \in B_k$, $L_{b,k}((s_1, \ldots, s_k)) = true$,
then K and L hold.

□

The abbreviations IB, IS1, and IS2 stand for induction base, induction step 1 and induction step 2, respectively.

Note 3.12. In concrete applications of Principle 3.11 throughout the book, the sets $B_k, k \geq 0$ will form a hierarchy, i.e., $B_k \subseteq B_{k+1}$ for every $k \geq 0$ (see Sect. 2.2). Hence, for any $k \geq 0$ and $b \in B_k$, it also holds that, for every $l \geq k$, $b \in B_l$. Moreover, for every $k \geq 0$ and $b \in B_k$, the predicates $L_{b,k}$ will have the following property: for every $l \geq k$ and sequence $s_1, \ldots, s_l$ of trees in T_Σ,

$$L_{b,k}((s_1, \ldots, s_k)) = L_{b,l}((s_1, \ldots, s_l)).$$

This, and the fact that the arity of a predicate will always be clear from the context, namely from the specification of its arguments, encourages us to drop the index k from $L_{b,k}$ in concrete applications of Principle 3.11. Thereby our formula becomes more readable. □

We can now prove by simultaneous induction that the derivation relation of a top-down tree transducer is terminating on the set of sentential forms. In the proof we will show how Principle 3.11 is applied to this concrete case. However, later, we shall leave such details to the reader.

Lemma 3.13. The derivation relation $\Rightarrow_T$ is terminating on $SF(T)$.

Proof. We prove that, for every sentential form φ, there exists an integer g_φ such that, for every derivation $\varphi = \varphi_0, \varphi_1, \ldots, \varphi_m$ starting from φ, we have $m \leq g_\varphi$.

To see this, we prove by simultaneous induction that K and L hold, where K and L are defined in the following way. Let

- $A = Q$ and
- for every $q \in Q$, let K_q be the predicate over T_Σ defined such that, for every $s \in T_\Sigma$, $K_q(s) = true$, if and only if there is an integer $g_{q,s}$ such that the derivations starting from $q(s)$ are not longer than $g_{q,s}$.
- Let $K = \{K_q \mid q \in Q\}$.

Moreover,

- for every $k \geq 0$, let $B_k = RHS(Q, \Delta, k)$ and
- for every $\xi \in RHS(Q, \Delta, k)$, let L_ξ be the predicate over $(T_\Sigma)^k$ defined such that, for every $\omega = (s_1, \ldots, s_k) \in (T_\Sigma)^k$, $L_\xi(\omega) = true$ if and only if there is an integer $g_{\xi,\omega}$ such that the derivations starting from $\xi[x_1 \leftarrow s_1, \ldots, x_k \leftarrow s_k]$ are not longer than $g_{\xi,\omega}$.
- Finally, let $L = \{L_\xi \mid \xi \in RHS(Q, \Delta, k) \text{ for some } k \geq 0\}$.

First we prove IB. As we noted after Def. 3.1, $RHS(Q, \Delta, 0) = T_\Delta$ and thus $\xi \in T_\Delta$. Moreover $\omega = ()$. Then $g_{\xi,()} = 0$ is suitable, since the only derivation starting from ξ is the sequence ξ.

Next we prove IS1. Let $s = \sigma(s_1, \ldots, s_k)$, for some $\sigma \in \Sigma^{(k)}$ and $s_1, \ldots, s_k \in T_\Sigma$, and let $\omega = (s_1, \ldots, s_k)$. Moreover, let $\varphi_0, \varphi_1, \ldots, \varphi_m$ be a derivation sequence starting from $q(s)$. We observe that, by the definition of $\Rightarrow_T$, $\varphi_1 = \xi[x_1 \leftarrow s_1, \ldots, x_k \leftarrow s_k]$, where ξ is the right-hand side of the (q, σ) rule in R. On the other hand, $\varphi_1, \ldots, \varphi_m$ is a derivation of length $m - 1$ starting from φ_1. Thus, by the induction hypothesis on L, there exists an integer $g_{\xi,\omega}$, such that $m - 1 \leq g_{\xi,\omega}$, hence, for $g_{q,s} = 1 + g_{\xi,\omega}$, we have $m \leq g_{q,s}$.

The IS2 is proved by induction on the structure of ξ.

(i) Let $\xi = q(x_i)$, for some $q \in Q$ and $1 \leq i \leq k$. Then $\xi[x_1 \leftarrow s_1, \ldots, x_k \leftarrow s_k] = q(s_i)$ and thus every derivation starting from $\xi[x_1 \leftarrow s_1, \ldots, x_k \leftarrow s_k]$ in fact starts from $q(s_i)$. Hence, by the induction hypothesis on K, $g_{\xi,\omega} = g_{q,s_i}$ is suitable for our purpose.

(ii) Now let $\xi = \delta(\xi_1, \ldots, \xi_l)$, for some $\delta \in \Delta^{(l)}$, $l \geq 0$, and $\xi_1, \ldots, \xi_l \in RHS(Q, \Delta, k)$. Then we have $\xi[x_1 \leftarrow s_1, \ldots, x_k \leftarrow s_k] = \delta(\xi_1[x_1 \leftarrow s_1, \ldots, x_k \leftarrow s_k], \ldots, \xi_l[x_1 \leftarrow s_1, \ldots, x_k \leftarrow s_k])$ and, by induction hypothesis on ξ, for every $1 \leq i \leq l$, there exists an integer $g_{\xi_i,\omega}$ such that the derivations starting from $\xi_i[x_1 \leftarrow s_1, \ldots, x_k \leftarrow s_k]$ are not longer than $g_{\xi_i,\omega}$. On the other hand, we note that, for every derivation

$$
\begin{aligned}
& \xi[x_1 \leftarrow s_1, \ldots, x_k \leftarrow s_k] \\
= \; & \delta(\xi_1[x_1 \leftarrow s_1, \ldots, x_k \leftarrow s_k], \ldots, \xi_l[x_1 \leftarrow s_1, \ldots, x_k \leftarrow s_k]) \\
= \; & \varphi_0, \varphi_1, \ldots, \varphi_m
\end{aligned}
$$

starting from $\xi[x_1 \leftarrow s_1, \ldots, x_k \leftarrow s_k]$, and $0 \leq i \leq m$, we have $\varphi_i = \delta(\zeta_{i1}, \ldots, \zeta_{il})$, for some sentential forms $\zeta_{i1}, \ldots, \zeta_{il} \in SF(T)$. Moreover, by the definition of $\Rightarrow_T$, for each $0 \leq i \leq m - 1$, there is exactly one $1 \leq j \leq l$ such that $\zeta_{ij} \Rightarrow_T \zeta_{(i+1)j}$ and $\zeta_{ij'} = \zeta_{(i+1)j'}$, for every other j' with $1 \leq j' \neq j \leq l$. Therefore $g_{\xi,\omega} = \sum_{i=1}^{l} g_{\xi_i,\omega}$ is suitable. (Note, that in the particular case $l = 0$, we define $g_{\xi,\omega} = 0$).

We now finish the proof of the lemma. Let φ be an arbitrary sentential form. Then, by Lemma 3.7, there exist $k \geq 0$, $\xi \in RHS(Q, \Delta, k)$, and a sequence $\omega = (s_1, \ldots, s_k) \in (T_\Sigma)^k$ with $\varphi = \xi[x_1 \leftarrow s_1, \ldots, x_k \leftarrow s_k]$. We have proved that L holds, hence any derivation starting from φ is not longer than $g_{\xi,\omega}$, hence we have $g_\varphi = g_{\xi,\omega}$. □

We observe that, in the proof by simultaneous induction that K and L hold, the proof of the items IB and IS2 can usually be performed in one step. This is because IB can be incorporated into IS2 by extending IS2 to the case $k = 0$. Actually, IS2 in the case $k = 0$ means that, for every $b \in B_0$, $L_b(()) = true$, which is exactly IB. Therefore we will now usually incorporate the proof of IB in that of IS2.

Lemma 3.14. The derivation relation $\Rightarrow_T$ is confluent on $SF(T)$.

Proof. By Lemmas 3.9 and 3.13, $\Rightarrow_T$ is locally confluent and terminating, respectively. Hence, by Lemma 2.5, $\Rightarrow_T$ is also confluent. □

The following easy technical lemma is only a preparation for showing that the normal form of every sentential form is in T_Δ.

Lemma 3.15. Let $\delta \in \Delta^{(l)}$, $l \geq 0$, and $\xi_1, \ldots, \xi_l \in SF(T)$. If $\xi_1, \ldots, \xi_l \in SF(T)$ have normal forms, then

$$nf(\Rightarrow_T, \delta(\xi_1, \ldots, \xi_l)) = \delta(nf(\Rightarrow_T, \xi_1), \ldots, nf(\Rightarrow_T, \xi_l)).$$

Proof. Our statement follows from the fact that a rule can only be applied to a sentential form in such a place where a state occurs. Hence, the symbol δ at the root of $\delta(\xi_1, \ldots, \xi_l)$ remains unchanged during the derivation steps. □

The following theorem is the basis of defining the tree transformation induced by a top-down tree transducer.

Theorem 3.16. For every sentential form $\varphi \in SF(T)$, the normal form $nf(\Rightarrow_T, \varphi)$ exists. Moreover, $nf(\Rightarrow_T, \varphi) \in T_\Delta$.

Proof. By Lemmas 3.13 and 3.14, the relation $\Rightarrow_T$ is confluent and terminating on $SF(T)$. Then, by Corollary 2.7, the normal form $nf(\Rightarrow_T, \varphi)$ exists, for every $\varphi \in SF(T)$.

In order to prove that $nf(\Rightarrow_T, \varphi) \in T_\Delta$, we define the predicates K and L and then show by simultaneous induction that they hold.

K : For every $q \in Q$ and $s = \sigma(s_1, \ldots, s_k) \in T_\Sigma$, it holds that $nf(\Rightarrow_T, q(s)) \in T_\Delta$.

L : For every $k \geq 0$, $\xi \in RHS(Q, \Delta, k)$, and $s_1, \ldots, s_k \in T_\Sigma$, it holds that $nf(\Rightarrow_T, \xi[x_1 \leftarrow s_1, \ldots, x_k \leftarrow s_k]) \in T_\Delta$. (Here we use that, by Lemma 3.7, $\xi[x_1 \leftarrow s_1, \ldots, x_k \leftarrow s_k] \in SF(T)$.)

First we show IS1. Let ξ be the right-hand side of the (q, σ)-rule in R. Then, by the definition of $\Rightarrow_T$, we have $q(s) \Rightarrow_T \xi[x_1 \leftarrow s_1, \ldots, x_k \leftarrow s_k]$ and, by Corollary 2.8, it holds that $nf(\Rightarrow_T, q(s)) = nf(\Rightarrow_T, \xi[x_1 \leftarrow s_1, \ldots, x_k \leftarrow$

$s_k]$). Since, by the induction hypothesis on L, $nf(\Rightarrow_T, \xi[x_1 \leftarrow s_1, \ldots, x_k \leftarrow s_k]) \in T_\Delta$, we get that $nf(\Rightarrow_T, q(s))$ is also in T_Δ.

The proof of IS2 (and IB) can be done by induction on the structure of ξ.

(i) Let $\xi = p(x_i)$, for some $p \in Q$ and $1 \leq i \leq k$. Then $\xi[x_1 \leftarrow s_1, \ldots, x_k \leftarrow s_k] = p(s_i)$ and thus $nf(\Rightarrow_T, \xi[x_1 \leftarrow s_1, \ldots, x_k \leftarrow s_k]) = nf(\Rightarrow_T, p(s_i))$. Since, by the induction hypothesis on K, $nf(\Rightarrow_T, p(s_i)) \in T_\Delta$, we also have $nf(\Rightarrow_T, \xi[x_1 \leftarrow s_1, \ldots, x_k \leftarrow s_k]) \in T_\Delta$.

(ii) Now let $\xi = \delta(\xi_1, \ldots, \xi_l)$, for some $\delta \in \Delta^{(l)}$ with $l \geq 0$. Then, $\xi[x_1 \leftarrow s_1, \ldots, x_k \leftarrow s_k] = \delta(\xi_1[x_1 \leftarrow s_1, \ldots, x_k \leftarrow s_k], \ldots, \xi_l[x_1 \leftarrow s_1, \ldots, x_k \leftarrow s_k])$ and thus, by Lemma 3.15,

$$
\begin{aligned}
&nf(\Rightarrow_T, \xi[x_1 \leftarrow s_1, \ldots, x_k \leftarrow s_k]) = \\
&\delta(nf(\Rightarrow_T, \xi_1[x_1 \leftarrow s_1, \ldots, x_k \leftarrow s_k]), \ldots \\
&\ldots, nf(\Rightarrow_T, \xi_l[x_1 \leftarrow s_1, \ldots, x_k \leftarrow s_k])).
\end{aligned}
$$

On the other hand, by induction hypothesis on ξ, for every $1 \leq i \leq l$, we have $nf(\Rightarrow_T, \xi_i[x_1 \leftarrow s_1, \ldots, x_k \leftarrow s_k]) \in T_\Delta$, and thus $nf(\Rightarrow_T, \xi[x_1 \leftarrow s_1, \ldots, x_k \leftarrow s_k]) \in T_\Delta$.

Now our lemma follows from the fact that L holds and that, by Lemma 3.7, each sentential form φ can be written in the form $\xi[x_1 \leftarrow s_1, \ldots, x_k \leftarrow s_k]$, for some $k \geq 0$, $\xi \in RHS(Q, \Delta, k)$, and $s_1, \ldots, s_k \in T_\Sigma$. □

We can now define the tree transformation induced by a top-down tree transducer. In fact, it is a mapping that takes every input tree $s \in T_\Sigma$ to the normal form of the initial sentential form $q_0(s)$.

Definition 3.17. Let $q \in Q$ be a state of T. Then

(a) The *tree transformation induced by T with q* is the mapping $\tau_{T,q}^{der} : T_\Sigma \rightarrow T_\Delta$ such that, for every $s \in T_\Sigma$, we define $\tau_{T,q}^{der}(s) = nf(\Rightarrow_T, q(s))$.

(b) The *tree transformation induced by T* is the mapping $\tau_T^{der} = \tau_{T,q_0}^{der}$. □

We call a tree transformation a *top-down tree transformation* if it can be induced by a top-down tree transducer. Moreover, the class of all top-down tree transformations is denoted by TOP.

By Theorem 3.16, the normal form of a sentential form φ is in T_Δ. On the other hand, if φ is a sentential form and, for some $s \in T_\Delta$, we have $\varphi \Rightarrow_T^* s$, then clearly s is irreducible and hence it is a normal form of φ. In conclusion this means that $\varphi \Rightarrow_T^* s$, for some $s \in T_\Delta$, if and only if s is a normal form of φ. Therefore the tree transformation induced by T can alternatively be written in the following form, which is generally used in the theory of tree transformation:

$$\tau_T^{der} = \{(r, s) \in T_\Sigma \times T_\Delta \mid q_0(r) \Rightarrow_T^* s\}.$$

To finish this section, we give an example of a tree transformation.

Example 3.18. It can be easily seen that the tree transformation induced by T' in Example 3.3 is

$$\tau_{T'}^{der} = \{(\alpha, \alpha)\} \cup \{(\gamma^n(\alpha), \delta(\gamma^{n-1}(\beta_1), \gamma^{n-1}(\beta_2))) \mid n \geq 1\}.$$

In the rest of the book we will call this tree transformation $\tau_{chain-to-chains}$ (Fig. 3.3). □

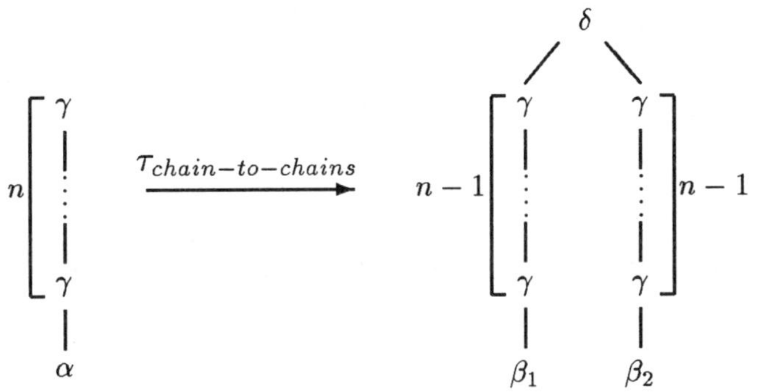

Fig. 3.3. The tree transformation $\tau_{chain-to-chains}$

3.3 Characterization of Top-Down Tree Transformations

The aim of this section is to introduce certain inductively defined functions for a top-down tree transducer which will be very close to the tree transformation induced by it and will serve as both a precise and convenient tool for proving statements concerning top-down tree transformations.

First we present the principle of definition by simultaneous induction which can be viewed as an extension of the principle of definition by structural induction (see Chap. 2). Then, using this new principle, we define two families of functions concerning a top-down tree transducer and show that they can also be expressed by normal forms of sentential forms. Hence they also define the tree transformation induced by a top-down tree transducer.

We start with the following definition.

Definition 3.19. Let Σ be a ranked alphabet and let C be a set. Let A be a set and, for every $a \in A$, let $f_a : T_\Sigma \to C$ be a function. Moreover, for every $k \geq 0$, let B_k be a set and, for every $b \in B_k$, let $g_{b,k} : (T_\Sigma)^k \to C$ be a function.

Let

$$F = \{f_a \mid a \in A\}$$

and let

$$G = \{g_{b,k} \mid b \in B_k \text{ for some } k \geq 0\}.$$

We say that F is defined if, for every $a \in A$ and $s \in T_\Sigma$, $f_a(s)$ is defined. Similarly, we say that G is defined if, for every $k \geq 0, b \in B_k$ and $s_1, \ldots, s_k \in T_\Sigma$, $g_{b,k}((s_1, \ldots, s_k))$ are defined. □

Definition by simultaneous induction is a method to define simultaneously families of functions like F and G. This method generalizes the principle of definition by structural induction (Principle 2.11). The principle of definition by simultaneous induction is as follows.

Principle 3.20. Let F and G be two families of functions as in Def. 3.19. The *principle of definition by simultaneous induction* is:

If
IB: for every $b \in B_0$, the value $g_{b,0}(())$ is defined, and
IS1: for every $a \in A$, $k \geq 0$, and $s_1, \ldots, s_k \in T_\Sigma$, if for every $b \in B_k$, $g_{b,k}((s_1, \ldots, s_k))$ is defined, then, for every $\sigma \in \Sigma^{(k)}$, $f_a(\sigma(s_1, \ldots, s_k))$ is defined, and
IS2: for every $k \geq 1$ and $s_1, \ldots, s_k \in T_\Sigma$, if for every $a_1, \ldots, a_k \in A$, $f_{a_1}(s_1), \ldots, f_{a_k}(s_k)$ are defined, then, for every $b \in B_k$, $g_{b,k}((s_1, \ldots, s_k))$ is defined,
then F and G are defined. □

The abbreviations IB, IS1, and IS2 stand for induction base, induction step 1, and induction step 2, respectively.

Similarly to the principle of proof by simultaneous induction, when we define F and G by simultaneous induction, the items IB and IS2 can usually be incorporated into one step. Namely, IB can be incorporated into IS2 by extending IS2 to the case $k = 0$. In fact, IS2 in the case $k = 0$ means that, for every $b \in B_0$, $g_{b,0}(())$ is defined, which is exactly IB.

Note 3.21. Similarly to Principle 3.11 in concrete applications of Principle 3.20, the sets $B_k, k \geq 0$ will form a hierarchy. Moreover, for every $k \geq 0$ and $b \in B_k$, the functions $g_{b,k}$ will have the property that, for every $l \geq k$ and sequence $s_1, \ldots, s_l$ of trees in T_Σ,

$$g_{b,k}((s_1, \ldots, s_k)) = g_{b,l}((s_1, \ldots, s_l)).$$

The arity of a function $g_{b,k}$ will always be clear from the context, namely from the specification of its arguments, therefore we drop the index k from $g_{b,k}$ in concrete applications of Principle 3.20.

Using this principle, we define two families of functions concerning a top-down tree transducer as follows.

Definition 3.22. Let $T = (Q, \Sigma, \Delta, q_0, R)$ be a top-down tree transducer. We define families

$$F_T = \{\tau^{ind}_{T,q} : T_\Sigma \to T_\Delta \mid q \in Q\}$$

and

$$G_T = \{\tau^{ind}_{T,\xi} : (T_\Sigma)^k \to T_\Delta \mid k \geq 0, \xi \in RHS(Q, \Delta, k)\}$$

of functions by simultaneous induction as follows.

IB: For every $\xi \in RHS(Q, \Delta, 0)$, let $\tau^{ind}_{T,\xi}(()) = \xi$.

IS1: For every $\sigma \in \Sigma^{(k)}$ with $k \geq 0$, $s_1, \ldots, s_k \in T_\Sigma$, and $q \in Q$, we define $\tau^{ind}_{T,q}(\sigma(s_1, \ldots, s_k)) = \tau^{ind}_{T,\xi}((s_1, \ldots, s_k))$, where $\xi = rhs(q, \sigma)$.

IS2: For every $k \geq 1$, $\xi \in RHS(Q, \Delta, k)$, and $\omega = (s_1, \ldots, s_k) \in (T_\Sigma)^k$, we define $\tau^{ind}_{T,\xi}(\omega)$ by induction on the structure of ξ as follows:

(i) If $\xi = p(x_i)$ for some $p \in Q$ and $1 \leq i \leq k$, then $\tau^{ind}_{T,\xi}(\omega) = \tau^{ind}_{T,p}(s_i)$.

(ii) If $\xi = \delta(\xi_1, \ldots, \xi_l)$ for some $\delta \in \Delta^{(l)}$ with $l \geq 0$ and $\xi_1, \ldots, \xi_l \in RHS(Q, \Delta, k)$, then $\tau^{ind}_{T,\xi}(\omega) = \delta(\tau^{ind}_{T,\xi_1}(\omega), \ldots, \tau^{ind}_{T,\xi_l}(\omega))$. □

We note that in the particular case $\xi \in T_\Delta$, it holds that $\tau^{ind}_{T,\xi}(\omega) = \xi$, whatever k is. This is always the case, in particular if $k = 0$. Therefore, for every $s = \sigma \in \Sigma^{(0)}$ and $q \in Q$, it holds that $\tau^{ind}_{T,q}(s) = \xi$, where $\xi = rhs(q, \sigma) \in RHS(Q, \Delta, 0) = T_\Delta$.

In the following we give an example for a computation of the functions τ^{ind}.

Example 3.23. Let T' be the tree transducer defined in Example 3.3. We compute $\tau^{ind}_{T',q_0}(\gamma^2(\alpha))$.

$$\begin{aligned}
\tau^{ind}_{T',q_0}(\gamma^2(\alpha)) &= \tau^{ind}_{T',\delta(q_1(x_1),q_2(x_1))}((\gamma(\alpha)))\\
&= \delta(\tau^{ind}_{T',q_1(x_1)}((\gamma(\alpha))), \tau^{ind}_{T',q_2(x_1)}((\gamma(\alpha))))\\
&= \delta(\tau^{ind}_{T',q_1}(\gamma(\alpha)), \tau^{ind}_{T',q_2}(\gamma(\alpha)))\\
&= \delta(\tau^{ind}_{T',\gamma(q_1(x_1))}((\alpha)), \tau^{ind}_{T',\gamma(q_2(x_1))}((\alpha)))\\
&= \delta(\gamma(\tau^{ind}_{T',q_1(x_1)}((\alpha))), \gamma(\tau^{ind}_{T',q_2(x_1)}((\alpha))))\\
&= \delta(\gamma(\tau^{ind}_{T',q_1}(\alpha)), \gamma(\tau^{ind}_{T',q_2}(\alpha)))\\
&= \delta(\gamma(\tau^{ind}_{T',\beta_1}(())), \gamma(\tau^{ind}_{T',\beta_2}(())))\\
&= \delta(\gamma(\beta_1), \gamma(\beta_2))
\end{aligned}$$

□

We now show that the functions τ^{der}, defined by normal forms, and τ^{ind}, defined by simultaneous induction, for a top-down tree transducer T coincide.

Since the proof will be performed by simultaneous induction, we also require the missing function $\tau^{der}_{T,\xi}$.

Definition 3.24. Let $T = (Q, \Sigma, \Delta, q_0, R)$ be a top-down tree transducer, $k \geq 0, \xi \in RHS(Q, \Delta, k)$. We define the function $\tau^{der}_{T,\xi} : (T_\Sigma)^k \to T_\Delta$ for every $\omega = (s_1, \ldots, s_k) \in (T_\Sigma)^h$ by $\tau^{der}_{T,\xi}(\omega) = nf(\Rightarrow_T, \xi[x_1 \leftarrow s_1, \ldots, x_k \leftarrow s_k])$. □

Now the proof of the equality of τ^{ind} and τ^{der} follows. We recall from Lemma 3.7 that $\xi[x_1 \leftarrow s_1, \ldots, x_k \leftarrow s_k] \in SF(T)$.

Theorem 3.25. Let $T = (Q, \Sigma, \Delta, q_0, R)$ be a top-down tree transducer. Then the predicates K and L hold, where

K : For every $q \in Q$ and $s \in T_\Sigma$, $\tau^{der}_{T,q}(s) = \tau^{ind}_{T,q}(s)$.
L : For every $\xi \in RHS(Q, \Delta, k)$ with $k \geq 0$ and $\omega = (s_1, \ldots, s_k) \in (T_\Sigma)^k$, $\tau^{der}_{T,\xi}(\omega) = \tau^{ind}_{T,\xi}(\omega)$.

Proof. We prove by simultaneous induction and start by proving IS1. Let $q \in Q, s = \sigma(s_1, \ldots, s_k)$, for some $\sigma \in \Sigma^{(k)}$, $k \geq 0$, and $s_1, \ldots, s_k \in T_\Sigma$. Moreover, let $\omega = (s_1, \ldots, s_k)$. First we note that, by Def. 3.4, it holds that $q(s) \Rightarrow_T \xi[x_1 \leftarrow s_1, \ldots, x_k \leftarrow s_k]$, where $\xi = rhs(q, \sigma)$. Then we can compute as follows:

$$\begin{array}{rcll}
\tau^{der}_{T,q}(s) & = & nf(\Rightarrow_T, q(s)) & \text{(by Def. 3.17)} \\
& = & nf(\Rightarrow_T, \xi[x_1 \leftarrow s_1, \ldots & \\
& & \ldots, x_k \leftarrow s_k]) & \text{(by Corollary 2.8)} \\
& = & \tau^{der}_{T,\xi}(\omega) & \text{(by Def. 3.24)} \\
& = & \tau^{ind}_{T,\xi}(\omega) & \text{(by induction hypothesis on } L) \\
& = & \tau^{ind}_{T,q}(s) & \text{(by Def. 3.22).}
\end{array}$$

Now we show IS2 by induction on the structure of ξ.

(i) Let $\xi = p(x_i)$, for some $p \in Q$ and $1 \leq i \leq k$. Then $\xi[x_1 \leftarrow s_1, \ldots, x_k \leftarrow s_k] = p(s_i)$ and thus $nf(\Rightarrow_T, \xi[x_1 \leftarrow s_1, \ldots, x_k \leftarrow s_k]) = nf(\Rightarrow_T, p(s_i))$. Hence we obtain the following:

$$\begin{array}{rcll}
\tau^{der}_{T,\xi}(\omega) & = & nf(\Rightarrow_T, \xi[x_1 \leftarrow s_1, \ldots & \\
& & \ldots, x_k \leftarrow s_k]) & \text{(by Def. 3.24)} \\
& = & nf(\Rightarrow_T, p(s_i)) & \\
& = & \tau^{der}_{T,p}(s_i) & \text{(by Def. 3.17)} \\
& = & \tau^{ind}_{T,p}(s_i) & \text{(by induction hypothesis on } K) \\
& = & \tau^{ind}_{T,\xi}(\omega) & \text{(by Def. 3.22).}
\end{array}$$

(ii) Let now $\xi = \delta(\xi_1, \ldots, \xi_l)$, for some $\delta \in \Delta^{(l)}$ with $l \geq 0$, and $\xi_1, \ldots, \xi_l \in RHS(Q, \Delta, k)$. First we note that $\xi[x_1 \leftarrow s_1, \ldots, x_k \leftarrow s_k] = \delta(\xi_1[x_1 \leftarrow s_1, \ldots, x_k \leftarrow s_k], \ldots, \xi_l[x_1 \leftarrow s_1, \ldots, x_k \leftarrow s_k])$ and therefore

$$\begin{aligned} & nf(\Rightarrow_T, \xi[x_1 \leftarrow s_1, \ldots, x_k \leftarrow s_k]) = \\ & \qquad \delta(nf(\Rightarrow_T, \xi_1[x_1 \leftarrow s_1, \ldots, x_k \leftarrow s_k]), \ldots, \\ & \qquad nf(\Rightarrow_T, \xi_l[x_1 \leftarrow s_1, \ldots, x_k \leftarrow s_k])) \end{aligned}$$

Thus we get the following.

$$\begin{aligned} \tau^{der}_{T,\xi}(\omega) &= nf(\Rightarrow_T, \xi[x_1 \leftarrow s_1, \ldots, x_k \leftarrow s_k]) \\ & \qquad \text{(by Def. 3.24)} \\ &= \delta(nf(\Rightarrow_T, \xi_1[x_1 \leftarrow s_1, \ldots, x_k \leftarrow s_k]), \ldots \\ & \quad \ldots, nf(\Rightarrow_T, \xi_l[x_1 \leftarrow s_1, \ldots, x_k \leftarrow s_k])) \\ &= \delta(\tau^{der}_{T,\xi_1}(\omega), \ldots, \tau^{der}_{T,\xi_l}(\omega)) \quad \text{(by Def. 3.24)} \\ &= \delta(\tau^{ind}_{T,\xi_1}(\omega), \ldots, \tau^{ind}_{T,\xi_l}(\omega)) \quad \text{(by induction hypothesis on } L) \\ &= \tau^{ind}_{T,\xi}(\omega) \quad \text{(by Def. 3.22).} \end{aligned}$$

□

We have proved that the two function families defined by means of normal forms and by simultaneous induction are the same. Therefore, in the rest of the chapter we drop the superscripts *der* and *ind* when we refer to the functions related to a top-down tree transducer. However, it should be clear from the context which definition of the function we are working with.

3.4 Height Property

In this section we show that the height of the output tree $\tau_T(s)$ associated to an input tree s by a top-down tree transducer T is linearly bounded by the height of s.

We need the following technical lemma which is essentially the generalization of the observation made after Def. 3.22.

Lemma 3.26. Let $T = (Q, \Sigma, \Delta, q_0, R)$ be a top-down tree transducer. Then, for every $k \geq 0$, $\xi \in RHS(Q, \Delta, k)$, and $\omega = (s_1, \ldots, s_k) \in (T_\Sigma)^k$, it holds that $\tau_{T,\xi}(\omega) = \xi[q(x_j) \leftarrow \tau_{T,q}(s_j) \; ; \; q(x_j) \in Q(X_k)]$ where $Q(X_k)$ denotes the set $\{q(x_i) \mid q \in Q, 1 \leq i \leq k\}$.

Proof. We prove by structural induction on ξ.

(i) Let $\xi = p(x_i)$ for some $p \in Q$ and $1 \leq i \leq k$. Then $\tau_{T,\xi}(\omega) = \tau_{T,p}(s_i)$ by the definition of $\tau_{T,\xi}$. On the other hand,

$$\begin{aligned} \tau_{T,p}(s_i) &= p(x_i)[q(x_j) \leftarrow \tau_{T,q}(s_j) \; ; \; q(x_j) \in Q(X_k)] \\ & \quad \text{(by the definition of } \leftarrow) \\ &= \xi[q(x_j) \leftarrow \tau_{T,q}(s_j) \; ; \; q(x_j) \in Q(X_k)] \\ & \quad \text{(by } \xi = p(x_i)). \end{aligned}$$

(ii) Now let $\xi = \delta(\xi_1, \ldots, \xi_l)$, for some $\delta \in \Delta^{(l)}$ with $l \geq 0$, and $\xi_1, \ldots, \xi_l \in RHS(Q, \Delta, k)$. Then we have

$$\begin{array}{rcl} \tau_{T,\xi}(\omega) & = & \delta(\tau_{T,\xi_1}(\omega), \ldots, \tau_{T,\xi_l}(\omega)) \\ & & \text{(by Def. 3.22)} \\ & = & \delta(\xi_1[q(x_j) \leftarrow \tau_{T,q}(s_j)\ ;\ q(x_j) \in Q(X_k)], \ldots \\ & & \ldots, \xi_l[q(x_j) \leftarrow \tau_{T,q}(s_j)\ ;\ q(x_j) \in Q(X_k)]) \\ & & \text{(by induction hypothesis on } \xi) \\ & = & \delta(\xi_1, \ldots, \xi_l)[q(x_j) \leftarrow \tau_{T,q}(s_j)\ ;\ q(x_j) \in Q(X_k)] \\ & = & \xi[q(x_j) \leftarrow \tau_{T,q}(s_j)\ ;\ q(x_j) \in Q(X_k)]. \end{array}$$

□

We can now prove the relation mentioned earlier between the input and output trees of a top-down tree transducer.

Lemma 3.27. Let $T = (Q, \Sigma, \Delta, q_0, R)$ be a top-down tree transducer. There is an integer $c > 0$ such that, for every $s \in T_\Sigma$, $height(\tau_T(s)) \leq c \cdot height(s)$.

Proof. Let c be the maximal height of the right-hand sides of rules in R. We show that, for every $q \in Q$ and $s \in T_\Sigma$, $height(\tau_{T,q}(s)) \leq c \cdot height(s)$. We prove by induction on s.

(i) Let $s = \sigma \in \Sigma^{(0)}$ and let $\xi = rhs(q, \sigma)$. Then $\tau_{T,q}(s) = \xi$ and we obtain that $height(\tau_{T,q}(s)) = height(\xi) \leq c = c \cdot height(s)$.

(ii) Now let $s = \sigma(s_1, \ldots, s_k)$, for some $k \geq 1$, $\sigma \in \Sigma^{(k)}$, and $s_1, \ldots, s_k \in T_\Sigma$. Moreover, again let $\xi = rhs(q, \sigma)$. Then we can compute as follows, where again $Q(X_k)$ denotes the set $\{q(x_i) \mid q \in Q, 1 \leq i \leq k\}$.

$$\begin{array}{rcl} height(\tau_{T,q}(s)) & = & height(\xi[q(x_j) \leftarrow \tau_{T,q}(s_j)\,;\, q(x_j) \in Q(X_k)]) \\ & & \text{(by Lemma 3.26)} \\ & \leq & height(\xi) + \\ & & max\{height(\tau_{T,q}(s_j)) \mid q(x_j) \in Q(X_k)\} \\ & & \text{(by the definition of } height) \\ & \leq & c + max\{height(\tau_{T,q}(s_j)) \mid q(x_j) \in Q(X_k)\} \\ & & \text{(by the definition of } c) \\ & \leq & c + max\{c \cdot height(s_j) \mid 1 \leq j \leq k\} \\ & & \text{(by induction hypothesis)} \\ & = & c \cdot (1 + max\{height(s_j) \mid 1 \leq j \leq k\}) \\ & = & c \cdot height(s). \end{array}$$

□

3.5 Subclasses of *TOP*

In this section we introduce and study restricted versions of top-down tree transducers. These will induce certain subclasses of *TOP*. Actually, we impose three kinds of restrictions and obtain the concepts of a homomorphism tree transducer, a linear top-down tree transducer and a superlinear top-down tree transducer. We give an example of the tree transformation induced by each of the restricted versions. Finally we compare the tree transformation power of the versions mentioned by giving an inclusion diagram of the tree transformation classes which they induce.

We start with defining the restrictions.

Definition 3.28. Let $T = (Q, \Sigma, \Delta, q_0, R)$ be a top-down tree transducer. Then

(a) T is a *homomorphism tree transducer* if it has only one state, i.e., if $Q = \{q_0\}$. A tree transformation is called a *homomorphism tree transformation* if it can be induced by a homomorphism tree transducer. The class of all homomorphism tree transformations is denoted by HOM.

(b) T is *linear* if, for every rule $q(\sigma(x_1, \ldots, x_k)) \to \xi$ in R and every $1 \leq i \leq k$, the variable x_i occurs at most once in ξ. A tree transformation is called a *linear top-down tree transformation* if it can be induced by a linear top-down tree transducer. The class of all linear top-down tree transformations is denoted by *l-TOP*.

(c) T is *superlinear* if it is linear and, for every $\sigma \in \Sigma^{(k)}$ with $k \geq 0$ and $q, q' \in Q$ with $q \neq q'$, $Var_X(rhs(q, \sigma)) \cap Var_X(rhs(q', \sigma)) = \emptyset$. Informally, this property means that for every $\sigma \in \Sigma^{(k)}$ with $k \geq 0$ and every $1 \leq i \leq k$, there is at most one $q \in Q$ such that x_i occurs in $rhs(q, \sigma)$. A tree transformation is called *superlinear top-down tree transformation* if it can be induced by a superlinear top-down tree transducer. The class of all superlinear top-down tree transformations is denoted by *sl-TOP*. □

The attributes linear and superlinear can also be applied to homomorphism tree transducers. Thus we also have classes *l-HOM* and *sl-HOM*.

In the following we give examples for the tree transformations induced by each of the above restricted top-down tree transducers. First we present a homomorphism tree transducer.

Example 3.29. Let the top-down tree transducer $T = (\{q\}, \Sigma, \Delta, q, R)$ be defined as follows:

- $\Sigma = \{\gamma^{(1)}, \alpha^{(0)}\}$, $\Delta = \{\sigma^{(2)}, \alpha^{(0)}\}$,
- R consists of the rules
  1) $q(\gamma(x_1)) \to \sigma(q(x_1), q(x_1))$,
  2) $q(\alpha) \to \alpha$.

Then obviously T is a homomorphism tree transducer, hence $\tau_T \in HOM$. A derivation by T looks as follows:

$$\begin{array}{llllll} q(\gamma^2(\alpha)) & \Rightarrow_T & \sigma(q(\gamma(\alpha)), q(\gamma(\alpha))) & \Rightarrow_T & \sigma(\sigma(q(\alpha), q(\alpha)), q(\gamma(\alpha))) \\ & \Rightarrow_T^2 & \sigma(\sigma(\alpha, \alpha), q(\gamma(\alpha))) & \Rightarrow_T^3 & \sigma(\sigma(\alpha, \alpha), \sigma(\alpha, \alpha)). \end{array}$$

Now we explicitly give τ_T. For this, let us denote the fully balanced tree over Δ of height n by s_n, i.e.,

(i) let $s_1 = \alpha$, and
(ii) let $s_{n+1} = \sigma(s_n, s_n)$, for every $n \geq 1$.

Then we can prove by induction on n that, for every $n \geq 0$, we have $\tau_T(\gamma^n(\alpha)) = s_{n+1}$:

(i) If $n = 0$, then $q(\gamma^0(\alpha)) = q(\alpha) \Rightarrow_T \alpha = s_1$, hence $\tau_T(\gamma^0(\alpha)) = s_1$.

(ii) Let $n > 0$. Then $q(\gamma^n(\alpha)) \Rightarrow_T \sigma(q(\gamma^{n-1}(\alpha)), q(\gamma^{n-1}(\alpha)))$ and, by induction hypothesis and Theorem 3.25, $q(\gamma^{n-1}(\alpha)) \Rightarrow_T^* s_n$. Hence we get $q(\gamma^n(\alpha)) \Rightarrow_T^* \sigma(s_n, s_n) = s_{n+1}$, meaning that $\tau_T(\gamma^n(\alpha)) = s_{n+1}$.

Thus the tree transformation induced by T is

$$\tau_T = \{(\gamma^n(\alpha), s_{n+1}) \mid n \geq 0\},$$

(Fig. 3.4). We call this tree transformation τ_{double}. □

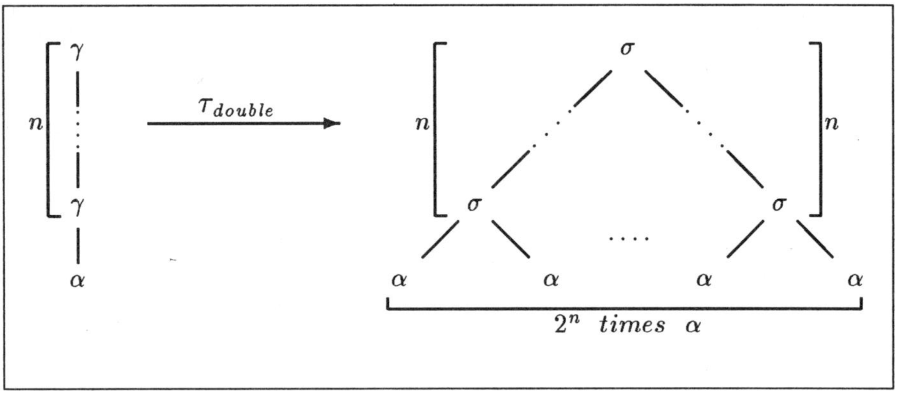

Fig. 3.4. The tree transformation τ_{double}

The following example shows a linear top-down tree transducer.

Example 3.30. Consider the top-down tree transducer
$T = (\{q_1, q_2\}, \Sigma, \Delta, q_1, R)$, where

- $\Sigma = \{\gamma^{(1)}, \alpha^{(0)}\}$, $\Delta = \{\gamma_1^{(1)}, \gamma_2^{(1)}, \alpha^{(0)}\}$,
- R is the set of the rules
  1) $q_1(\gamma(x_1)) \rightarrow \gamma_1(q_2(x_1))$,
  2) $q_2(\gamma(x_1)) \rightarrow \gamma_2(q_1(x_1))$,
  3) $q_1(\alpha) \rightarrow \alpha$,
  4) $q_2(\alpha) \rightarrow \alpha$.

Then T is a linear top-down tree transducer and hence $\tau_T \in l\text{-}TOP$.

At the same time, T is not superlinear because x_1 appears in both $rhs(q_1,\gamma)$ and $rhs(q_2,\gamma)$. A derivation for T looks as follows:

$$\begin{array}{lllll} q_1(\gamma^3(\alpha)) & \Rightarrow_T & \gamma_1(q_2(\gamma^2(\alpha))) & \Rightarrow_T & \gamma_1(\gamma_2(q_1(\gamma(\alpha)))) \\ & \Rightarrow_T & \gamma_1(\gamma_2(\gamma_1(q_2(\alpha)))) & \Rightarrow_T & \gamma_1(\gamma_2(\gamma_1(\alpha))). \end{array}$$

Again, we are going to specify the tree transformation induced by T. We define the trees $s_{1,m}, s_{2,m}$, for each $m \geq 1$, by induction on m as follows:

(i) Let $s_{1,1} = s_{2,1} = \alpha$.
(ii) For $m \geq 1$, let $s_{1,m+1} = \gamma_1(s_{2,m})$ and $s_{2,m+1} = \gamma_2(s_{1,m})$.

Then, it should be clear that the tree transformation induced by T is

$$\tau_T = \{(r, s_{1,m}) \mid r \in T_\Sigma, m = height(r)\}.$$

The above tree transformation will be called $\tau_{alternate}$, (Fig. 3.5). □

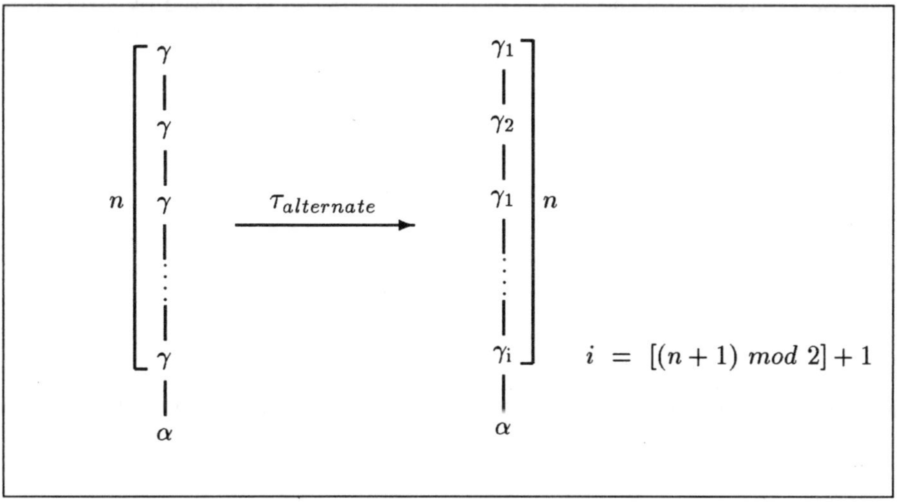

Fig. 3.5. The tree transformation $\tau_{alternate}$

The third example is a superlinear top-down tree transducer.

Example 3.31. Let $T = (\{q_1, q_2\}, \Sigma, \Delta, q_1, R)$ be the top-down tree transducer, where

- $\Sigma = \{\delta^{(2)}, \alpha^{(0)}\}$, $\Delta = \{1^{(1)}, 2^{(1)}, \alpha^{(0)}\}$
- R is the set of the four rules
  1) $q_1(\delta(x_1,x_2)) \to 1(q_2(x_1))$,
  2) $q_2(\delta(x_1,x_2)) \to 2(q_1(x_2))$,

3) $q_1(\alpha) \to \alpha$,
4) $q_2(\alpha) \to \alpha$.

Clearly T is a superlinear top-down tree transducer and thus $\tau_T \in$ *sl-TOP*.

First we present a derivation by T:

$$q_1(\delta(\delta(\alpha,\alpha),\alpha)) \Rightarrow_T 1(q_2(\delta(\alpha,\alpha))) \Rightarrow_T 1(2(q_1(\alpha))) \Rightarrow 1(2(\alpha)).$$

Next we give τ_T. We define, for $i = \{1,2\}$, the i-path $w_i \in T_\Delta$ of a tree $r \in T_\Sigma$, by structural induction as follows:

(i) If $r = \alpha$, then $w_i = \alpha$.
(ii) Let $r = \delta(r_1, r_2)$, where $r_1, r_2 \in T_\Sigma$, and let the i-path of r_j be $v_{i,j}$, for $j = \{1,2\}$. Then we define $w_1 = 1v_{2,1}$ and $w_2 = 2v_{1,2}$.

Intuitively, the i-path of a tree $r = \delta(r_1, r_2) \in T_\Sigma$ can be obtained by taking the i-th branch at the root δ of r and then taking the $(i \bmod(2) + 1)$-path of the subtree r_i we reached. (For example, let $r = \delta(\delta(\alpha,\alpha),\alpha)$. Then the 1-path of r is $12(\alpha)$ and the 2-path of it is $2(\alpha)$.) It can be proved by structural induction that, for each $r \in T_\Sigma$, we have $\tau_{T,q_i}(r)$ as the i-path of r. Hence we obtain

$$\tau_T = \{(r, w) \mid r \in T_\Sigma \text{ and } w \text{ is the 1-path of } r\}.$$

We will call this tree transformation τ_{zigzag} (Fig. 3.6). □

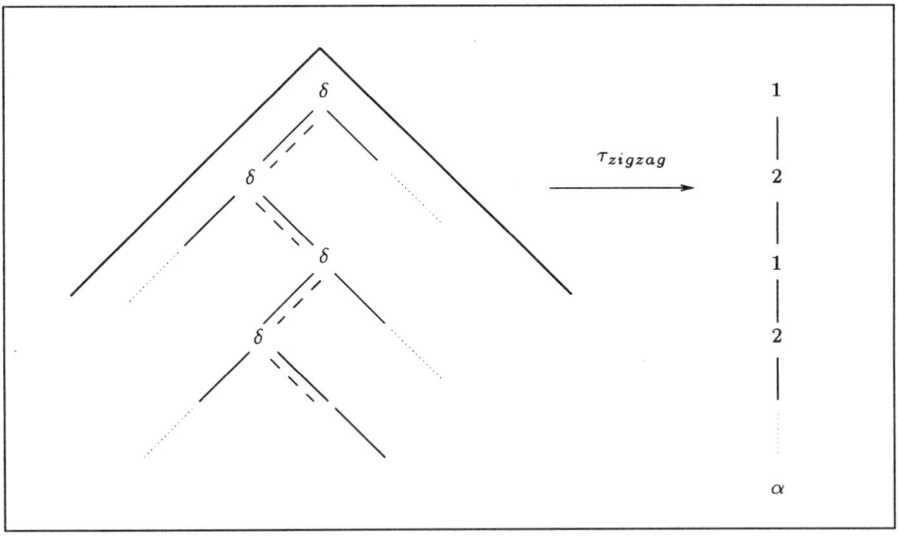

Fig. 3.6. The tree transformation τ_{zigzag}

We devote the rest of this section to comparing the tree transformational capacity of the variants of top-down tree transducers we have defined up to now. We perform the comparison by giving the inclusion diagram of the set of the six classes *TOP*, *l-TOP*, *sl-TOP*, *HOM*, *l-HOM*, and *sl-HOM* above, with respect to inclusion. We construct the inclusion diagram with the method presented in Sect. 2.2, i.e., we first give a conjecture for the inclusion diagram and then prove that it is the inclusion diagram. We recall that the proof involves determining the minimal set of inequalities that are necessary to prove that the conjectured diagram is the inclusion diagram.

We start by giving some inclusions and one equality which are immediate consequences of Def. 3.28.

Corollary 3.32.

$$\begin{array}{lrcl} 1) & HOM & \subseteq & TOP \\ 2) & l\text{-}TOP & \subseteq & TOP \\ 3) & sl\text{-}TOP & \subseteq & l\text{-}TOP \\ 4) & l\text{-}HOM & \subseteq & HOM \\ 5) & sl\text{-}HOM & \subseteq & sl\text{-}TOP \\ 6) & sl\text{-}HOM & = & l\text{-}HOM. \end{array}$$

Proof. The inclusions 1) - 5) and $sl\text{-}HOM \subseteq l\text{-}HOM$ hold by definition. We only have to prove the converse of the last one.

Therefore, take a linear homomorphism tree transducer $T = (\{q_0\}, \Sigma, \Delta, q_0, R)$ and an input symbol $\sigma \in \Sigma^{(k)}$ with $k \geq 0$. Then T is obviously superlinear because there are no two different states in $\{q_0\}$, see (c) of Def. 3.28. Hence $l\text{-}HOM \subseteq sl\text{-}HOM$, too. □

We now give a conjecture for the inclusion diagram. We conjecture that the inclusion diagram of the set of the six classes we consider is that which can be seen in Fig. 3.7. We now prove that this is true.

We first note that every inclusion which is shown by the diagram has already been proved in Corollary 3.32. Moreover, we observe that the inequalities which are necessary to prove are as follows:

$$\begin{array}{lcl} TOP - HOM & \neq & \emptyset \\ HOM - sl\text{-}HOM & \neq & \emptyset \\ TOP - l\text{-}TOP & \neq & \emptyset \\ l\text{-}TOP - sl\text{-}TOP & \neq & \emptyset \\ sl\text{-}TOP - sl\text{-}HOM & \neq & \emptyset \\ HOM - sl\text{-}TOP & \neq & \emptyset \\ HOM - l\text{-}TOP & \neq & \emptyset \\ sl\text{-}TOP - HOM & \neq & \emptyset \\ l\text{-}TOP - HOM & \neq & \emptyset. \end{array}$$

Then it is straightforward to see that the minimal inequalities are the following.

$$\begin{array}{lcl} HOM - l\text{-}TOP & \neq & \emptyset \\ sl\text{-}TOP - HOM & \neq & \emptyset \\ l\text{-}TOP - sl\text{-}TOP & \neq & \emptyset \end{array}$$

Next we prove the first one.

Lemma 3.33. *$HOM - l\text{-}TOP \neq \emptyset$.*

Proof. We prove by contradiction that the tree transformation τ_{double} appearing in Example 3.29 cannot be induced by any linear top-down tree transducer. Therefore, assume that $\tau_{double} = \tau_{T'}$, for some linear top-down tree transducer $T' = (Q, \Sigma, \Delta, q_0, R')$. Take an arbitrary element of T_Σ, which, of course, has the form $\gamma^n(\alpha)$, for some $n \geq 0$. Since $\tau_{double} = \tau_{T'}$, there is the derivation $q_0(\gamma^n(\alpha)) \Rightarrow_{T'} s_{n+1}$. Suppose that the mth step of this derivation is the first such one in which the right-hand side of the rule applied for γ is not of the form $p(x_1)$, for some $p \in Q$. Then the involved derivation can be shown as

$$q_0(\gamma^n(\alpha)) \Rightarrow_{T'}^{m-1} p(\gamma^{n-m+1}(\alpha)) \Rightarrow_{T'} \xi[x_1 \leftarrow \gamma^{n-m}(\alpha)] \Rightarrow_{T'}^{*} s_{n+1},$$

where $p \in Q$, $p(x_1)$ is the right-hand side of the rule applied in the $(m-1)$th step and $p(\gamma(x_1)) \to \xi$ is the rule applied in the mth step.

Next we show that ξ contains at least one occurrence of α. Actually, either $\xi = \alpha$ or ξ contains at least one occurrence of σ. In the first case our statement obviously holds. Considering the second case we observe, on the one hand, that σ has rank 2 and, on the other, that x_1 can occur at most once in ξ because T' is linear. However these two conditions mean that ξ should contain at least one occurrence of α in the second case, too.

So there is at least one occurrence of α in ξ anyway. Our next observation is that the lengths of all paths in ξ leading from its root to occurrences of α must be the same (specifically they must be $n+1$, because the length of all such paths in s_{n+1} is $n+1$).

Finally, let $n' > n$ and consider the derivation

$$q_0(\gamma^{n'}(\alpha)) \Rightarrow_{T'}^{m-1} p(\gamma^{n'-m+1}(\alpha)) \Rightarrow_{T'} \xi[x_1 \leftarrow \gamma^{n'-m}(\alpha)] \Rightarrow_{T'}^{*} s_{n'+1}.$$

Then there will be paths of length $n+1$ in $s_{n'+1}$ leading from its root to occurrences of α, because there are such paths in ξ, due to our observation. However, this is a contradiction because the length of every such path in $s_{n'+1}$ must be $n'+1 > n+1$.

Hence τ_{double} cannot be induced by a linear top-down tree transducer. □

Corollary 3.34. *$HOM - sl\text{-}TOP \neq \emptyset$.*

Proof. The statement is an obvious consequence of Corollary 3.32 and Lemma 3.33. □

We now prove the second minimal inequality.

Lemma 3.35. *sl-TOP* $-$ *HOM* $\neq \emptyset$.

Proof. We show that the tree transformation τ_{zigzag}, see Example 3.31, cannot be induced by any homomorphism tree transducer.

Again, we prove by contradiction. Assume that there is a homomorphism tree transducer $T' = (\{p\}, \Sigma, \Delta, p, R')$, such that $\tau_{zigzag} = \tau_{T'}$. Moreover, let $R' = \{p(\delta(x_1, x_2)) \to \xi, p(\alpha) \to \zeta\}$.

Concerning the form of ξ, three different cases are possible, namely, either

- $\xi \in T_\Delta$, or
- x_1 occurs in ξ but x_2 does not, or
- x_2 occurs in ξ but x_1 does not.

(Note that x_1 and x_2 cannot both occur in ξ, because all symbols in Δ have rank at most 1.)

We show that all cases are impossible.

Case 1: $\xi \in T_\Delta$. Then, for every input tree $r = \delta(r_1, r_2) \in T_\Sigma$, we have $p(r) = p(\delta(r_1, r_2)) \Rightarrow_{T'} \xi$, showing that

$$\tau_{T'} = \{(\alpha, \zeta)\} \cup \{(r, \xi) \mid r = \delta(r_1, r_2), \text{ for some } r_1, r_2 \in T_\Sigma\}.$$

However this contradicts $\tau_{zigzag} = \tau_{T'}$, see Example 3.31.

Case 2: x_1 occurs in ξ but x_2 does not. Then, obviously $\xi = vp(x_1)$, for some $v \in (\Delta^{(1)})^*$. Now, take an input tree $r = \delta(\delta(\alpha, r_1), \alpha)$, where $r_1 \in T_\Sigma$ is an arbitrary tree. For this r, we have

$$p(r) = p(\delta(\delta(\alpha, r_1), \alpha)) \Rightarrow_{T'} vp(\delta(\alpha, r_1)) \Rightarrow_{T'} vvp(\alpha) \Rightarrow_{T'} vv\zeta$$

meaning that $\tau_{T'}(r) = vv\zeta$ does not depend on r_1. On the other hand, as was shown in Example 3.31, $\tau_{zigzag}(r) = 12w$, where w is the 1-path of r_1, which of course does depend on r_1. This contradicts $\tau_{zigzag} = \tau_{T'}$.

Case 3: x_2 occurs in ξ but x_1 does not. In this case $\xi = vp(x_2)$, for some $v \in (\Delta^{(1)})^*$. Take an input tree $r = \delta(r_1, \alpha)$, where $r_1 \in T_\Sigma$ is arbitrary. Then we can compute as follows

$$p(r) = p(\delta(r_1, \alpha)) \Rightarrow_{T'} vp(\alpha) \Rightarrow_{T'} v\zeta,$$

which means that $\tau_{T'}(r) = v\zeta$, for every $r_1 \in T_\Sigma$. On the other hand, $\tau_{zigzag}(r) = 1w$, where w is the 2-path of r_1, which contradicts $\tau_{zigzag} = \tau_{T'}$.

With this we have proved that τ_{zigzag} cannot be induced by any homomorphism tree transducer. □

Corollary 3.36. $l\text{-}TOP - HOM \neq \emptyset$.

Proof. The statement immediately follows from Corollary 3.32 and Lemma 3.35. □

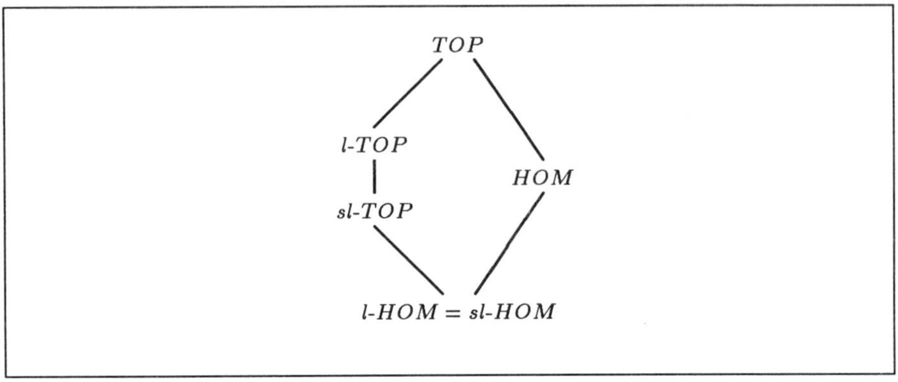

Fig. 3.7. The inclusion diagram of TOP and all its subclasses

We now state and prove that the conjectured diagram is the inclusion diagram of the six tree transformation classes involved.

Theorem 3.37. The diagram in Fig. 3.7 is the inclusion diagram of the tree transformation classes TOP, $l\text{-}TOP$, $sl\text{-}TOP$, HOM, $l\text{-}HOM$, and $sl\text{-}HOM$.

Proof. We have already proved all the inclusions and two inequalities which are necessary to verify our statement.

The only element which remains to be proved is the inequality $l\text{-}TOP - sl\text{-}TOP \neq \emptyset$. This will follow from two lemmas of the forthcoming section. In fact, in Lemma 3.41 we will prove that $l\text{-}TOP$ is closed with respect to composition, while, in Lemma 3.43, we will show that $sl\text{-}TOP$ is not closed under composition. Hence $l\text{-}TOP$ and $sl\text{-}TOP$ cannot be equal. □

Closing this section, we present a lemma which expresses that identical tree transformations can be induced already by the simplest top-down tree transducers we have considered. In the forthcoming sections we will use this lemma.

Lemma 3.38. $ID \subseteq sl\text{-}HOM$, i.e., for every ranked alphabet Σ, the identical tree transformation $Id(T_\Sigma)$ is in $sl\text{-}HOM$.

Proof. Let Σ be an arbitrary ranked alphabet. Consider the top-down tree transducer $T = (\{q\}, \Sigma, \Sigma, q, R)$ where R is the smallest set of rules such that for every $\sigma \in \Sigma^{(k)}$ with $k \geq 0$ the rule

$$q(\sigma(x_1,\ldots,x_k)) \to \sigma(q(x_1),\ldots,q(x_k))$$

is in R. Then we can prove that for every $s \in T_\Sigma$, $\tau_T(s) = s$, hence $\tau_T = Id(T_\Sigma)$. Moreover, T is a linear homomorphism tree transducer and hence, by Corollary 3.32, it is in *sl-HOM*, too. □

3.6 Composition and Decomposition Results

In this section composition and decomposition results concerning *TOP* and its subclasses will be proved. We speak about a composition theorem when it is proved that the composition of two (or sometimes more) tree transformation classes is the subclass of a single tree transformation class. Conversely, if a tree transformation class is a subclass of the composition of some other tree transformation classes, we speak about decomposition results because in this case every tree transformation of that class can be decomposed into tree transformations in the other classes. In both cases equality can hold. This gives a kind of characterization of the single class which is on either side of that equation.

In the first theorem we show that *TOP* is closed under composition, i.e., that the composition of two top-down tree transformations can always be induced by a top-down tree transducer.

Theorem 3.39. $TOP \circ TOP = TOP$.

Proof. By Lemma 3.38 and Corollary 3.32, $ID \subseteq TOP$. Hence, by Lemma 2.14, $TOP \subseteq TOP \circ TOP$.

The proof of $TOP \circ TOP \subseteq TOP$ is more technical. It requires to construct, for any two top-down tree transducers $T_1 = (Q_1, \Sigma, \Delta, q_1, R_1)$ and $T_2 = (Q_2, \Delta, \Omega, q_2, R_2)$, a third top-down tree transducer T such that $\tau_{T_1} \circ \tau_{T_2} = \tau_T$.

First we modify T_2 by introducing the top-down tree transducer T_2', so that T_2' can process the right-hand sides of rules in R_1. Therefore we define $mx = \max\{k \mid \Sigma^{(k)} \neq \emptyset\}$ and the ranked alphabets

$$Q_1(X_{mx}) = \{q(x_i)^{(0)} \mid q \in Q_1, 1 \leq i \leq mx\}$$

and

$$Q_2 \times Q_1(X_{mx}) = \{(p,q)(x_i)^{(0)} \mid (p,q) \in Q_2 \times Q_1, 1 \leq i \leq mx\}.$$

We note that the ranked alphabets $Q_1(X_{mx})$ and $Q_2 \times Q_1(X_{mx})$ contain only elements of rank 0 and that strings like $q(x_i)^{(0)}$ and $(p,q)(x_i)^{(0)}$ denote individual symbols here.

Then we put $T_2' = (Q_2, \Delta \cup Q_1(X_{mx}), \Omega \cup Q_2 \times Q_1(X_{mx}), q_2, R_2')$, where $R_2' = R_2 \cup \{p(q(x_i)) \to (p,q)(x_i) \mid p \in Q_2 \text{ and } q(x_i) \in Q_1(X_{mx})\}$.

Next we prove the following statement, which shows the connection between T_2 and T_2'.

Statement. For every $p \in Q_2$, $\xi \in RHS(Q_1, \Delta, k)$ with $0 \le k \le mx$, and family $\{t_{\bar{q},j} \in T_\Delta \mid \bar{q} \in Q_1, 1 \le j \le k\}$ of trees, we have

$$\begin{aligned} & \tau_{T_2,p}(\xi[\bar{q}(x_j) \leftarrow t_{\bar{q},j}\,; \bar{q}(x_j) \in Q_1(X_k)]) \\ = \quad & \tau_{T_2',p}(\xi)[(\bar{p},\bar{q})(x_j) \leftarrow \tau_{T_2,\bar{p}}(t_{\bar{q},j})\,;\, (\bar{p},\bar{q})(x_j) \in Q_2 \times Q_1(X_k)]. \end{aligned}$$

We prove this statement by structural induction on ξ.

(i) Let $\xi = q(x_i)$, for some $q \in Q_1$ and let $1 \le i \le k$. Then

$$\begin{aligned} & \tau_{T_2,p}(\xi[\bar{q}(x_j) \leftarrow t_{\bar{q},j}\,; \bar{q}(x_j) \in Q_1(X_k)]) \\ = \quad & \tau_{T_2,p}(t_{q,i}) \\ = \quad & (p,q)(x_i)[(\bar{p},\bar{q})(x_j) \leftarrow \tau_{T_2,\bar{p}}(t_{\bar{q},j})\,;\, (\bar{p},\bar{q})(x_j) \in Q_2 \times Q_1(X_k)] \\ = \quad & \tau_{T_2',p}(\xi)[(\bar{p},\bar{q})(x_j) \leftarrow \tau_{T_2,\bar{p}}(t_{\bar{q},j})\,;\, (\bar{p},\bar{q})(x_j) \in Q_2 \times Q_1(X_k)]. \end{aligned}$$

(ii) Let $\xi = \delta(\xi_1, \ldots, \xi_l)$, for some $\delta \in \Delta^{(l)}$ with $l \ge 0$, and $\xi_1, \ldots, \xi_l \in RHS(Q_1, \Delta, k)$. Moreover, let $\bar{\xi}$ be the right-hand side of the (p, δ)-rule in R_2. Then we can compute as follows:

$$\begin{aligned} & \tau_{T_2,p}(\xi[\bar{q}(x_j) \leftarrow t_{\bar{q},j}\,;\, \bar{q}(x_j) \in Q_1(X_k)]) \\ = \quad & \tau_{T_2,p}(\delta(\xi_1[\bar{q}(x_j) \leftarrow t_{\bar{q},j}\,;\, \bar{q}(x_j) \in Q_1(X_k)], \ldots \\ & \ldots, \xi_l[\bar{q}(x_j) \leftarrow t_{\bar{q},j}\,;\, \bar{q}(x_j) \in Q_1(X_k)])) \\ = \quad & \tau_{T_2,\bar{\xi}}(\xi_1[\bar{q}(x_j) \leftarrow t_{\bar{q},j}\,;\, \bar{q}(x_j) \in Q_1(X_k)], \ldots \\ & \ldots, \xi_l[\bar{q}(x_j) \leftarrow t_{\bar{q},j}\,;\, \bar{q}(x_j) \in Q_1(X_k)]) \\ & \text{(by Def. 3.22)} \\ = \quad & \bar{\xi}[\tilde{q}(x_i) \leftarrow \tau_{T_2,\tilde{q}}(\xi_i[\bar{q}(x_j) \leftarrow t_{\bar{q},j}\,;\, \bar{q}(x_j) \in Q_1(X_k)])\,;\, \tilde{q}(x_i) \in Q_1(X_k)] \\ & \text{(by Lemma 3.26)} \\ = \quad & \bar{\xi}[\tilde{q}(x_i) \leftarrow \tau_{T_2',\tilde{q}}(\xi_i)[(\bar{p},\bar{q})(x_j) \leftarrow \\ & \tau_{T_2,\bar{p}}(t_{\bar{q},j}); (\bar{p},\bar{q})(x_j) \in Q_2 \times Q_1(X_k)]\,;\, \tilde{q}(x_i) \in Q_1(X_k)] \\ & \text{(by induction hypothesis)} \\ = \quad & \bar{\xi}[\tilde{q}(x_i) \leftarrow \tau_{T_2',\tilde{q}}(\xi_i)\,;\, \tilde{q}(x_i) \in Q_1(X_k)][(\bar{p},\bar{q})(x_j) \leftarrow \\ & \tau_{T_2,\bar{p}}(t_{\bar{q},j}); (\bar{p},\bar{q})(x_j) \in Q_2 \times Q_1(X_k)] \\ & \text{(by the associative property of } \leftarrow) \\ = \quad & \tau_{T_2',\bar{\xi}}(\xi_1, \ldots, \xi_l)[(\bar{p},\bar{q})(x_j) \leftarrow \tau_{T_2,\bar{p}}(t_{\bar{q},j})\,;\, (\bar{p},\bar{q})(x_j) \in Q_2 \times Q_1(X_k)] \\ & \text{(by Lemma 3.26)} \\ = \quad & \tau_{T_2',p}(\xi)[(\bar{p},\bar{q})(x_j) \leftarrow \tau_{T_2,\bar{p}}(t_{\bar{q},j})\,;\, (\bar{p},\bar{q})(x_j) \in Q_2 \times Q_1(X_k)] \\ & \text{(by Def. 3.22).} \end{aligned}$$

□

We now define the top-down tree transducer $T = (Q_2 \times Q_1, \Sigma, \Omega, (q_2, q_1), R)$, where R is the smallest set of rules satisfying the following condition. If the rule $q(\sigma(x_1, \ldots, x_k)) \to \xi$ is in R_1, for some

$q \in Q_1, \sigma \in \Sigma^{(k)}$ with $k \geq 0$ and $\xi \in RHS(Q_1, \Delta, k)$, then, for every $p \in Q_2$, the rule

$$(p,q)(\sigma(x_1,\ldots,x_k)) \to \tau_{T_2',p}(\xi)$$

is in R.

We note that T_2' can work on the right-hand sides of rules in R_1 because, for every $0 \leq k \leq mx$, we have $RHS(Q_1, \Delta, k) \subseteq T_{\Delta \cup Q_1(X_{mx})}$. Moreover, it can be proved by an easy induction that, for every $\xi \in RHS(Q_1, \Delta, k)$ with $0 \leq k \leq mx$ and $p \in Q_2$, it holds that $\tau_{T_2',p}(\xi) \in RHS(Q_2 \times Q_1, \Omega, k)$. Hence the definition of R is syntactically correct.

To see that $\tau_T = \tau_{T_1} \circ \tau_{T_2}$, we prove that the predicates K and L hold by simultaneous induction.

K : For every $p \in Q_2, q \in Q_1$, and $s \in T_\Sigma$, it holds that $\tau_{T_2,p}(\tau_{T_1,q}(s)) = \tau_{T,(p,q)}(s)$.
L : For every $0 \leq k \leq mx$, $\xi \in RHS(Q_1, \Delta, k)$, $p \in Q_2$, and $\omega = (s_1,\ldots,s_k) \in (T_\Sigma)^k$, we have $\tau_{T_2,p}(\tau_{T_1,\xi}(\omega)) = \tau_{T,\xi'}(\omega)$, where $\xi' = \tau_{T_2',p}(\xi)$.

First we prove IS1. Let $s = \sigma(s_1,\ldots,s_k) \in T_\Sigma$, for some $k \geq 0$, $\sigma \in \Sigma^{(k)}$, and $s_1,\ldots,s_k \in T_\Sigma$. Moreover, let ξ be the right-hand side of the (q,σ)-rule in R_1 and let $\omega = (s_1,\ldots,s_k)$. Then we have

$$\begin{aligned} \tau_{T_2,p}(\tau_{T_1,q}(s)) &= \tau_{T_2,p}(\tau_{T_1,\xi}(\omega)) && \text{(by Def. 3.22)} \\ &= \tau_{T,\xi'}(\omega) && \text{(by induction hypothesis on } L\text{)}, \end{aligned}$$

where $\xi' = \tau_{T_2',p}(\xi)$. On the other hand, by the definition of T, ξ' is the right-hand side of the $((p,q),\sigma)$-rule in R, and hence $\tau_{T,\xi'}(\omega) = \tau_{T,(p,q)}(s)$.

We now prove IS2. We compute as follows.

$$\begin{aligned} & \tau_{T_2,p}(\tau_{T_1,\xi}(\omega)) \\ = \; & \tau_{T_2,p}(\xi[\bar{q}(x_j) \leftarrow \tau_{T_1,\bar{q}}(s_j)\,;\, \bar{q}(x_j) \in Q_1(X_k)]) \\ & \text{(by Lemma 3.26)} \\ = \; & \tau_{T_2',p}(\xi)[(\bar{p},\bar{q})(x_j) \leftarrow \tau_{T_2,\bar{p}}(\tau_{T_1,\bar{q}}(s_j))\,;\,(\bar{p},\bar{q})(x_j) \in Q_2 \times Q_1(X_k)] \\ & \text{(by the Statement above)} \\ = \; & \tau_{T_2',p}(\xi)[(\bar{p},\bar{q})(x_j) \leftarrow \tau_{T,(\bar{p},\bar{q})}(s_j)\,;\,(\bar{p},\bar{q})(x_j) \in Q_2 \times Q_1(X_k)] \\ & \text{(by induction hypothesis on } K\text{)} \\ = \; & \tau_{T,\xi'}(\omega) \\ & \text{(by Lemma 3.26)} \end{aligned}$$

where $\xi' = \tau_{T_2',p}(\xi)$.

Now, since K holds for $p = q_2$ and $q = q_1$, we obtain $\tau_T = \tau_{T_1} \circ \tau_{T_2}$. With this we have proved that $TOP \circ TOP \subseteq TOP$. □

Next we show that HOM is also closed under composition.

Corollary 3.40. $HOM \circ HOM = HOM$.

Proof. By Lemma 3.38 and by Lemma 3.32, we have $ID \subseteq HOM$. Hence, by Lemma 2.14, $HOM \subseteq HOM \circ HOM$.

To prove the converse inclusion, we make the following observation. If in the proof of Theorem 3.39, the top-down tree transducers T_1 and T_2 have the only states q_1 and q_2, respectively, then T will have the only state (q_2, q_1), that is, T is also a homomorphism tree transducer. Hence, by $\tau_T = \tau_{T_1} \circ \tau_{T_2}$, we get $HOM \circ HOM \subseteq HOM$. □

We now show that l-TOP is also closed under composition.

Corollary 3.41. $l\text{-}TOP \circ l\text{-}TOP = l\text{-}TOP$.

Proof. By Lemma 3.38 and by Lemma 3.32, we have $ID \subseteq l\text{-}TOP$. Hence, by Lemma 2.14, $l\text{-}TOP \subseteq l\text{-}TOP \circ l\text{-}TOP$.

The exact proof of $l\text{-}TOP \circ l\text{-}TOP \subseteq l\text{-}TOP$ is more complex, although intuitively it can be seen easily.

We prove that if, in the proof of Theorem 3.39, the tree transducers T_1 and T_2 are linear, then T constructed from them is also linear. In other words, having the notations of Theorem 3.39, we should prove that, for every $(p, q) \in Q_2 \times Q_1$, $\sigma \in \Sigma^{(k)}$ with $k \geq 0$, and $1 \leq i \leq k$, the variable x_i appears at most once in $rhs((p, q), \sigma)$. This is the consequence of the two following, slightly more general statements.

Let $p \in Q_2$, $\xi \in RHS(Q_1, \Delta, k)$ with $k \geq 0$ and $1 \leq i \leq k$. Then the following two statements hold.

(a) Variables occurring in $\tau_{T_2',p}(\xi)$ also occur in ξ.

(b) If T_2 is linear and x_i occurs at most once in ξ, then it occurs also at most once in $\tau_{T_2',p}(\xi)$.

The proof of both statements is performed by structural induction on ξ.

(i) Let $\xi = q(x_j)$, for some $q \in Q_1$ and $1 \leq j \leq k$. Then $\tau_{T_2',p}(\xi) = (p, q)(x_j)$, hence both (a) and (b) hold independently whether T_2' is linear or not.

(ii) Let $\xi = \delta(\xi_1, \ldots, \xi_l)$, for some $l \geq 0$, $\delta \in \Delta^{(l)}$, and $\xi_1, \ldots, \xi_l \in RHS(Q_1, \Delta, k)$. Let, moreover, $\bar{\xi}$ be the right-hand side of the (p, δ)-rule in R_2'. Then we have

$$\begin{aligned} \tau_{T_2',p}(\xi) &= \tau_{T_2',p}(\delta(\xi_1, \ldots, \xi_l)) = \tau_{T_2',\bar{\xi}}((\xi_1, \ldots, \xi_l)) \\ &= \bar{\xi}[\bar{p}(x_j) \leftarrow \tau_{T_2',\bar{p}}(\xi_j)\,;\, \bar{p}(x_j) \in Q_2(X_k)]. \qquad (*) \end{aligned}$$

By the induction hypothesis, variables occurring in $\tau_{T_2',\bar{p}}(\xi_j)$ also occur in ξ_j, for every $\bar{p} \in Q_2$ and $1 \leq j \leq l$. Hence statement (a) holds.

Now suppose that T_2 is linear and x_i occurs at most once in ξ. Then there is at most one $1 \leq j \leq l$, such that x_i occurs in ξ_j. Moreover, by the induction hypothesis, if x_i occurs in ξ_j at most once, then it occurs at most once in $\tau_{T_2',\bar{p}}(\xi_j)$, for every $\bar{p} \in Q_2$.

Since, for every $\bar{p} \in Q_2$ and $1 \leq n \leq l$, variables occurring in $\tau_{T_2',\bar{p}}(\xi_n)$ also occur in ξ_n, there is at most one $1 \leq n \leq l$, such that x_i occurs in $\tau_{T_2',\bar{p}}(\xi_n)$.

On the other hand, T_2' is linear, because T_2 is, hence for every $1 \le n \le l$, a subtree of the form $\bar{p}(x_n)$ occurs at most once in $\bar{\xi}$. Then it can be seen from (*) that x_i occurs at most once in $\tau_{T_2',p}(\xi)$, finishing the proof of (a) and (b).

Now suppose that both T_1 and T_2 are linear. Let $(p, q) \in Q_2 \times Q_1$ and $\sigma \in \Sigma^{(k)}$, for some $k \ge 0$. Then, for every $1 \le i \le k$, x_i occurs at most once in the right-hand side ξ of the (q, σ)-rule in R_1. Moreover, by the above statement, for every $1 \le i \le k$, the variable x_i occurs at most once in $\tau_{T_2',p}(\xi)$, which is, by the definition of T, the right-hand side of the $((p, q), \sigma)$-rule in R. Hence T is linear. □

We are going to show that superlinear top-down tree transformations are not closed under composition. This will be the consequence of the following stronger result.

Lemma 3.42. *l-HOM* $\circ$ *sl-TOP* $-$ *sl-TOP* $\neq \emptyset$.

Proof. We prove this in the following way. We define a linear homomorphism tree transducer T_1 and then we show that $\tau_{T_1} \circ \tau_T \notin$ *sl-TOP*, where T is the superlinear top-down tree transducer introduced in Example 3.31.

To this end, let $T_1 = (\{p_1\}, \Sigma_1, \Sigma, p_1, R_1)$, where

- $\Sigma_1 = \{\delta^{(2)}, \gamma^{(1)}, \alpha^{(0)}\}$ and $\Sigma = \{\delta^{(2)}, \alpha^{(0)}\}$ and
- R_1 is the set consisting of the three rules
  - $p_1(\delta(x_1, x_2)) \to \delta(p_1(x_1), p_1(x_2))$,
  - $p_1(\gamma(x_1)) \to p_1(x_1)$,
  - $p_1(\alpha) \to \alpha$.

It should be obvious that T_1 induces the tree transformation $\tau_{T_1} : T_{\Sigma_1} \to T_\Sigma$, where, for $t \in T_{\Sigma_1}$, we have $\tau_{T_1}(t) = t_1$ if and only if t_1 is obtained from t by deleting every occurrence of γ from t.

Then, the mapping $\tau_{T_1} \circ \tau_T : T_{\Sigma_1} \to T_\Delta$ can be explained as follows. For every $t \in T_{\Sigma_1}$, $(\tau_{T_1} \circ \tau_T)(t) = t'$, if and only if t' is obtained from t in the following way: first form t_1 by deleting all occurrences of γ from t, then take the 1-path of t_1 (see Example 3.31).

We prove $\tau_{T_1} \circ \tau_T \notin$ *sl-TOP* by contradiction, i.e., we assume that $\tau_{T_1} \circ \tau_T = \tau_{T'}$ for some superlinear top down tree transducer $T' = (Q', \Sigma_1, \Delta, q_0, R')$. We study which kind of rules are in R'.

Our first observation is that the rule $q_0(\alpha) \to \alpha$ should be in R', because $\tau_{T'}(\alpha) = \alpha$.

Next we observe that $\tau_{T'}(\gamma(\alpha)) = \alpha$. This implies that either $q_0(\gamma(x_1)) \to \alpha \in R'$ or there is a state $q \in Q'$ such that the rules $q_0(\gamma(x_1)) \to q(x_1)$ and $q(\alpha) \to \alpha$ are in R'.

It is easy to see that the first case is impossible. Otherwise, if $q_0(\gamma(x_1)) \to \alpha$ is in R', then, for every $t \in T_\Sigma$, we have $q_0(\gamma(t)) \Rightarrow_{T'} \alpha$, yielding that $\tau_{T'}(\gamma(t)) = \alpha$. However this is a contradiction because $\tau_{T'}(\gamma(t))$ is the 1-path of t, by $\tau_{T_1} \circ \tau_T = \tau_{T'}$.

Therefore $q_0(\gamma(x_1)) \to q(x_1)$ and $q(\alpha) \to \alpha$ are in R' for some $q \in Q'$. We show that necessarily $q = q_0$. For this, we consider the (q, γ)-rule, which, by the definition of a top-down tree transducer, must be in R'. Let $\xi = rhs(q, \gamma)$. We obtain easily that ξ cannot be in T_Δ. If $\xi \in T_\Delta$, then for every $t \in T_\Sigma$, we get $q_0(\gamma^2(t)) \Rightarrow_{T'} q(\gamma(t)) \Rightarrow_{T'} \xi$ yielding that $\tau_{T'}(\gamma^2(t)) = \xi$, for every $t \in T_\Sigma$. This cannot hold because $\tau_{T'}(\gamma^2(t))$ is the 1-path of t. Thus x_1 should occur in ξ. However, since T' is superlinear, x_1 can only occur in both $rhs(q_0, \gamma)$ and $rhs(q, \gamma)$ in the case $q = q_0$. Hence we conclude that $q_0(\gamma(x_1)) \to q_0(x_1)$ is also in R'.

Next we observe that $\tau_{T'}(\delta(\alpha, \alpha)) = 1\alpha$ and that $\tau_{T'}(\delta(\delta(\alpha, \alpha), \alpha)) = 12\alpha$. It is an easy exercise to show that these two equations and the fact that T' is superlinear imply that $q_0(\delta(x_1, x_2)) \to 1p(x_1)$ must be in R', for some $p \in Q'$. We show now that $p \neq q_0$. Otherwise, if $p = q_0$, then we have

$$q_0(\delta(\delta(\alpha, \alpha), \alpha)) \Rightarrow_{T'} 1q_0(\delta(\alpha, \alpha)) \Rightarrow_{T'} 11q_0(\alpha) \Rightarrow_{T'} 11\alpha$$

contradicting $\tau_{T'}(\delta(\delta(\alpha, \alpha), \alpha)) = 12\alpha$.

Hence $p \neq q_0$. Then x_1 cannot occur in $rhs(p, \gamma)$, because T' is superlinear and, as we already have seen, it occurs in $rhs(q_0, \gamma)$. Consequently, $p(\gamma(x_1)) \to \xi$ is in R', for some $\xi \in T_\Delta$.

However, we show that this is again impossible. If $\xi \in T_\Delta$, then, for every $t \in T_\Sigma$, we have $q_0(\delta(\gamma(t), \alpha)) \Rightarrow_{T'} 1p(\gamma(t)) \Rightarrow_{T'} 1\xi$. On the other hand, $\tau_{T'}(\delta(\gamma(t), \alpha)) = 1w$, where w is the 2-path of t, which is a contradiction.

With this we proved that $\tau_{T_1} \circ \tau_T$ cannot be induced by any superlinear top-down tree transducer. □

Now we have obtained the following result.

Theorem 3.43. The class *sl-TOP* is not closed under composition.

Proof. By Lemma 3.42 and by Corollary 3.32 we have $sl\text{-}TOP \circ sl\text{-}TOP - sl\text{-}TOP \neq \emptyset$. □

We will now prove two characterizations of *TOP* in terms of the composition of its proper subclasses. In the proof of the first such theorem we will need to specify the right-hand sides of the rules of a top-down tree transducer in a more detailed form. Therefore we establish the following lemma.

Lemma 3.44. Let $T = (Q, \Sigma, \Delta, q_0, R)$ be a top-down tree transducer, $k \geq 0$, and let $\xi \in RHS(Q, \Delta, k)$. Assume, for every $1 \leq i \leq k$, the number of occurrences of x_i in ξ to be n_i. Let $n = n_1 + \ldots + n_k$. Then there exists a context $t \in C_{\Delta,n}$ and, for every $1 \leq i \leq k$, there exist states $q_{i1}, \ldots, q_{in_i} \in Q$, such that

$$\begin{aligned} \xi \;=\; & t[z_1 \leftarrow q_{11}(x_1), \ldots, z_{n_1} \leftarrow q_{1n_1}(x_1), \ldots \\ & \ldots, z_{u_k+1} \leftarrow q_{k1}(x_k), \ldots, z_{u_k+n_k} \leftarrow q_{kn_k}(x_k)], \end{aligned}$$

where $u_i = n_1 + \ldots + n_{i-1}$ for every $2 \leq i \leq k$. (The statement is visualized in Fig. 3.8. Note that, for $k = 0$, it is clear that $n = 0$ and $t = \xi$.)

Proof. Although the proof is rather technical, it can easily be performed by an induction on the structure of ξ. Therefore we omit the details. □

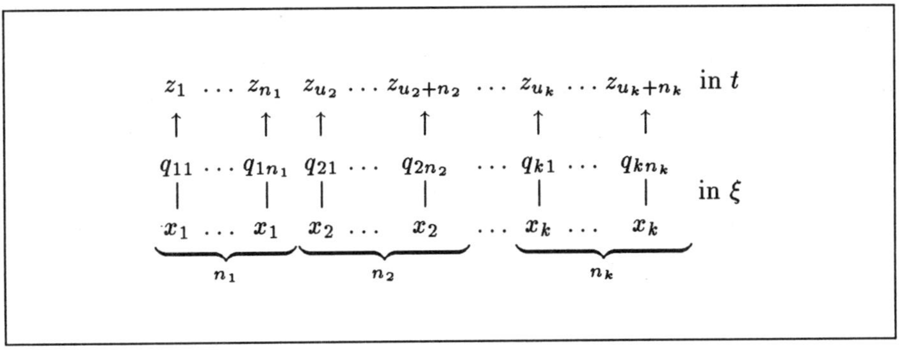

Fig. 3.8. Detail of variables in ξ

The next theorem is a characterization of top-down tree transformations stating that top-down tree transformations are exactly those tree transformations which can be obtained as the composition of a homomorphism tree transformation and a linear top-down tree transformation.

Theorem 3.45. $TOP = HOM \circ l\text{-}TOP$.

Proof. The inclusion $HOM \circ l\text{-}TOP \subseteq TOP$ can be obtained by the following calculation:

$$\begin{array}{lcll} HOM \circ l\text{-}TOP & \subseteq & TOP \circ TOP & \text{(by Corollary 3.32)} \\ & = & TOP & \text{(by Theorem 3.39).} \end{array}$$

We show that the converse inclusion also holds. Therefore, let $T = (Q, \Sigma, \Delta, q_0, R)$ be a top-down tree transducer. We construct a homomorphism tree transducer $T_1 = (\{p\}, \Sigma, \Sigma', p, R_1)$ and a linear top-down tree transducer $T_2 = (Q_2, \Sigma', \Delta, q_2, R_2)$ such that $\tau_T = \tau_{T_1} \circ \tau_{T_2}$.

Therefore let l be the number of rules in R. Let us fix an arbitrary linear order $r_1, \ldots, r_l$ on the rules in R. Moreover, for every $i \geq 1$ and $1 \leq j \leq l$, let $n(i,j)$ be the number of the occurrences of x_i's in the jth rule r_j. Let

$$n = max\{n(i,j) \mid 1 \leq j \leq l, i \geq 1\}.$$

We first define T_1. Let Σ' be the smallest ranked alphabet such that, for every $k \geq 0$, $\Sigma'^{(k \cdot n)} = \{\sigma' \mid \sigma \in \Sigma^{(k)}\}$. Moreover, let R_1 be the smallest set of rules such that, for every $k \geq 0$ and $\sigma \in \Sigma^{(k)}$, the rule

$$p(\sigma(x_1,\ldots,x_k)) \to \sigma'(\overbrace{p(x_1),\ldots,p(x_1)}^{n},\ldots,\overbrace{p(x_k),\ldots,p(x_k)}^{n})$$

is in R_1, where every variable occurs n times on the right-hand side. (Note that, for $k=0$, we have that $p(\sigma) \to \sigma'$ is in R_1.)

We now define T_2. Let $Q_2 = Q$, $q_2 = q_0$, and let R_2 be the set constructed in the following way. For every $1 \le j \le l$, let the jth rule r_j in R be constructed as

$$r_j = q(\sigma(x_1,\ldots,x_k)) \to \xi. \quad (*)$$

For brevity, let $n_1 = n(1,j),\ldots,n_k = n(k,j)$ and $\bar{n} = n_1 + \ldots + n_k$. By Lemma 3.44, there exists a context $t \in C_{\Delta,\bar{n}}$ such that

$$\begin{aligned} \xi \;=\; & t[z_1 \leftarrow q_{11}(x_1),\ldots,z_{n_1} \leftarrow q_{1n_1}(x_1),\ldots \\ & \ldots,z_{u_k+1} \leftarrow q_{k1}(x_k),\ldots,z_{u_k+n_k} \leftarrow q_{kn_k}(x_k)] \end{aligned}$$

where u_i is now equal to $(i-1)\cdot n$.

Then, put the rule

$$\begin{aligned} q(\sigma'(x_1,\ldots,x_{k\cdot n})) \quad & \to t[z_1 \leftarrow q_{11}(x_1),\ldots,z_{n_1} \leftarrow q_{1n_1}(x_{n_1}),\ldots \\ & \ldots,z_{u_k+1} \leftarrow q_{k1}(x_{(k-1)\cdot n+1}),\ldots \\ & \ldots,z_{u_k+n_k} \leftarrow q_{kn_k}(x_{(k-1)\cdot n+n_k})] \end{aligned}$$

in R_2. (By the note we made in Lemma 3.44, in the case $k=0$, we have $\xi \in T_\Delta$ and $q(\sigma') \to \xi \in R_2$.)

Intuitively, the homomorphism tree transducer T_1 handles the task of multiplying variables of T. Since T_1 does not know how many times a variable must be copied, it makes n copies of every variable. This ensures that every subtree of an input tree to T_1 is multiplied sufficiently many times. Then, T_2 takes as many copies of a subtree as T has produced and operates on them in the same way as T did. The process by which T_2 chooses the necessary copies out of the n copies produced by T_1 is visualized in Fig. 3.9.

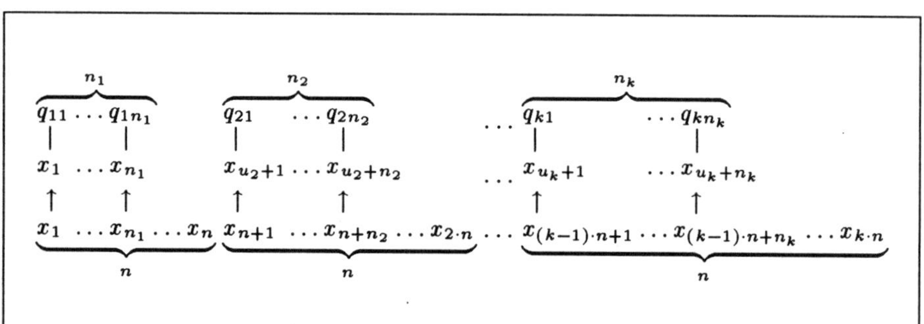

Fig. 3.9. T_2 chooses the copies it needs out of the n copies produced by T_1

To show that $\tau_T = \tau_{T_1} \circ \tau_{T_2}$, it is sufficient to prove that, for every $s \in T_\Sigma$ and $q \in Q$, we have $\tau_{T,q}(s) = \tau_{T_2,q}(\tau_{T_1,p}(s))$. We prove this by structural induction on s. In order to avoid complex formulas, we put $s' = \tau_{T_1,p}(s)$.

Now let $s = \sigma \in \Sigma_0$. Then $\tau_{T,q}(s) = \xi$, where $\xi = rhs(q,\sigma)$. On the other hand, $s' = \sigma'$ and $q(\sigma') \rightarrow \xi$ is a rule in R_2. Hence $\tau_{T_2,q}(s') = \xi$, too.

Now let $s = \sigma(s_1, \ldots, s_k)$, where $k \geq 0$, $\sigma \in \Sigma^{(k)}$, and $s_1, \ldots, s_k \in T_\Sigma$. First compute $\tau_{T,q}(s)$. Assume that the (q,σ)-rule in R has the form as in $(*)$ and that all notations we defined for $(*)$ are valid for the rule involved. Then we have

$$\begin{aligned}
\tau_{T,q}(s) &= \tau_{T,q}(\sigma(s_1,\ldots,s_k)) = \xi[\bar{q}(x_j) \leftarrow \tau_{T,\bar{q}}(s_j)\,;\, \bar{q}(x_j) \in Q(X_k)] \\
&= t[z_1 \leftarrow \tau_{T,q_{11}}(s_1), \ldots, z_{n_1} \leftarrow \tau_{T,q_{1n_1}}(s_1), \ldots \\
&\quad \ldots, z_{u_k+1} \leftarrow \tau_{T,q_{k1}}(s_k), \ldots, z_{u_k+n_k} \leftarrow \tau_{T,q_{kn_k}}(s_k)].
\end{aligned}$$

We now compute $\tau_{T_2,q}(s')$. We first observe that

$$s' = \sigma'(s'_1, \ldots, s'_1, \ldots, s'_k, \ldots, s'_k),$$

where s'_i occurs n times, for every $1 \leq i \leq k$. Hence

$$\begin{aligned}
\tau_{T_2,q}(s') &= \tau_{T_2,q}(\sigma'(s'_1,\ldots,s'_1,\ldots,s'_k,\ldots,s'_k)) \\
&= t[z_1 \leftarrow \tau_{T_2,q_{11}}(s'_1), \ldots, z_{n_1} \leftarrow \tau_{T_2,q_{1n_1}}(s'_1), \ldots \\
&\quad \ldots, z_{u_k+1} \leftarrow \tau_{T_2,q_{k1}}(s'_k), \ldots, z_{u_k+n_k} \leftarrow \tau_{T_2,q_{kn_k}}(s'_k)].
\end{aligned}$$

Now, by the induction hypothesis, for every $1 \leq i \leq k$ and $1 \leq j \leq n_i$, $\tau_{T,q_{ij}}(s_i) = \tau_{T_2,q_{ij}}(s'_i)$ and thus $\tau_{T,q}(s) = \tau_{T_2,q}(s')$. This completes the proof. □

Our next result is a decomposition result. It says that every linear top-down tree transformation can be obtained as the decomposition of a homomorphism tree transformation and a superlinear top-down tree transformation.

Lemma 3.46. *l-TOP ⊆ HOM ∘ sl-TOP.*

Proof. Let $T = (Q, \Sigma, \Delta, q_0, R)$ be a linear top-down tree transducer. Assume that R contains n rules. We define a homomorphism tree transducer $T_1 = (\{p\}, \Sigma, \Sigma', p, R_1)$ and a superlinear top-down tree transducer $T_2 = (Q_2, \Sigma', \Delta, q_2, R_2)$ such that $\tau_T = \tau_{T_1} \circ \tau_{T_2}$.

To this end, let $r_1, \ldots, r_n$ be an arbitrary, but fixed order of the rules in R.

We first define T_1. Let Σ' be the smallest ranked alphabet, for which $\Sigma'^{(k\cdot n)} = \{\sigma' \mid \sigma \in \Sigma^{(k)}\}$, for every $k \geq 0$.

Then, let R_1 be the smallest set of rules satisfying the following condition. For every $k \geq 0$ and $\sigma \in \Sigma_k$, let the rule

$$p(\sigma(x_1,\ldots,x_k)) \rightarrow \sigma'(\overbrace{p(x_1),\ldots,p(x_1)}^{n},\ldots,\overbrace{p(x_k),\ldots,p(x_k)}^{n})$$

be in R_1. (Note that x_i occurs n times in $rhs(p, \sigma)$, for every $1 \le i \le k$).

We now define T_2. Let $Q_2 = Q$, $q_2 = q_0$, and let R_2 be constructed as follows. For every $1 \le i \le n$, consider the rule r_i. Assume that it has the form $q(\sigma(x_1, \ldots, x_k)) \to \xi$ for some $k \ge 0$ and $\sigma \in \Sigma^{(k)}$. Then put the rule $q(\sigma'(x_1, \ldots, x_{k \cdot n})) \to \xi'$ in R_2, where the tree ξ' is obtained from ξ by substituting x_j in ξ by $x_{(j-1)n+i}$, for every $1 \le j \le k$. (More formally, we let $\xi' = \xi[x_1 \leftarrow x_i, x_2 \leftarrow x_{n+i}, \ldots, x_k \leftarrow x_{(k-1) \cdot n+i}]$.)

We show that T_2 is superlinear. Since T is linear and, during the above construction, different variables in the right-hand side of a rule in R are substituted by different ones, T_2 is certainly linear. Moreover, take a (q, σ')-rule and a (q', σ')-rule from R_2, where $\sigma' \in \Sigma'^{(k \cdot n)}$, for some $k \ge 0$, and $q, q' \in Q_2$ are different states. Suppose that the (q, σ)-rule is the ith rule and the (q', σ)-rule is the jth rule in the sequence $r_1, \ldots, r_n$. Obviously, $i \ne j$. Then, by the construction of R_2, $Var_X(rhs(q, \sigma')) \subseteq \{x_i, x_{n+i}, \ldots, x_{(k-1)n+i}\}$ and $Var_X(rhs(q', \sigma')) \subseteq \{x_j, x_{n+j}, \ldots, x_{(k-1)n+j}\}$. Since, by $i \ne j$,

$$\{x_i, x_{n+i}, \ldots, x_{(k-1)n+i}\} \cap \{x_j, x_{n+j}, \ldots, x_{(k-1)n+j}\} = \emptyset,$$

we obtain that $Var_X(rhs(q, \sigma')) \cap Var_X(rhs(q', \sigma')) = \emptyset$. Hence, T_2 is superlinear.

To prove that $\tau_T = \tau_{T_1} \circ \tau_{T_2}$, it is enough to show that the following equivalence holds. For every $s \in T_\Sigma$ and $q \in Q$, $\tau_{T,q}(s) = \tau_{T_2,q}(\tau_{T_1,p}(s))$. The proof can be performed by an induction on the structure of s. We leave it for an exercise. □

To finish this section we give another characterization of top-down tree transformations in terms of homomorphism tree transformations and superlinear tree transformations.

Corollary 3.47. *TOP* = *HOM* ∘ *sl-TOP*.

Proof.

	TOP	
=	*HOM* ∘ *l-TOP*	(by Lemma 3.45)
⊆	*HOM* ∘ *HOM* ∘ *sl-TOP*	(by Lemma 3.46)
=	*HOM* ∘ *sl-TOP*	(by Corollary 3.40)
⊆	*HOM* ∘ *l-TOP*	(by Corollary 3.32)
=	*TOP*	(by Lemma 3.45)

□

3.7 Composition Semigroup Generated by *TOP*, *l-TOP*, and *HOM*

The motivation of this section

We have so far considered five fundamental classes of top-down tree transformations: *TOP*, *l-TOP*, *sl-TOP*, *HOM* and *l-HOM*. We have given their inclusion diagram in Sect. 3.5. Moreover we have obtained several composition results and inclusions, proved in the Corollaries 3.32–3.47. We now observe that the knowledge we possess is sufficient to state the equality of, or the inclusion between, some further tree transformation classes which can be obtained by composition from the five fundamental classes. For example, let us take the tree transformation classes $TOP \circ HOM \circ HOM \circ l\text{-}TOP$ and $HOM \circ TOP$. Then we can easily prove that they are equal, that is to say,

$$TOP \circ HOM \circ HOM \circ l\text{-}TOP = HOM \circ TOP.$$

That is, we can show that both sides of the equality are equal to *TOP* by computing as follows:

$$\begin{array}{lll} & TOP \circ HOM \circ HOM \circ l\text{-}TOP & \\ = & TOP \circ HOM \circ l\text{-}TOP & \text{(by Corollary 3.40)} \\ = & TOP \circ TOP & \text{(by Theorem 3.45)} \\ = & TOP & \text{(by Theorem 3.39)} \end{array}$$

and

$$\begin{array}{lll} & TOP & \\ \subseteq & HOM \circ TOP & \text{(by Lemma 2.14 and Lemma 3.38)} \\ \subseteq & TOP \circ TOP & \text{(by Corollary 3.32)} \\ = & TOP & \text{(by Theorem 3.39),} \end{array}$$

proving also that $HOM \circ TOP = TOP$.

On the other hand, let us consider, for example, the tree transformation classes $HOM \circ l\text{-}TOP$ and $l\text{-}TOP \circ HOM$. We know that, by Theorem 3.45, $HOM \circ l\text{-}TOP = TOP$. Moreover, by Corollary 3.32 and Theorem 3.39, $l\text{-}TOP \circ HOM \subseteq TOP \circ TOP = TOP$. Hence we obtain

$$l\text{-}TOP \circ HOM \subseteq HOM \circ l\text{-}TOP.$$

However, at this point, we do not yet know whether proper inclusion or equality holds between these classes.

This motivates us to deal with the following problem in this section. We would like to (and will) give an algorithm which can determine, given *any two expressions* built up with composition from some of our fundamental tree transformation classes, which relation out of $\subset, \supset, =$ and $\bowtie$ holds between them (recall that $\bowtie$ stands for the incomparability relationship).

We structure the discussion as follows. First we abstract from the concrete choice of the fundamental classes, i.e., we consider instead an arbitrary finite set H, of tree transformation classes, and we present a general method which can construct an algorithm with the following property: given any two expressions built up by composition from the elements of H, the algorithm can decide which of the relations holds between them. Then we apply the general method to a concrete choice of H consisting of four of our fundamental tree transformation classes. More precisely we carry out the two following main procedures, I and II, in this section.

I. *We give a general method which, given a finite set H of tree transformation classes and given two tree transformation classes $U_1 \circ \cdots \circ U_m$ and $V_1 \circ \cdots \circ V_n$, such that, for every $1 \leq i \leq m$ and $1 \leq j \leq n$, $U_i, V_j \in H$, determines which one of the following four conditions holds:*

$$\begin{array}{rrcl} \text{(i)} & U_1 \circ \cdots \circ U_m & = & V_1 \circ \cdots \circ V_n, \\ \text{(ii)} & U_1 \circ \cdots \circ U_m & \subset & V_1 \circ \cdots \circ V_n, \\ \text{(iii)} & V_1 \circ \cdots \circ V_n & \subset & U_1 \circ \cdots \circ U_m\ , \\ \text{(iv)} & U_1 \circ \cdots \circ U_m & \bowtie & V_1 \circ \cdots \circ V_n, \end{array}$$

II. *Then we apply this general method to the concrete set $H = \{TOP, l\text{-}TOP, HOM\}$.* (We note that application to the whole set of our fundamental tree transformation classes would lead too far and exceed the scope of this book.)

Presentation of the general method

In this subsection we present the general method. The necessary notions of universal algebra were given in Sects. 2.1 and 2.5. However, in order to form the method, we need some more notions and notations as follows.
Suppose that H is already given. We consider two semigroups in terms of H:

- the free semigroup H^+ with the operation of concatenation (denoted here by $\cdot$) and
- the semigroup

$$[H] = \{U_1 \circ \cdots \circ U_m \mid m \geq 1, U_i \in H, \text{for every } 1 \leq i \leq m\}$$

generated by H with the operation of composition (denoted by $\circ$).

We also consider the homomorphism $|\ | : H^+ \rightarrow [H]$ which is the unique extension of the identity mapping over H and for which, therefore,

$$|U_1 \cdot \ldots \cdot U_m| = U_1 \circ \cdots \circ U_m,$$

for all elements $U_1, \ldots, U_m \in H$. In this way, strings over H, i.e., elements of H^+, represent tree transformation classes in $[H]$. Finally, let us denote the kernel of $|\ |$ by θ.

We can now present the general method as follows. Before studying it, the reader is advised to consult Sect. 2.1 about the concept of a set of representatives.

GENERAL METHOD:

(a) Give a string rewrite system S over H that presents $[H]$, i.e., give a finite set $S \subseteq H^* \times H^*$ for which $H^+/\Leftrightarrow_S^* \cong [H]$, and such that $\Leftrightarrow_S^* = \theta$.

(b) Give a set REP of representatives for the congruence classes of $\Leftrightarrow_S^*$.

(c) Give the inclusion diagram of the subset $|REP| = \{|u| \mid u \in REP\}$ of $[H]$, i.e., the set of tree transformation classes represented by the representatives.

(d) Give an algorithm that, for every string $w \in H^+$, computes a representative $u \in REP$ such that $w \Leftrightarrow_S^* u$.

Next we show that if we can carry out the tasks described by (a)–(d) for the set H, we have also obtained an algorithm that has the desired property.

Lemma 3.48. Suppose that the tasks (a)–(d) of the general method have been accomplished for H. Then there is an algorithm that determines, given two arbitrary tree transformation classes $U_1 \circ \cdots \circ U_m$ and $V_1 \circ \cdots \circ V_n$, where $U_i, V_j \in H$, for every $1 \leq i \leq m$ and $1 \leq j \leq n$, which of the conditions (i)–(iv) (of the previous subsection) holds.

Proof. By (a) and (b), $\Leftrightarrow_S^* = \theta$ and REP is a set of representatives for $\Leftrightarrow_S^*$, respectively. Thus, REP is a set of representatives for the congruence classes of θ. (Hence, $|REP| = [H]$, too.)

The algorithm works as follows. Let us take the tree transformation classes $U_1 \circ \cdots \circ U_m$ and $V_1 \circ \cdots \circ V_n$ and form the words $U_1 \cdot \ldots \cdot U_m$ and $V_1 \cdot \ldots \cdot V_n$, respectively. First, by algorithm (d), compute the representatives u and v such that

$$U_1 \cdot \ldots \cdot U_m \Leftrightarrow_S^* u \text{ and } V_1 \cdot \ldots \cdot V_n \Leftrightarrow_S^* v.$$

Then, since $\Leftrightarrow_S^* = \theta$, we also have

$$U_1 \circ \ldots \circ U_m = |u| \text{ and } V_1 \circ \ldots \circ V_n = |v|.$$

Thus, one of the conditions (i)–(iv) holds for the tree transformation classes $U_1 \circ \cdots \circ U_m$ and $V_1 \circ \cdots \circ V_n$ if and only if the corresponding condition

$$\begin{array}{llcl}
\text{(i')} & |u| & = & |v|, \\
\text{(ii')} & |u| & \subset & |v|, \\
\text{(iii')} & |v| & \subset & |u|, \\
\text{(iv')} & |u| & \bowtie & |v|
\end{array}$$

holds for $|u|$ and $|v|$. However, having the inclusion diagram from (c), we can read from the diagram which of the conditions (i')–(iv') holds. □

We note that the general method only fixes a series of tasks; and the application of the method means that the tasks should be implemented for an actual monoid. However, it does not guarantee any result for a concrete application because during the implementation of a task we may come across difficulties that we cannot solve. Still, the method was successfully applied for several monoids - see the bibliographic notes at the end of this chapter.

An application of the general method

In this subsection we apply the general method to a concrete instance of H, namely, to the set

$$H = \{TOP, l\text{-}TOP, HOM\}.$$

Having applied the general method to such concrete instances of H, our experience has been that it is useful to implement the tasks (a)–(c) of the general method by performing the following five steps.

STEP 1. In this step we produce the string rewrite system S, which is our candidate for presenting $[H]$. Let S consist of eight rules:

$$\begin{array}{lllll}
1) & TOP \cdot TOP & \to & TOP \\
2) & TOP \cdot HOM & \to & TOP \\
3) & HOM \cdot TOP & \to & TOP \\
4) & HOM \cdot l\text{-}TOP & \to & TOP \\
5) & l\text{-}TOP \cdot l\text{-}TOP & \to & l\text{-}TOP \\
6) & TOP \cdot l\text{-}TOP & \to & TOP \\
7) & l\text{-}TOP \cdot TOP & \to & TOP \\
8) & HOM \cdot HOM & \to & HOM.
\end{array}$$

Since S should be such that $\theta = \Leftrightarrow_S^*$, our candidate S should fulfill this condition. However in this step, we can prove only the inclusion $\Leftrightarrow_S^* \subseteq \theta$. The converse inclusion will be proved only in *STEP 5.*

Lemma 3.49. $\Leftrightarrow_S^* \subseteq \theta$

Proof. First we show that for every $(u \to v) \in S$, we have $|u| = |v|$, or equivalently $u\theta v$. In words we say that elements of S are *valid in* $[H]$.

We observe that Theorem 3.39, Theorem 3.45, Corollary 3.41, and Corollary 3.40 establish that rules 1), 4), 5) and 8) are valid is S, respectively.

Next we prove that 2) is also valid. This can easily be done by computing as follows.

$$\begin{array}{lll} & TOP & \\ \subseteq & TOP \circ HOM & \text{(by Lemma 2.14)} \\ \subseteq & TOP \circ TOP & \text{(by Corollary 3.32)} \\ = & TOP & \text{(by Theorem 3.39),} \end{array}$$

which establishes that 2) is also valid in S. The validity of the rules 3), 6) and 7) can be seen in a similar way, so we omit the detailed proof.

Now we can prove that $\Rightarrow_S \subseteq \theta$. In fact, let $w, z \in H^+$ such that $w \Rightarrow_S z$. Then, by the definition of $\Rightarrow_S$, there are strings x and y in H^* and there is a rule $(u \to v) \in S$ so that $w = xuy$ and $z = xvy$. Then we can compute as follows:

$$\begin{array}{lll} & |w| & \\ = & |xuy| & \\ = & |x| \circ |u| \circ |y| & \text{(because } |\;| \text{ is a homomorphism)} \\ = & |x| \circ |v| \circ |y| & \text{(because } |u| = |v|) \\ = & |xvy| & \text{(because } |\;| \text{ is a homomorphism)} \\ = & |z|, & \end{array}$$

proving that $w\theta z$. Analogously, we can prove that $\Rightarrow_S^{-1} \subseteq \theta$, which yields that $\Leftrightarrow_S \subseteq \theta$ also. Finally, we obtain $\Leftrightarrow_S^* \subseteq \theta^* = \theta$, proving the lemma. □

STEP 2. In this step we propose our candidate for the set of representatives of the congruence classes of $\Leftrightarrow_S^*$. Namely, we conjecture that

$$REP = \{TOP, HOM, l\text{-}TOP, l\text{-}TOP \cdot HOM\}$$

is this set. However, we can prove that only in *STEP 5.*

STEP 3. In this step we give the inclusion diagram of the tree transformation classes represented by the elements of REP, i.e., of the set $|REP| = \{|u| \mid u \in REP\}$.

In determining the inclusion diagram of $|REP|$, we follow the method described in Sect. 2.2. That is, we conjecture that the inclusion diagram of $|REP|$ is the diagram in Fig. 3.10. Then with the method described in Sect. 2.2, we determine the sets of formal inclusions and formal inequalities whose validity is necessary and sufficient to prove that our conjecture is correct. The formal inclusions are:

$$\begin{array}{lll} HOM & \subseteq & l\text{-}TOP \circ HOM \\ l\text{-}TOP & \subseteq & l\text{-}TOP \circ HOM \\ l\text{-}TOP \circ HOM & \subseteq & TOP. \end{array}$$

The formal inequalities are:

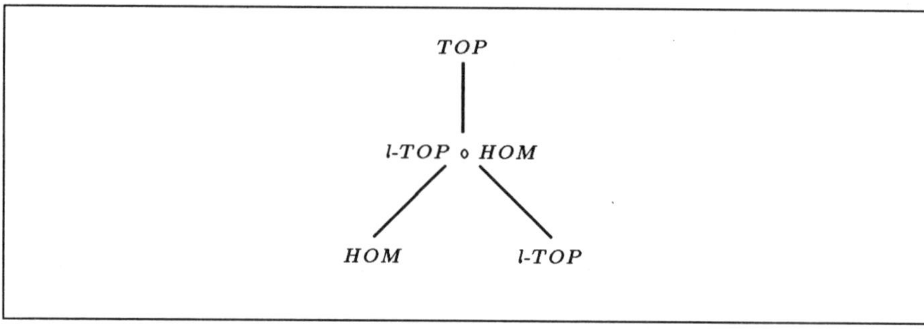

Fig. 3.10. The inclusion diagram of $|REP|$

$$\begin{array}{lcl} TOP - l\text{-}TOP \circ HOM & \neq & \emptyset \\ l\text{-}TOP \circ HOM - HOM & \neq & \emptyset \\ l\text{-}TOP \circ HOM - l\text{-}TOP & \neq & \emptyset \\ HOM - l\text{-}TOP & \neq & \emptyset \\ l\text{-}TOP - HOM & \neq & \emptyset, \end{array}$$

moreover, one can easily see that the minimal formal inequalities are:

$$\begin{array}{lcl} TOP - l\text{-}TOP \circ HOM & \neq & \emptyset \\ HOM - l\text{-}TOP & \neq & \emptyset \\ l\text{-}TOP - HOM & \neq & \emptyset. \end{array}$$

According to the method, we now prove all inclusions and inequalities which will also prove that our conjecture is the inclusion diagram of $|REP|$. The second and third minimal formal inequalities have been proved in Lemma 3.33 and Corollary 3.36, respectively. Thus it remains to prove the first minimal inequality.

Lemma 3.50. *$TOP - l\text{-}TOP \circ HOM \neq \emptyset$.*

Proof. We show that the tree transformation $\tau_{chain-to-chains}$, induced by the top-down tree transducer appearing in Example 3.3, is not in $l\text{-}TOP \circ HOM$.

We prove by contradiction, that is, we assume that $\tau_{chain-to-chains} = \tau_{T_1} \circ \tau_{T_2}$, where $T_1 = (Q_1, \Sigma, \Omega, q_1, R_1)$ is a linear top-down tree transducer and $T_2 = (\{q\}, \Omega, \Delta, q, R_2)$ is a homomorphism tree transducer.

We observe a fact which can be easily proved by the reader. Let $\xi \in RHS(Q_1, \Omega, 1)$ such that x_1 occurs exactly once in ξ. Then there exist a context $t \in C_{\Omega,1}$ and a state $p \in Q_1$ such that $\xi = t[p(x_1)]$. The proof can be done by structural induction on ξ.

Thus, since T_1 is linear, the following condition holds for the γ-rules in R_1. If $p(\gamma(x_1)) \to \xi$ is a γ-rule in R_1, then either $\xi \in T_\Omega$ or there is a context $t \in C_{\Omega,1}$ and a state $p' \in Q_1$ such that $\xi = t[p'(x_1)]$.

Now we prove a statement concerning computations carried out by T_1. For this define

$$k_1 = max\{height(\xi) \,|\, \xi \text{ is the right-hand side of some rule in } R_1\}.$$

Statement 1. Let $p \in Q_1$ and $n \geq 0$ be arbitrary. Then there is an integer $0 \leq m \leq n$, and there are contexts $t_1, \ldots, t_m \in C_{\Omega,1}$ and a tree $t_{m+1} \in T_\Omega$ such that

- x_1 occurs exactly once in any of $t_1, \ldots, t_m$,
- for every $1 \leq i \leq m+1$, $height(t_i) \leq k_1$, and
- $\tau_{T_1,p}(\gamma^n(\alpha)) = t_1[t_2 \ldots [t_{m+1}] \ldots]$.

Proof. We prove by induction on n.

For $n = 0$, we put $m = 0$ and $t_1 = rhs(p, \alpha)$. Then we have $\tau_{T_1,p}(\gamma^n(\alpha)) = \tau_{T_1,p}(\alpha) = t_1$. Moreover, $t_1 \in T_\Omega$ and $height(t_1) \leq k_1$, hence this case is proved.

Now let $n \geq 1$ and let $\xi = rhs(p, \gamma)$. (Note that $\xi \in RHS(Q_1, \Omega, 1)$.) Then, by the definition of $\tau_{T_1,p}$,

$$\tau_{T_1,p}(\gamma^n(\alpha)) = \tau_{T_1,\xi}(\gamma^{n-1}(\alpha)).$$

Now two cases are possible.

Case 1: $\xi \in T_\Omega$

Then we put $m = 0$ and $t_1 = \xi$. Obviously, $height(t_1) \leq k_1$. Then we have finished, because, by the note made after Def. 3.22, we have $\tau_{T_1,\xi}(\gamma^{n-1}(\alpha)) = \xi = t_1$.

Case 2: The variable x_1 occurs exactly once in ξ. Then, there is a context $t_1 \in C_{\Omega,1}$ and a state $p' \in Q_1$, such that $\xi = t_1[p'(x_1)]$ and $height(t_1) \leq k_1$. Moreover,

$$\begin{array}{rcl} \tau_{T_1,p}(\gamma^n(\alpha)) & = & \tau_{T_1,\xi}(\gamma^{n-1}(\alpha)) \text{ (by the definition of } \tau_{T_1,p}) \\ & = & \xi[p'(x_1) \leftarrow \tau_{T_1,p'}(\gamma^{n-1}(\alpha))] \text{ (by Lemma 3.7)} \\ & = & t_1[\tau_{T_1,p'}(\gamma^{n-1}(\alpha))]. \text{ (by the definition of } t_1) \end{array}$$

Now, by the induction hypothesis on n, there is an integer $0 < m' < n - 1$ and there are contexts $t'_1, \ldots, t'_{m'} \in C_{\Omega,1}$ and a tree $t'_{m'+1} \in T_\Omega$, such that $\tau_{T_1,p'}(\gamma^{n-1}(\alpha)) = t'_1[\cdots[t'_{m'+1}]\cdots]$. Moreover, for every $1 \leq i \leq m' + 1$, $height(t'_i) \leq k_1$. Then, for the integer $m = m' + 1$ and trees $t_1, t_2 = t'_1, \ldots, t_m = t'_{m'}, t_{m+1} = t'_{m'+1}$, we have

$$\tau_{T_1,p}(\gamma^n(\alpha)) = t_1[t_2 \ldots [t_{m+1}] \ldots].$$

This ends the proof of Statement 1.

Statement 2. Let $n \geq 0$ and $s \in T_\Omega$ be such that $\tau_{T_1}(\gamma^n(\alpha)) = s$. Let s' be a subtree of s and assume that $s' = \sigma(s_1, \ldots, s_l)$, for some $\sigma \in \Omega$ and $s_1, \ldots, s_l \in T_\Omega$. Then $height(s_i) > k_1$ can hold for at most one $1 \leq i \leq l$.

Proof. By Statement 1, $s = t_1[t_2 \ldots [t_{m+1}] \ldots]$, where the trees $t_1, \ldots, t_{m+1}$ have the properties described there. Then there is an integer $1 \leq j \leq m+1$, such that all but one of the trees $s_1, \ldots, s_l$ should be a subtree of t_j. Hence the heights of all but one of them should be at most k_1.

Now we can finish the proof of the lemma. For this, define

$$k_2 = max\{height(\xi) \mid \xi \text{ is the right-hand side of some rule in } R_2\}$$

and let $n > k_2(k_1 + 1) + 1$.

By our indirect assumption, there exists a tree $s \in T_\Omega$, such that $\tau_{T_1}(\gamma^n(\alpha)) = s$ and $\tau_{T_2}(s) = \delta(\gamma^{n-1}(\beta_1), \gamma^{n-1}(\beta_2))$. This latter equation means that $q(s) \Rightarrow^*_{T_2} \delta(\gamma^{n-1}(\beta_1), \gamma^{n-1}(\beta_2))$. Let us show this derivation in detail. There is a place where δ is derived, hence we can write

$$\begin{aligned} q(s) &\Rightarrow^*_{T_2} \quad q(s') = q(\sigma(s_1, \ldots, s_m)) \Rightarrow_{T_2} \delta(\gamma^{u_1}(q(s_i)), \gamma^{u_2}(q(s_j))) \\ &\Rightarrow^*_{T_2} \quad \delta(\gamma^{u_1}\gamma^{l_1}(\beta_1), \gamma^{u_2}\gamma^{l_2}(\beta_2)) = \delta(\gamma^{n-1}(\beta_1), \gamma^{n-1}(\beta_2)), \end{aligned}$$

where $m \geq 0$, $\sigma \in \Omega^{(m)}$, $s' = \sigma(s_1, \ldots, s_m) \in T_\Omega$ is a subtree of s, and $q(\sigma(x_1, \ldots, x_m)) \to \delta(\gamma^{u_1}(q(x_i)), \gamma^{u_2}(q(x_j)))$ is the (q, σ)-rule in R_2. Our first observation is that $i \neq j$. Actually, $i = j$ would imply $\gamma^{l_1}(\beta_1) = \gamma^{l_2}(\beta_2)$, which is impossible because $\beta_1 \neq \beta_2$. Then, by Statement 2, one of the trees s_i and s_j, let us say s_i, satisfies the condition $height(s_i) \leq k_1$. Then we have $l_1 \leq k_1 k_2$. We obtain

$$n - 1 = u_1 + l_1 \leq k_2 + k_1 \cdot k_2 = k_2(k_1 + 1),$$

that is, $n \leq k_2(k_1 + 1) + 1$ contradicting the choice of n. With this we have proved that $\tau_{chain-to-chains}$ cannot be the composition of a linear top-down tree transformation and a homomorphism tree transformation. □

We can now prove that our conjecture for the inclusion diagram of $|REP|$ is correct.

Lemma 3.51. The diagram in Fig. 3.10 is the inclusion diagram of the set $|REP| = \{|u| \mid u \in REP\}$.

Proof. We have estabilshed in Lemma 3.33, Lemma 3.50 and Corollary 3.36 that $HOM - l\text{-}TOP \neq \emptyset$, $TOP - l\text{-}TOP \circ HOM \neq \emptyset$ and $l\text{-}TOP - HOM \neq \emptyset$ hold. At the same time it is an easy exercise to show that all inclusions shown by the diagram also hold. Hence the lemma is proved. □

STEP 4. In this step we prove the following.

Lemma 3.52. For every $w \in H^+$, an element $u \in REP$ can be given such that $w \Rightarrow^*_S u$.

Proof. We prove the lemma by induction on $length(w)$.

Let $length(w) = 1$, implying that $w \in H$. Then we also have $w \in REP$, by the definition of REP. Hence, for $u = w$, we have $u \in REP$ and $w \Rightarrow^*_S u$.

Now suppose that $length(w) > 1$, which means that $w = x \cdot U$, for some $x \in H^+$ and $U \in H$. Moreover, by the induction hypothesis on $length(w)$, a representative $v \in REP$ can effectively be computed, such that $x \Rightarrow_S^* v$. Then, regarding v, four main cases are possible and three subcases can appear concerning U within each of the four. In each case, we compute u.

Case 1: $v = TOP$. Then $w = x \cdot U \Rightarrow_S^* v \cdot U = TOP \cdot U$.

- Subcase $U = TOP$. Then we have $u = TOP$, because $TOP \cdot U = TOP \cdot TOP \Rightarrow_S TOP$, by rule 1).
- Subcase $U = l\text{-}TOP$. Then $u = TOP$, because $TOP \cdot U = TOP \cdot l\text{-}TOP \Rightarrow_S TOP$, by rule 6).
- Subcase $U = HOM$. Then $u = TOP$, because $TOP \cdot U = TOP \cdot HOM \Rightarrow_S TOP$, by rule 2).

Case 2: $v = l\text{-}TOP$. Then $w = x \cdot U \Rightarrow_S^* v \cdot U = l\text{-}TOP \cdot U$.

- Subcase $U = TOP$. Then we have $u = TOP$, because $l\text{-}TOP \cdot U = l\text{-}TOP \cdot TOP \Rightarrow_S TOP$, by rule 7).
- Subcase $U = l\text{-}TOP$. Then we have $u = l\text{-}TOP$, because $l\text{-}TOP \cdot U = l\text{-}TOP \cdot l\text{-}TOP \Rightarrow_S l\text{-}TOP$, by rule 5).
- Subcase $U = HOM$. Then we have $u = l\text{-}TOP \cdot HOM$, because $l\text{-}TOP \cdot U = l\text{-}TOP \cdot HOM$ and there is no rule for $l\text{-}TOP \cdot HOM$.

Case 3: $v = HOM$. Then $w = x \cdot U \Rightarrow_S^* v \cdot U = HOM \cdot U$.

- Subcase $U = TOP$. Then we have $u = TOP$, because $HOM \cdot U = HOM \cdot TOP \Rightarrow_S TOP$, by rule 3).
- Subcase $U = l\text{-}TOP$. Then we have $u = TOP$, because $HOM \cdot U = HOM \cdot l\text{-}TOP \Rightarrow_S TOP$, by rule 4).
- Subcase $U = HOM$. Then we have $u = HOM$, because $HOM \cdot U = HOM \cdot HOM \Rightarrow_S HOM$, by rule 8).

Case 4: $v = l\text{-}TOP \cdot HOM$. Then $w = x \cdot U \Rightarrow_S^* v \cdot U = l\text{-}TOP \cdot HOM \cdot U$.

- Subcase $U = TOP$. Then $u = TOP$, because $l\text{-}TOP \cdot HOM \cdot U = l\text{-}TOP \cdot HOM \cdot TOP \Rightarrow_S l\text{-}TOP \cdot TOP$, by rule 3), and $l\text{-}TOP \cdot TOP \Rightarrow_S TOP$, by rule 7).
- Subcase $U = l\text{-}TOP$. Then $u = TOP$, because $l\text{-}TOP \cdot HOM \cdot U = l\text{-}TOP \cdot HOM \cdot l\text{-}TOP \Rightarrow_S l\text{-}TOP \cdot TOP$, by rule 4), and $l\text{-}TOP \cdot TOP \Rightarrow_S TOP$, by rule 7).
- Subcase $U = HOM$. Then $u = l\text{-}TOP \cdot HOM$, because $l\text{-}TOP \cdot HOM \cdot U = l\text{-}TOP \cdot HOM \cdot HOM \Rightarrow_S l\text{-}TOP \cdot HOM$, by rule 8).

This finishes the proof of our lemma. □

STEP 5. In this step we can prove that the tasks (a)–(d) of the general method have been performed for our particular choice of H.

Lemma 3.53. The tasks (a)–(d) have been performed for

$$H = \{TOP, l\text{-}TOP, HOM\}.$$

Proof. First we prove that $\Leftrightarrow_S^* = \theta$. By Lemma 3.49, $\Leftrightarrow_S^* \subseteq \theta$. We now show that the converse inclusion also holds. Let $w, w' \in H^+$ such that $w\theta w'$. By Lemma 3.52 there are elements $u, u' \in REP$, such that $w \Rightarrow_S^* u$ and $w' \Rightarrow_S^* u'$. Since $\Leftrightarrow_S^* \subseteq \theta$, we also have $\Rightarrow_S^* \subseteq \theta$, and hence $w\theta u$ and $w'\theta u'$. Recalling that θ is a congruence, we obtain $u\theta u'$. However, by Lemma 3.51, two elements of REP represent the same tree transformation class in $[H]$ if and only if they are equal, hence $u = u'$. Therefore $w \Leftrightarrow_S^* w'$ also holds. Thus, by $[H] \cong H^+/\theta$ and $\theta = \Leftrightarrow_S^*$, we get $[H] \cong H^+/\Leftrightarrow_S^*$. Hence task (a) has been completed for H.

Next we prove that REP is a set of representatives for the congruence classes of $\Leftrightarrow_S^*$. Let w be an element of an arbitrary congruence class of $\Leftrightarrow_S^*$. Then, by Lemma 3.52, there is an element $u \in REP$ such that $w \Rightarrow_S^* u$, hence u is in the $\Leftrightarrow_S^*$-class of w. Now assume that the elements $u, u' \in REP$ are both in the $\Leftrightarrow_S^*$ class of w. Then, by $\Leftrightarrow_S^* = \theta$, we have $u\theta u'$, hence, by Lemma 3.51, we have $u = u'$. This shows that task (b) and, again by Lemma 3.51, task (c) have been completed for H.

Finally, Lemma 3.52 implies that task (d) has been performed. □

Now we can prove the main result of this section.

Theorem 3.54. Let $H = \{TOP, l\text{-}TOP, HOM\}$. Then there is an algorithm that decides, given two arbitrary tree transformation classes $U_1 \circ \cdots \circ U_m$ and $V_1 \circ \cdots \circ V_n$, where $U_i, V_j \in H$, for every $1 \le i \le m$ and $1 \le j \le n$, which one of the conditions

$$\begin{array}{rrcl}
\text{(i)} & U_1 \circ \cdots \circ U_m & = & V_1 \circ \cdots \circ V_n, \\
\text{(ii)} & U_1 \circ \cdots \circ U_m & \subset & V_1 \circ \cdots \circ V_n, \\
\text{(iii)} & V_1 \circ \cdots \circ V_n & \subset & U_1 \circ \cdots \circ U_m, \\
\text{(iv)} & U_1 \circ \cdots \circ U_m & \bowtie & V_1 \circ \cdots \circ V_n,
\end{array}$$

holds.

Proof. By Lemma 3.53, the tasks appearing in the description of the general method have been carried out. Hence, by Lemma 3.48 the algorithm exists.
□

An application of the algorithm

We show by an example how our algorithm works. Let us take the two tree transformation classes

$$l\text{-}TOP \circ HOM \circ l\text{-}TOP \circ l\text{-}TOP \text{ and } l\text{-}TOP \circ l\text{-}TOP \circ HOM \circ HOM$$

which are constructed by composition from the elements of H.

First, by Lemma 3.48, we have to compute the representatives u and v of the words *l-TOP* · *HOM* · *l-TOP* · *l-TOP* and *l-TOP* · *l-TOP* · *HOM* · *HOM*, respectively. We do this by applying the algorithm described in Lemma 3.52.

An easy way to apply the algorithm of Lemma 3.52 is to prepare a table from which entries can be readily obtained later. For $H = \{TOP, l\text{-}TOP, HOM\}$, the table looks as follows.

	TOP	*l-TOP*	*HOM*
TOP	*TOP*	*TOP*	*TOP*
l-TOP	*TOP*	*l-TOP*	*l-TOP* ∘ *HOM*
HOM	*TOP*	*TOP*	*HOM*
l-TOP ∘ *HOM*	*TOP*	*TOP*	*l-TOP* ∘ *HOM*

This table, denoted by T, has the following properties. Its rows are labeled by elements of REP and its columns are labeled by elements of H. Moreover, for $v \in REP$ and $U \in H$, the entry $T(v, U)$ at (v, U) is also an element of REP such that $v \cdot U \Rightarrow_S^* T(v, U)$.

Now, if there is a $w = U_1 \cdot U_2 \cdot \ldots \cdot U_n \in H^+$, the following small program computes $u \in REP$, for which $w \Rightarrow_S^* u$.

$u := U_1$;
for $i = 2$ **to** n **do**
  $u := T(u, U_i)$

By running this program on $w = l\text{-}TOP \cdot HOM \cdot l\text{-}TOP \cdot l\text{-}TOP$, we get

$$w = l\text{-}TOP \cdot HOM \cdot l\text{-}TOP \cdot l\text{-}TOP \Rightarrow_S^* TOP,$$

implying that

$$l\text{-}TOP \circ HOM \circ l\text{-}TOP \circ l\text{-}TOP = TOP.$$

Similarly, running the program on $w = l\text{-}TOP \cdot l\text{-}TOP \cdot HOM \cdot HOM$, we get

$$l\text{-}TOP \cdot l\text{-}TOP \cdot HOM \cdot HOM \Rightarrow_S^* l\text{-}TOP \cdot HOM,$$

which implies that

$$l\text{-}TOP \circ l\text{-}TOP \circ HOM \circ HOM = l\text{-}TOP \circ HOM.$$

From the inclusion diagram appearing in Lemma 3.51, we can see that $l\text{-}TOP \circ HOM \subset TOP$. Hence also

$$l\text{-}TOP \circ l\text{-}TOP \circ HOM \circ HOM \subset l\text{-}TOP \circ HOM \circ l\text{-}TOP \circ l\text{-}TOP.$$

Presenting $[H]$ by a terminating and confluent rewrite system

In this subsection we show that the string rewrite system S which consists of the rules 1) through 8), is terminating and confluent as well (Sect. 2.4). Moreover we have chosen REP in such a way that $REP = IRR(H, S)$ holds, where we recall that $IRR(H, S)$ denotes the set of irreducible strings over H with respect to S (Sect. 2.6). This makes it easier to compute, for every $w \in H^+$, the representative $u \in REP$ with $w \Rightarrow_S^* u$. Then we can more easily compute the representative of a string representing a tree transformation class. First we prove the two initial statements (in the opposite order) and then we discuss their importance.

Lemma 3.55. $REP = IRR(H, S)$

Proof. Recall that $REP = \{TOP, HOM, l\text{-}TOP, l\text{-}TOP \cdot HOM\}$. The fact that $REP \subseteq IRR(H, S)$ can easily be seen as follows. The length of the left-hand side of any rule in S is 2. Therefore none of them is applicable to TOP, HOM and $l\text{-}TOP$, hence these are in $IRR(H, S)$. Moreover, there is no rule with a left-hand side $l\text{-}TOP \cdot HOM$ in S either, thus $l\text{-}TOP \cdot HOM \in IRR(H, S)$ also.

To prove the converse inclusion, let, for $n \geq 1$,

$$IRR_n = \{w \in IRR(H, S) \mid length(w) = n\}.$$

We show that $IRR_n \subseteq REP$, for every $n \geq 1$.

Obviously $IRR_1 \subseteq REP$, because $IRR_1 = \{TOP, l\text{-}TOP, HOM\}$. Next we observe that $IRR_2 = \{l\text{-}TOP \cdot HOM\}$ because, as can be seen immediately, for any elements $U, V \in H$ with $U \neq l\text{-}TOP$ and $V \neq HOM$, $U \cdot V$ is the left-hand side of a rule in S, hence it cannot be in IRR_2. Therefore $IRR_2 \subseteq REP$, too. Now we show that $IRR_3 = \emptyset$, which arises from the following observation. We observe that each element of IRR_3, if there are any, should be either of the form $U \cdot w$ or of the form $w \cdot U$, for some $w \in IRR_2$ and $U \in H$, therefore either of the form $U \cdot l\text{-}TOP \cdot HOM$ or of the form $l\text{-}TOP \cdot HOM \cdot U$. However, both $U \cdot l\text{-}TOP$ and $HOM \cdot U$ are the left-hand sides of a rule in S, whatever U is, therefore neither $U \cdot l\text{-}TOP \cdot HOM$ nor $l\text{-}TOP \cdot HOM \cdot U$ can be irreducible. Finally, $IRR_3 = \emptyset$ implies that $IRR_n = \emptyset$, for every $n \geq 3$, and thus $IRR_n \subseteq REP$, for every $n \geq 3$.

This also proves $IRR(H, S) \subseteq REP$. □

Lemma 3.56. S is terminating and confluent.

Proof. The fact that S is terminating immediately follows from $length(u) > length(v)$, for every rule $u \to v \in S$. Hence, $length(w) > length(z)$ for every $w, z \in H^+$ with $w \Rightarrow_S z$. By Lemma 3.55, $IRR(H, S) = REP$. Moreover, we saw in the proof of Lemma 3.53, that each congruence class of $\Leftrightarrow_S^*$ contains exactly one element of REP and thus exactly one irreducible element with respect to $\Rightarrow_S$. Hence, by Lemma 2.6, S is confluent as well. □

The importance of Lemmas 3.55 and 3.56 is evident in that when computing the representative in REP for an arbitrary string $w \in H^+$, we do not have to use the algorithm described in Lemma 3.52.

Instead, we can do the following. Given a string $w \in H^+$, we compute an irreducible element $u \in IRR(H,S)$ so that $w \Rightarrow_S^* u$, by applying rules in S in an arbitrary order as far as we can. The process will stop because S is terminating. On the other hand the same irreducible u will be reached in any case, because S is confluent. Since $IRR(H,S) = REP$, we also have $u \in REP$.

For example, take again the two tree transformation classes

$$l\text{-}TOP \circ HOM \circ l\text{-}TOP \circ l\text{-}TOP \text{ and } l\text{-}TOP \circ l\text{-}TOP \circ HOM \circ HOM.$$

Then we have

$$\begin{array}{lll} & l\text{-}TOP \cdot HOM \cdot l\text{-}TOP \cdot l\text{-}TOP & \\ \Rightarrow_S & l\text{-}TOP \cdot TOP \cdot l\text{-}TOP & \text{(by rule 4)} \\ \Rightarrow_S & l\text{-}TOP \cdot TOP & \text{(by rule 6)} \\ \Rightarrow_S & TOP & \text{(by rule 7),} \end{array}$$

but also

$$\begin{array}{lll} & l\text{-}TOP \cdot HOM \cdot l\text{-}TOP \cdot l\text{-}TOP & \\ \Rightarrow_S & l\text{-}TOP \cdot HOM \cdot l\text{-}TOP & \text{(by rule 5)} \\ \Rightarrow_S & l\text{-}TOP \cdot TOP & \text{(by rule 4)} \\ \Rightarrow_S & TOP & \text{(by rule 7).} \end{array}$$

Similarly, we obtain that $l\text{-}TOP \cdot l\text{-}TOP \cdot HOM \cdot HOM \Rightarrow_S^* l\text{-}TOP \cdot HOM$. Since $l\text{-}TOP \circ HOM \subset TOP$, the same inclusion relation holds between the original classes, too, as we have seen already.

3.8 Bibliographic Notes

The concept of a top-down tree transducer was introduced in [Rou69, Rou70b] and [Tha70]. Both papers consider top-down tree transducers as formal models which are suitable for studying properties of the syntax directed translation method. Then fundamental properties of top-down tree transducers were more thoroughly investigated in [Eng75], [Bak78], and [Bak79]; special top-down tree transducers, and composition and decomposition problems were considered mainly. The results of Sect. 3.5 come from these papers. Lemma 3.9 also follows from Lemma 3.1 of [Hue80], because there are no critical pairs arising from the rules of a deterministic top-down tree transducer.

The top-down tree transducer is enriched with a regular look-ahead capacity in [Eng77], which increases its transformational power and yields nice closure properties. In the papers [FV89a], [FV89b], and [FV89c] deterministic top-down tree transducers with regular look-ahead were also considered. The connection between top-down tree transducers and the theory of L-systems was studied in the papers [Eng76] and [ERS80]. In [Eng80] a collection of problems, some of them still unsolved even now, was presented. The work [Eng81] is a study of the connection between the syntax-directed translation method and its formal models, among which top-down tree transducers were also considered. In [Eng82a] it was proved that surface languages of top-down tree transducers form a hierarchy with respect to composition. Leading on from this result, composition of (nondeterministic) top-down tree transformations also forms a hierarchy with respect to the number of components appearing in the composition. The papers [Eng78] and [Eng82b] also deal with top-down tree transducers. A good, though not complete, collection of the results appearing in the above papers can be found with a uniform terminology in the handbook [GS84].

A relatively new area in the theory of tree transducers is looking for semigroups generated by tree transformation classes in which the inclusion (and hence the equivalence) of tree transformation classes is decidable. Such investigations were started in [FV87b]. Then, in the works [FV87a], [FV88a],[FV88b], [FV90], [FV91], and [Fül91] a method was developed for finding an algorithm that can decide the inclusion in certain semigroups of tree transformation classes. Also in these papers the method was successfully applied for two semigroups. A summary of these results can be found in [FV92b]. Further applications of the method occur in [SV93], [GV96], [DF96], and [DF98]. We have also applied this method in Sect. 3.7 of this book.

4. Macro Tree Transducers

In this chapter we introduce another formal model of syntax-directed semantics: the macro tree transducer, which is an abstraction from denotational semantics. It is more powerful than a top-down tree transducer in the sense that it can handle context information (as discussed in Sect. 1.1). In fact, macro tree transducers handle the context information in an implicit way, i.e., there are parameters given to the states in which context information can be assembled and modified. In this way, macro tree transducers can be considered as a straightforward extension of top-down tree transducers.

The structure of this chapter is as follows.

1. Basic definitions
2. Induced tree transformation
3. Characterization of macro tree transformations
4. Height property
5. Composition and decomposition results
6. Bibliographic Notes

4.1 Basic Definitions

Since context information is handled implicitly by having additional parameters, we introduce a new set of variables to take over the role of these parameters. Let $Y = \{y_1, y_2, \ldots\}$ be a set of variables disjoint from X (introduced in Sect. 2.7) and, for every $n \geq 1$, let $Y_n = \{y_1, y_2, \ldots y_n\}$. In fact, we will sometimes call the elements of Y *context parameters*.

As in the case of top-down tree transducers, we first define the set of right-hand sides, elements of which may be the right-hand sides of rules of a macro tree transducer. The new phenomenon with respect to right-hand sides of rules of top-down tree transducers is the possibility of nesting states in the parameter positions of other states.

Definition 4.1. Let Q and Δ be disjoint ranked alphabets and let $k, n \geq 0$ be integers. The set $RHS(Q, \Delta, k, n)$ of *right-hand sides over Q, Δ, k, and n* is the smallest subset RHS of $T_{Q\cup\Delta}(X_k \cup Y_n)$ satisfying the following conditions:

(i) For every $p \in Q^{(l+1)}$ with $l \geq 0$, $1 \leq i \leq k$, and $\xi_1, \ldots, \xi_l \in RHS$, the term $p(x_i, \xi_1, \ldots, \xi_l)$ is in RHS.
(ii) For every $\delta \in \Delta^{(l)}$ with $l \geq 0$ and $\xi_1, \ldots, \xi_l \in RHS$, the term $\delta(\xi_1, \ldots, \xi_l)$ is in RHS.
(iii) $Y_n \subseteq RHS$. □

In case (i) we say that "ξ_j occurs in the j-th parameter position of p."

In the definition of a macro tree transducer, we also allow the initial state to have parameters. Since parameters carry context information, it is clear intuitively that the concrete values for the parameters of the initial state have to be given explicitly. In effect, these values form the environment in which the macro tree transducer operates.

Definition 4.2. A *macro tree transducer* is a tuple $M = (Q, \Sigma, \Delta, q_0, R, E)$, where

- Q is a ranked alphabet, called the set of *states*; every state has at least rank 1.
- Σ and Δ are ranked alphabets with $Q \cap (\Sigma \cup \Delta) = \emptyset$, called the *input alphabet* and the *output alphabet*, respectively.
- $q_0 \in Q^{(r+1)}$ with $r \geq 0$ is a designated state, called the *initial state*.
- R is a finite set of *rules* such that, for every $q \in Q^{(n+1)}$ with $n \geq 0$ and $\sigma \in \Sigma^{(k)}$ with $k \geq 0$, there is exactly one rule of the form

$$q(\sigma(x_1, \ldots, x_k), y_1, \ldots, y_n) \rightarrow \xi$$

  in R where $x_1, \ldots, x_k \in X$, $y_1, \ldots, y_n \in Y$, and $\xi \in RHS(Q, \Delta, k, n)$.
- E is a tuple $(t_1, \ldots, t_r)$ where $t_1, \ldots, t_r \in T_\Delta$ and $r + 1 = rank(q_0)$; E is called the *environment*.

□

By the following note, the formal model of the top-down tree transducer can be retrieved in an easy way from the formal model of the macro tree transducer.

Note 4.3. If Q is a unary ranked alphabet, then M is a top-down tree transducer. In fact, since the environment is the empty sequence, it can be dropped. □

The rule $q(\sigma(x_1, \ldots, x_k), y_1, \ldots, y_n) \rightarrow \xi$ is called the (q, σ)-rule of M; the variables $y_1, \ldots, y_n$ carry context information for the input subtree $\sigma(x_1, \ldots, x_k)$; ξ is called its *right-hand side* which is also denoted by $rhs(q, \sigma)$. Thus the expression $\xi = rhs(q, \sigma)$ reads "ξ is the right-hand side of the (q, σ)-rule." Every (q, σ)-rule for some $q \in Q$ is also called a *σ-rule*.

Example 4.4. Let the macro tree transducer $M_{bal} = (Q, \Sigma, \Delta, q, R, E)$ be defined by

- $Q = \{q^{(2)}\}$
- $\Sigma = \Delta = \{\sigma^{(2)}, \alpha^{(0)}\}$
- R consists of two rules:
  1) $q(\sigma(x_1, x_2), y_1) \to q(x_2, q(x_1, y_1))$
  2) $q(\alpha, y_1) \to \sigma(y_1, y_1)$
- $E = (\alpha)$.

For instance, the first rule of M_{bal} is called the (q, σ)-rule. Its right-hand side, i.e., the term $q(x_2, q(x_1, y_1))$, is also referred to as $rhs(q, \sigma)$. In particular, the term $q(x_1, y_1)$ is nested in the parameter position of the outermost occurrence of q.

Later we will see that M_{bal} takes as input a tree s over Σ and produces as output the fully balanced tree of height $k + 1$ where k is the number of leaves of s (see Example 3.29 for the definition of fully balanced trees). In the next section we will return to this example and state in a more formal way the input-output behavior of M_{bal}. □

4.2 Induced Tree Transformation

Now, for an arbitrary macro tree transducer M, we will define the tree transformation which is induced by M. We proceed in the same way as in Chap. 3: first, we define the derivation relation of M; second, we prove that the derivation relation of M is locally confluent and terminating; and third, roughly speaking, we define the transformation of an input tree s as the normal form of the term $q_0(s, y_1, \ldots, y_n)$ (note that this normal form exists because of Corollary 2.7). Clearly, in the third step, we have to observe the environment of M.

The following definitions and lemmas concern an arbitrary, but fixed, macro tree transducer $M = (Q, \Sigma, \Delta, q_0, R, E)$.

First, let us define the derivation relation of M (see also Fig. 4.1). We should point out that the derivation relation is a binary relation over $T_{Q \cup \Sigma \cup \Delta}(Y)$ and not just over $T_{Q \cup \Sigma \cup \Delta}$, i.e., the context parameters y_i's may occur in the terms which are derived, and in that situation they are considered as nullary symbols. This slight extension is just a technical convenience for subsequent proofs.

Definition 4.5. The *derivation relation induced by* M is the binary relation $\Rightarrow_M$ over the set $T_{Q \cup \Sigma \cup \Delta}(Y)$ such that, for every $\varphi, \psi \in T_{Q \cup \Sigma \cup \Delta}(Y)$, we define $\varphi \Rightarrow_M \psi$, if

- there is a context $\beta \in C_{Q \cup \Sigma \cup \Delta \cup Y, 1}$
- there is a state $q \in Q^{(n+1)}$ with $n \geq 0$
- there is a $\sigma \in \Sigma^{(k)}$ for some $k \geq 0$
- there are trees $s_1, \ldots, s_k \in T_\Sigma$
- there are trees $\varphi_1, \ldots, \varphi_n \in T_{Q \cup \Sigma \cup \Delta}(Y)$

such that
$\varphi = \beta[q(\sigma(s_1, \ldots, s_k), \varphi_1, \ldots, \varphi_n)]$ and
$\psi = \beta[\xi[x_1 \leftarrow s_1, \ldots, x_k \leftarrow s_k][y_1 \leftarrow \varphi_1, \ldots, y_n \leftarrow \varphi_n]]$ where
$\xi = rhs(q, \sigma)$. □

Note that $\xi[x_1 \leftarrow s_1, \ldots, x_k \leftarrow s_k][y_1 \leftarrow \varphi_1, \ldots, y_n \leftarrow \varphi_n] = \xi[x_1 \leftarrow s_1, \ldots, x_k \leftarrow s_k, y_1 \leftarrow \varphi_1, \ldots, y_n \leftarrow \varphi_n]$, i.e., the sequential substitution can also be performed simultaneously.

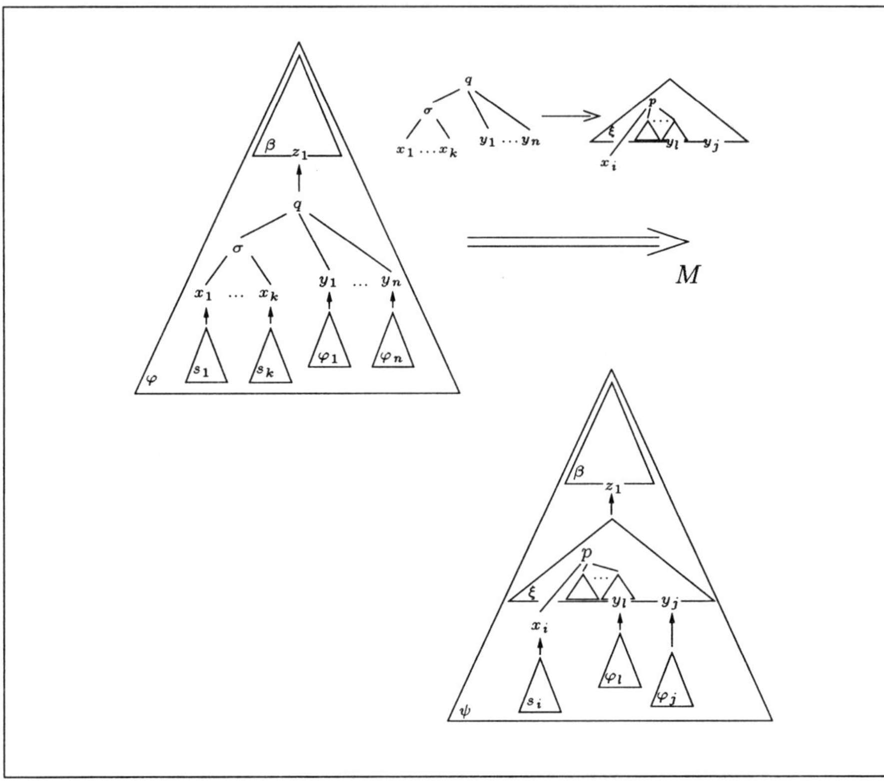

Fig. 4.1. A derivation step induced by M

Example 4.6. Let us consider the macro tree transducer M_{bal} of Example 4.4 and show a derivation step. Let

$$\varphi = q(\alpha, q(\sigma(\alpha, \alpha), \sigma(\alpha, y_1))).$$

Then there are a context $\beta = q(\alpha, z_1)$, a state q, an input symbol $\sigma \in \Sigma^{(2)}$, trees $s_1 = \alpha$, $s_2 = \alpha$, and $\varphi_1 = \sigma(\alpha, y_1)$ such that $\varphi = \beta[q(\sigma(s_1, s_2), \varphi_1)]$. Then, $\varphi \Rightarrow_{M_{bal}} \psi$, where

$$
\begin{aligned}
\psi &= \beta[rhs(q,\sigma)[x_1 \leftarrow s_1, x_2 \leftarrow s_2][y_1 \leftarrow \varphi_1]] \\
&= q(\alpha, z_1)[q(x_2, q(x_1, y_1))[x_1 \leftarrow \alpha, x_2 \leftarrow \alpha][y_1 \leftarrow \sigma(\alpha, y_1)]] \\
&= q(\alpha, z_1)[q(\alpha, q(\alpha, \sigma(\alpha, y_1)))] \\
&= q(\alpha, q(\alpha, q(\alpha, \sigma(\alpha, y_1))))
\end{aligned}
$$

We also show a sequence of derivation steps (see Fig. 4.2):

$$
\begin{aligned}
& q(\sigma(\alpha, \sigma(\alpha, \alpha)), y_1) \\
\Rightarrow_{M_{bal}}\ & q(\sigma(\alpha, \alpha), q(\alpha, y_1)) \\
& \text{(by rule (1) with } \beta = z_1) \\
\Rightarrow_{M_{bal}}\ & q(\alpha, q(\alpha, q(\alpha, y_1))) \\
& \text{(by rule (1) with } \beta = z_1) \\
\Rightarrow_{M_{bal}}\ & q(\alpha, q(\alpha, \sigma(y_1, y_1))) \\
& \text{(by rule (2) with } \beta = q(\alpha, q(\alpha, z_1))) \\
\Rightarrow_{M_{bal}}\ & q(\alpha, \sigma(\sigma(y_1, y_1), \sigma(y_1, y_1))) \\
& \text{(by rule (2) with } \beta = q(\alpha, z_1)) \\
\Rightarrow_{M_{bal}}\ & \sigma(\sigma(\sigma(y_1, y_1), \sigma(y_1, y_1)), \sigma(\sigma(y_1, y_1), \sigma(y_1, y_1))) \\
& \text{(by rule (1) with } \beta = z_1)
\end{aligned}
$$

□

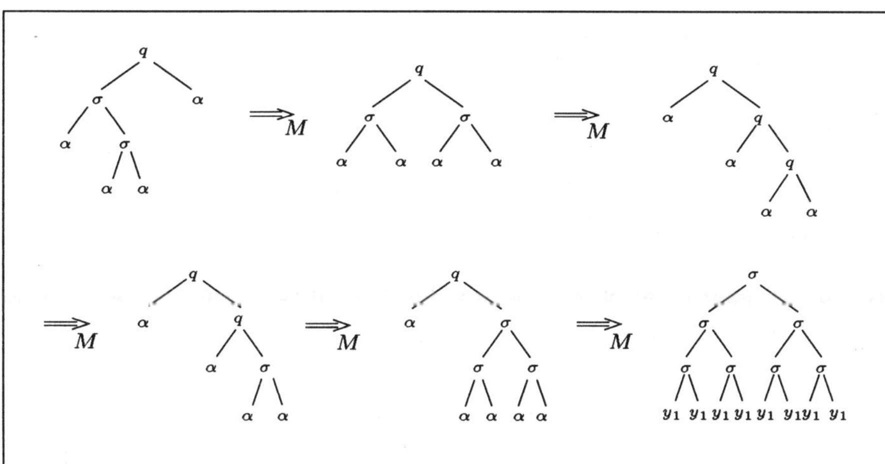

Fig. 4.2. A sequence of derivation steps of the macro tree transducer M_{bal}

Definition 4.7. The set of *sentential forms of* M, denoted by $SF(M)$, is the smallest subset SF of $T_{Q \cup \Sigma \cup \Delta}(Y)$ for which the following conditions hold.

(i) For every $p \in Q^{(l+1)}$ with $l \geq 0$, $s \in T_\Sigma$, and $\varphi_1, \ldots, \varphi_l \in SF$, the term $p(s, \varphi_1, \ldots, \varphi_l) \in SF$.

(ii) For every $\delta \in \Delta^{(l)}$ with $l \geq 0$ and $\varphi_1, \ldots, \varphi_l \in SF$, the term $\delta(\varphi_1, \ldots, \varphi_l)$ is in SF.
(iii) $Y \subseteq SF$. □

There is a relationship between sentential forms and right-hand sides which is similar to the situation for top-down tree transducers (cf. Lemma 3.7).

Lemma 4.8. (a) For every right-hand side $\xi \in RHS(Q, \Delta, k, n)$ with $k, n \geq 0$, $s_1, \ldots, s_k \in T_\Sigma$, and $\varphi_1, \ldots, \varphi_n \in SF(M)$, the term $\xi[x_1 \leftarrow s_1, \ldots, x_k \leftarrow s_k][y_1 \leftarrow \varphi_1, \ldots, y_n \leftarrow \varphi_n] \in SF(M)$.

(b) For every sentential form $\varphi \in SF(M)$, there are $k \geq 0$, $n \geq 0$, $\xi \in RHS(Q, \Delta, k, n)$, and $s_1, \ldots, s_k \in T_\Sigma$ such that $\varphi = \xi[x_1 \leftarrow s_1, \ldots, x_k \leftarrow s_k]$.

Proof. Statement (a) is proved by structural induction on ξ similar to the proof of the corresponding lemma for top-down tree transducers (cf. Lemma 3.7 (a)).

Statement (b) is proved by structural induction on φ again similar to the proof of Lemma 3.7 (b). In particular, if φ has the form $p(s, \varphi_1, \ldots, \varphi_l)$ for some state p, then the proof uses the same technique of index shift as in part (ii) of the proof of Lemma 3.7 (b). □

Also the proof of the closure of $SF(M)$ under $\Rightarrow_M$ can be seen as a straightforward generalization of the corresponding proof for top-down tree transducers (cf. Lemma 3.8).

Lemma 4.9. $SF(M)$ is closed under $\Rightarrow_M$.

Proof. We have to show that, for every $\varphi \in SF(M)$ and $\psi \in T_{Q \cup \Sigma \cup \Delta}(Y)$, if $\varphi \Rightarrow_M \psi$, then $\psi \in SF(M)$.

The proof is performed by structural induction in the same way as in the proof of Lemma 3.8. In particular, if φ has the form $p(s, \varphi_1, \ldots, \varphi_l)$, then we have to distinguish two cases: first, a rule is applied to the root of φ, and second, a rule is applied to one of the subtrees φ_i. In the first case, the proof is similar to part (i) of the proof of Lemma 3.8; in the second case, the proof is similar to part (ii) of the proof of Lemma 3.8. □

It follows from the previous lemma that $(SF(M), \Rightarrow_M)$ is a derivation system in the sense of Sect. 2.4. Next we shall prove that the derivation relation $\Rightarrow_M$ is locally confluent and terminating.

Lemma 4.10. The derivation relation $\Rightarrow_M$ is locally confluent on $SF(M)$.

Proof. Let φ, φ_1, and φ_2 be sentential forms of M such that $\varphi \Rightarrow_M \varphi_1$ and $\varphi \Rightarrow_M \varphi_2$. We show that there exists a sentential form $\psi \in SF(M)$ such that $\varphi_1 \Rightarrow_M^* \psi$ and $\varphi_2 \Rightarrow_M^* \psi$.

By Def. 4.5, for every i with $i = \{1, 2\}$,

- there is a context $\beta_i \in C_{Q\cup\Sigma\cup\Delta\cup Y,1}$,
- there is a state $q_i \in Q^{(n_i+1)}$ for some $n_i \geq 0$,
- there is a $\sigma_i \in \Sigma^{(k_i)}$ for some $k_i \geq 0$,
- there are trees $s_{i1}, \ldots, s_{ik_i} \in T_\Sigma$,
- there are trees $\varphi_{i1}, \ldots, \varphi_{in_i} \in T_{Q\cup\Sigma\cup\Delta}(Y)$

such that

$$\begin{aligned}
\varphi &= \beta_1[q_1(\sigma_1(s_{11}, \ldots, s_{1k_1}), \varphi_{11}, \ldots, \varphi_{1n_1})] \\
&= \beta_2[q_2(\sigma_2(s_{21}, \ldots, s_{2k_2}), \varphi_{21}, \ldots, \varphi_{2n_2})] \\
\varphi_1 &= \beta_1[\xi_1[x_1 \leftarrow s_{11}, \ldots, x_{k_1} \leftarrow s_{1k_1}][y_1 \leftarrow \varphi_{11}, \ldots, y_{n_1} \leftarrow \varphi_{1n_1}]], \\
\varphi_2 &= \beta_2[\xi_2[x_1 \leftarrow s_{21}, \ldots, x_{k_2} \leftarrow s_{2k_2}][y_1 \leftarrow \varphi_{21}, \ldots, y_{n_2} \leftarrow \varphi_{2n_2}]],
\end{aligned}$$

where, for every i with $i = \{1, 2\}$, the term $\xi_i = rhs(q_i, \sigma_i)$.

Let w_i be the occurrence of z_1 in β_i. The two cases in which

1) $w_1 = w_2$ and

2) w_1 and w_2 are disjoint paths, i.e., neither is a prefix of the other

are treated in the same way as in the proof of the corresponding lemma for top-down tree transducers (cf. Lemma 3.9)

Now assume that w_1 is a proper prefix of w_2. Then there is a j with $1 \leq j \leq n_1$ such that the term $q_2(\sigma_2(s_{21}, \ldots, s_{2k_2}), \varphi_{21}, \ldots, \varphi_{2n_2})$ occurs in φ_{1j}. Moreover, there is a v such that $w_2 = w_1 j v$. Assume that y_j occurs κ times in ξ_1. Let φ'_{1j} denote the term which is obtained from φ_{1j} by replacing the subterm at occurrence v by the term $\xi_2[x_1 \leftarrow s_{21}, \ldots, x_{k_2} \leftarrow s_{2k_2}][y_1 \leftarrow \varphi_{21}, \ldots, y_{n_2} \leftarrow \varphi_{2n_2}]$. Then define $\psi = \beta_1[\xi_1[x_1 \leftarrow s_{11}, \ldots, x_{k_1} \leftarrow s_{1k_1}]$ $[y_1 \leftarrow \varphi_{11}, \ldots, y_{j-1} \leftarrow \varphi_{1(j-1)}, y_j \leftarrow \varphi'_{1j}, y_{j+1} \leftarrow \varphi_{1(j+1)}, \ldots, y_{n_1} \leftarrow \varphi_{1n_1}]]$.

Clearly, $\varphi_1 \Rightarrow_M^\kappa \psi$ and $\varphi_2 \Rightarrow_M \psi$. □

Lemma 4.11. The derivation relation $\Rightarrow_M$ is terminating on $SF(M)$.

Proof. We first prove the following two statements by simultaneous induction (see Def. 3.10 and Principle 3.11).

K: For every $s \in T_\Sigma$ and $q \in Q^{(n+1)}$ with $n \geq 0$, there is a function $g_{q,s} : \mathbb{N}^n \to \mathbb{N}$ such that for every $\varphi_1, \ldots, \varphi_n \in SF(M)$, if the derivations starting from φ_i are not longer than c_i for every $1 \leq i \leq n$, then the derivations starting from $q(s, \varphi_1, \ldots, \varphi_n)$ are not longer than $g_{q,s}(c_1, \ldots, c_n)$.

L: For every $\xi \in RHS(Q, \Delta, k, n)$ with $k, n \geq 0$, and $\omega = (s_1, \ldots, s_k) \in (T_\Sigma)^k$, there is a function $g_{\xi,\omega} : \mathbb{N}^n \to \mathbb{N}$ such that for every $\varphi_1, \ldots, \varphi_n \in SF(M)$, if the derivations starting from φ_i are not longer than c_i for every $1 \leq i \leq n$, then the derivations starting from $\xi[x_1 \leftarrow s_1, \ldots, x_k \leftarrow s_k][y_1 \leftarrow \varphi_1, \ldots, y_n \leftarrow \varphi_n]$ are not longer than $g_{\xi,\omega}(c_1, \ldots, c_n)$.

Proof of IB: Let $\xi \in RHS(Q, \Delta, 0, n)$ and let $\omega = ()$. The proof is by structural induction on ξ.

(i) ξ cannot have the form $p(x_i, \xi_1, \ldots, \xi_l)$.

(ii) $\xi = \delta(\xi_1, \ldots, \xi_l)$ for some $\delta \in \Delta^{(l)}$ and $l \geq 0$. If $l > 0$, then define $g_{\xi,\omega}(c_1, \ldots, c_n) = g_{\xi_1,\omega}(c_1, \ldots, c_n) + \ldots + g_{\xi_l,\omega}(c_1, \ldots, c_n)$, and if $l = 0$, then define $g_{\xi,\omega}(c_1, \ldots, c_n) = 1$. Note that, by induction hypothesis on this structural induction, there are functions $g_{\xi_i,\omega} : \mathbb{N}^n \to \mathbb{N}$ such that the statement holds.

Now let $\varphi_1, \ldots, \varphi_n$ be arbitrary sentential forms in $SF(M)$. Assume that the derivations starting from φ_i are not longer than c_i. Then clearly, the derivations starting from ξ cannot be longer than $g_{\xi,\omega}(c_1, \ldots, c_n)$.

(iii) $\xi = y_j$ for some $y_j \in Y_n$. Then with $g_{\xi,\omega}(c_1, \ldots, c_n)$ defined as c_j, the statement follows immediately.

Proof of IS1: Let $s = \sigma(s_1, \ldots, s_k) \in T_\Sigma$, $q \in Q^{(n+1)}$, and $k, n \geq 0$. Then define $g_{q,s}(c_1, \ldots, c_n) = c_1 + \ldots + c_n + 1 + g_{rhs(q,\sigma),\omega}(c_1, \ldots, c_n)$ where $\omega = (s_1, \ldots, s_k)$.

An arbitrary derivation starting from $q(s, \varphi_1, \ldots, \varphi_n)$ looks either like

$$\begin{aligned} & q(s, \varphi_1, \ldots, \varphi_n) \\ \Rightarrow_M^* \quad & q(s, \varphi_1', \ldots, \varphi_n') \end{aligned}$$

and then, if the derivations starting from φ_i are not longer than c_i, it is not longer than $c_1 + \ldots + c_n$, or it looks like

$$\begin{aligned} & q(s, \varphi_1, \ldots, \varphi_n) \\ \Rightarrow_M^* \quad & q(s, \varphi_1', \ldots, \varphi_n') = q(\sigma(s_1, \ldots, s_k), \varphi_1', \ldots, \varphi_n') \\ \Rightarrow_M \quad & rhs(q, \sigma)[x_1 \leftarrow s_1, \ldots, x_k \leftarrow s_k][y_1 \leftarrow \varphi_1', \ldots, y_n \leftarrow \varphi_n'] \\ \Rightarrow_M^* \quad & \varphi. \end{aligned}$$

Then, by using the induction hypothesis on L, such a derivation has maximally the length $c_1 + \ldots + c_n + 1 + g_{rhs(q,\sigma),\omega}(c_1, \ldots, c_n)$.

Proof of IS2: Let $\xi \in RHS(Q, \Delta, k, n)$ and let $\omega = (s_1, \ldots, s_k)$. The proof is by structural induction where the cases (ii) and (iii) are proved in the same way as in the proof of IB. Now let ξ have the form $p(x_i, \xi_1, \ldots, \xi_l)$ for some state p. Then define

$$g_{\xi,\omega}(c_1, \ldots, c_n) = g_{p,s_i}(g_{\xi_1,\omega}(c_1, \ldots, c_n), \ldots, g_{\xi_l,\omega}(c_1, \ldots, c_n)).$$

By an argument which is similar to the proof of IS1, the statement follows.

This ends the proof of statements K and L.

Now let $\varphi \in SF(M)$. Then, by Lemma 4.8(b), there is a $\xi \in RHS(Q, \Delta, k, n)$ and there are $s_1, \ldots, s_k \in T_\Sigma$ such that $\varphi = \xi[x_1 \leftarrow s_1, \ldots, x_k \leftarrow s_k]$. Then, by L, there is a function $g_{\xi,\omega} : \mathbb{N}^n \to \mathbb{N}$ with $\omega = (s_1, \ldots, s_k)$ such that the derivations starting from φ are not longer than $g_{\xi,\omega}(0, \ldots, 0)$. Note that the derivations starting from y_j are not longer than zero steps. □

Lemma 4.12. The derivation relation $\Rightarrow_M$ is confluent on $SF(M)$.

Proof. This statement follows immediately from the facts that $\Rightarrow_M$ is locally confluent (Lemma 4.10) and terminating (Lemma 4.11), and from Newman's lemma (Lemma 2.5). □

Theorem 4.13. For every $\varphi \in SF(M)$, the normal form $nf(\Rightarrow_M, \varphi)$ exists.

Proof. The existence and uniqueness of the term $nf(\Rightarrow_M, \varphi)$ is guaranteed by the fact that $\Rightarrow_M$ is confluent and terminating on $SF(M)$ (Lemmas 4.12 and 4.11) and by Corollary 2.7. □

The following lemma will help us in the proof of the next theorem. It states that, when computing the normal form of a sentential form $q(s, \varphi_1, \ldots, \varphi_l)$, then this can be done by computing first the normal forms of the φ_i's and then the normal form of $nf(\Rightarrow_M, q(s, nf(\Rightarrow_M, \varphi_1), \ldots, nf(\Rightarrow_M, \varphi_n)))$.

Lemma 4.14. For every $q \in Q^{(n+1)}$, $n \geq 0$, $s \in T_\Sigma$, and $\varphi_1, \ldots, \varphi_n \in SF(M)$,

$$nf(\Rightarrow_M, q(s, \varphi_1, \ldots, \varphi_n)) = nf(\Rightarrow_M, q(s, nf(\Rightarrow_M, \varphi_1), \ldots, nf(\Rightarrow_M, \varphi_n)))$$

Proof. By Theorem 4.13, $\varphi_i \Rightarrow_M^* nf(\Rightarrow_M, \varphi_i)$, for every $1 \leq i \leq n$. Hence, $q(s, \varphi_1, \ldots, \varphi_n) \Rightarrow_M^* q(s, nf(\Rightarrow_M, \varphi_1), \ldots, nf(\Rightarrow_M, \varphi_n))$. Thus our lemma follows from Corollary 2.8. □

Theorem 4.15. For every $\varphi \in SF(M)$, $nf(\Rightarrow_M, \varphi) \in T_\Delta(Y)$.

Proof. $nf(\Rightarrow_M, \varphi) \in T_\Delta(Y)$ is verified by proving the following two statements by simultaneous induction.

K: For every $s \in T_\Sigma$ and $q \in Q^{(n+1)}$ with $n \geq 0$, the term $nf(\Rightarrow_M, q(s, y_1, \ldots, y_n)) \in T_\Delta(Y_n)$.

L: For every $\xi \in RHS(Q, \Delta, k, n)$ with $k, n \geq 0$, and $s_1, \ldots, s_k \in T_\Sigma$, the term $nf(\Rightarrow_M, \xi[x_1 \leftarrow s_1, \ldots, x_k \leftarrow s_k]) \in T_\Delta(Y_n)$.

Proof of IB: Let $\xi \in RHS(Q, \Delta, 0, n)$. Then it can be proved by structural induction over ξ that $nf(\Rightarrow_M, \xi) \in T_\Delta(Y_n)$. The case in which ξ has the form $\delta(\xi_1, \ldots, \xi_l)$ is proved by a corresponding proof to that for top-down tree transducers (cf. Lemma 3.16). Now let $\xi = y_j$ for some j with $1 \leq j \leq n$. Then $nf(\Rightarrow_M, y_j) = y_j \in T_\Delta(Y_n)$.

Proof of IS1: Let $s \in T_\Sigma$ and $q \in Q^{(n+1)}$ with $n \geq 0$. Then there are $k \geq 0$, $\sigma \in \Sigma^{(k)}$, and $s_1, \ldots, s_k \in T_\Sigma$ such that $s = \sigma(s_1, \ldots, s_k)$. Then,

$nf(\Rightarrow_M, q(s, y_1, \ldots, y_n)) = nf(\Rightarrow_M, q(\sigma(s_1, \ldots, s_k), y_1, \ldots, y_n)) = nf(\Rightarrow_M, rhs(q, \sigma)[x_1 \leftarrow s_1, \ldots, x_k \leftarrow s_k])$ by Corollary 2.8. By induction hypothesis on L, $nf(\Rightarrow_M, rhs(q,\sigma)[x_1 \leftarrow s_1, \ldots, x_k \leftarrow s_k]) \in T_\Delta(Y_n)$. Hence, $nf(\Rightarrow_M, q(s, y_1, \ldots, y_n)) \in T_\Delta(Y_n)$.

Proof of IS2: Let $\xi \in RHS(Q, \Delta, k, n)$, $k, n \geq 0$, and $s_1, \ldots, s_k \in T_\Sigma$. The proof is by structural induction on ξ where we represent $[x_1 \leftarrow s_1, \ldots, x_k \leftarrow s_k]$ by $[...]$.

(i) Let $\xi = q(x_i, \xi_1, \ldots, \xi_l)$ for some q, x_i, and $\xi_1, \ldots, \xi_l$. Then we represent $\Rightarrow_M$ by $\Rightarrow$ and the following holds:

$$\begin{aligned}
&nf(\Rightarrow, q(x_i, \xi_1, \ldots, \xi_l)[...]) \\
&= nf(\Rightarrow, q(s_i, \xi_1[...], \ldots, \xi_l[...])) \\
&= nf(\Rightarrow, q(s_i, nf(\Rightarrow, \xi_1[...]), \ldots, nf(\Rightarrow, \xi_l[...]))) \text{ (by Lemma 4.14)} \\
&= nf(\Rightarrow, q(s_i, y_1, \ldots, y_l)[y_1 \leftarrow nf(\Rightarrow, \xi_1[...]), \ldots, y_l \leftarrow nf(\Rightarrow, \xi_l[...])]).
\end{aligned}$$

By induction hypothesis on K, $nf(\Rightarrow_M, q(s_i, y_1, \ldots, y_l)) \in T_\Delta(Y_l)$. By induction hypothesis of this structural induction, for every $1 \leq j \leq l$, $nf(\Rightarrow_M, \xi_j[...]) \in T_\Delta(Y_n)$. Hence, $nf(\Rightarrow_M, q(s_i, y_1, \ldots, y_l)[y_1 \leftarrow nf(\Rightarrow_M, \xi_1[...]), \ldots, y_l \leftarrow nf(\Rightarrow_M, \xi_l[...])]) \in T_\Delta(Y_n)$.

(ii) and (iii) are proved in the same way as in the proof of IB.

Finally we can use the same method as in Lemma 4.11 to finish the proof. □

Now we can define the tree transformation τ_M^{der} induced by M similarly to Def. 3.17. Again, the superscript *der* indicates that the tree transformation is defined by means of the derivation relation.

Definition 4.16. Let $E = (t_1, \ldots, t_r)$ be the environment of M.

(a) Let $q \in Q^{(n+1)}$ with $n \geq 0$. The *tree transformation induced by M with q* is the function $\tau_{M,q}^{der} : T_\Sigma \to T_\Delta(Y_n)$ defined as follows: for every $s \in T_\Sigma$, define $\tau_{M,q}^{der}(s) = nf(\Rightarrow_M, q(s, y_1, \ldots, y_n))$.

(b) The *tree transformation induced by M* is the function $\tau_M^{der} : T_\Sigma \to T_\Delta$ defined by $\tau_M^{der}(s) = \tau_{M,q_0}^{der}(s)[y_1 \leftarrow t_1, \ldots, y_r \leftarrow t_r]$. □

We call a tree transformation a *macro tree transformation* if it can be induced by a macro tree transducer. The class of all macro tree transformations is denoted by MAC.

Example 4.17. The macro tree transducer M_{bal} of Example 4.4 computes the tree transformation $\tau_{M_{bal}}^{der}$ which takes a tree s over $\{\sigma^{(2)}, \alpha^{(0)}\}$ with k leaves as input and produces a fully balanced tree of height $k+1$ over the same ranked alphabet as output. To verify this statement, first observe that, for every s with k leaves

$$q(s, y_1) \Rightarrow^*_{M_{bal}} \underbrace{q(\alpha, \ldots q(\alpha,}_{k} y_1 \underbrace{) \ldots)}_{k}.$$

This observation can be proved by induction on k. Then, by starting from the innermost occurrence of $q(\alpha, \ldots)$ and using the (q, α)-rule $q(\alpha, y_1) \rightarrow \sigma(y_1, y_1)$ k times, the fully balanced tree of height $k+1$ over $\{\sigma^{(2)}, y_1^{(0)}\}$ is derived. After substituting α for y_1 according to Def. 4.16, we obtain the fully balanced tree of height $k+1$ over $\{\sigma^{(2)}, \alpha^{(0)}\}$. □

4.3 Characterization of Macro Tree Transformations

As in Section 3.3, we will give an inductive characterization of τ_M^{der} for every macro tree transducer M. Proofs of properties can then take advantage of this characterization. The main difference from the situation of top-down tree transducers (cf. Def. 3.22) is the possibility that states may occur nested in the right-hand sides of rules. Thus, the following definition will contain a kind of second-order substitution (in IS2(i)). Recall that M denotes an arbitrary, but fixed, macro tree transducer $(Q, \Sigma, \Delta, q_0, R, E)$.

Definition 4.18. We define the set

$$F_M = \{\tau_{M,q}^{ind} : T_\Sigma \rightarrow T_\Delta(Y) \mid q \in Q\}$$

and the set

$$G_M = \{\tau_{M,\xi}^{ind} : (T_\Sigma)^k \rightarrow T_\Delta(Y) \mid k \geq 0, \xi \in \bigcup_{n \geq 0} RHS(Q, \Delta, k, n)\}$$

of functions by simultaneous induction as follows.

IS1: For every $q \in Q$, $k \geq 0$, $\sigma \in \Sigma^{(k)}$, and $s_1, \ldots, s_k \in T_\Sigma$, we define

$$\tau_{M,q}^{ind}(\sigma(s_1, \ldots, s_k)) = \tau_{M,rhs(q,\sigma)}^{ind}((s_1, \ldots, s_k)).$$

IS2: For every $k, n \geq 0$, $\xi \in RHS(Q, \Delta, k, n)$, and $\omega = (s_1, \ldots, s_k) \in (T_\Sigma)^k$, we define $\tau_{M,\xi}^{ind}(\omega)$ inductively on the structure of ξ.

(i) If $\xi = p(x_i, \xi_1, \ldots, \xi_l)$ for some $p \in Q^{(l+1)}$ with $l \geq 0$, $1 \leq i \leq k$, and $\xi_1, \ldots, \xi_l \in RHS(Q, \Delta, k, n)$, then $\tau_{M,\xi}^{ind}(\omega) = \tau_{M,p}^{ind}(s_i)[y_j \leftarrow \tau_{M,\xi_j}^{ind}(\omega); 1 \leq j \leq l]$.

(ii) If $\xi = \delta(\xi_1, \ldots, \xi_l)$ for some $\delta \in \Delta^{(l)}$ with $l \geq 0$ and $\xi_1, \ldots, \xi_l \in RHS(Q, \Delta, k, n)$, then $\tau_{M,\xi}^{ind}(\omega) = \delta(\tau_{M,\xi_1}^{ind}(\omega), \ldots, \tau_{M,\xi_l}^{ind}(\omega))$.

(iii) If $\xi = y_j$, then $\tau_{M,\xi}^{ind}(\omega) = y_j$. □

Note that the base IB of the definition by simultaneous induction is part of the step IS2 (for $k = 0$; cf. Principle 3.20).

Example 4.19. Let us again consider the macro tree transducer M_{bal} of Example 4.4 and let us compute the value of the function $\tau^{ind}_{M_{bal},q}$ on the argument $s = \sigma(\alpha, \sigma(\alpha, \alpha))$. This computation should be compared with the derivation shown in Example 4.6; for ease of reference, we have indicated by an asterisk the lines in the following sequence of equalities which correspond to sentential forms in the derivation. We abbreviate M_{bal} by M.

$$\begin{aligned}
\tau^{ind}_{M,q}(s) &=^* \tau^{ind}_{M,q}(\sigma(\alpha, \sigma(\alpha, \alpha)))\\
&= \tau^{ind}_{M,rhs(q,\sigma)}((\alpha, \sigma(\alpha, \alpha)))\\
&= \tau^{ind}_{M,q(x_2,q(x_1,y_1))}((\alpha, \sigma(\alpha, \alpha)))\\
&= \tau^{ind}_{M,q}(\sigma(\alpha, \alpha))[y_1 \leftarrow \tau^{ind}_{M,q(x_1,y_1)}((\alpha, \sigma(\alpha, \alpha)))]\\
&= \tau^{ind}_{M,q}(\sigma(\alpha, \alpha))[y_1 \leftarrow \tau^{ind}_{M,q}(\alpha)[y_1 \leftarrow \tau^{ind}_{M,y_1}((\alpha, \sigma(\alpha, \alpha)))]]\\
&= \tau^{ind}_{M,q}(\sigma(\alpha, \alpha))[y_1 \leftarrow \tau^{ind}_{M,q}(\alpha)[y_1 \leftarrow y_1]]\\
&=^* \tau^{ind}_{M,q}(\sigma(\alpha, \alpha))[y_1 \leftarrow \tau^{ind}_{M,q}(\alpha)]\\
&= \tau^{ind}_{M,q(x_2,q(x_1,y_1))}((\alpha, \alpha))[y_1 \leftarrow \tau^{ind}_{M,q}(\alpha)]\\
&=^* \tau^{ind}_{M,q}(\alpha)[y_1 \leftarrow \tau^{ind}_{M,q}(\alpha)][y_1 \leftarrow \tau^{ind}_{M,q}(\alpha)]\\
&= \tau^{ind}_{M,q}(\alpha)[y_1 \leftarrow \tau^{ind}_{M,q}(\alpha)][y_1 \leftarrow \tau^{ind}_{M,\sigma(y_1,y_1)}(())]\\
&=^* \tau^{ind}_{M,q}(\alpha)[y_1 \leftarrow \tau^{ind}_{M,q}(\alpha)][y_1 \leftarrow \sigma(y_1, y_1)]\\
&= \tau^{ind}_{M,q}(\alpha)[y_1 \leftarrow \tau^{ind}_{M,\sigma(y_1,y_1)}(())][y_1 \leftarrow \sigma(y_1, y_1)]\\
&=^* \tau^{ind}_{M,q}(\alpha)[y_1 \leftarrow \sigma(y_1, y_1)][y_1 \leftarrow \sigma(y_1, y_1)]\\
&= \tau^{ind}_{M,\sigma(y_1,y_1)}(())[y_1 \leftarrow \sigma(y_1, y_1)][y_1 \leftarrow \sigma(y_1, y_1)]\\
&=^* \sigma(y_1, y_1)[y_1 \leftarrow \sigma(y_1, y_1)][y_1 \leftarrow \sigma(y_1, y_1)]\\
&= \sigma(\sigma(\sigma(y_1, y_1), \sigma(y_1, y_1)), \sigma(\sigma(y_1, y_1), \sigma(y_1, y_1)))
\end{aligned}$$

□

Lemma 4.20. For every $\xi \in T_\Delta(Y_n)$, $k \geq 0$, and $\omega \in (T_\Sigma)^k$, the equality $\tau^{ind}_{M,\xi}(\omega) = \xi$ holds.

Proof. This statement is proved easily by structural induction on ξ. □

In fact, it is also true for macro tree transducers that the functions defined inductively characterize the tree transformation which has been defined on the basis of the derivation relation. Clearly, $\tau^{der}_{M,q}$ corresponds to $\tau^{ind}_{M,q}$. In order that we also have a corresponding function for $\tau^{ind}_{M,\xi}$ with $\xi \in RHS(Q, \Delta, k, n)$, we define the function $\tau^{der}_{M,\xi} : (T_\Sigma)^k \to T_\Delta(Y)$ such that for every $\omega = (s_1, \ldots, s_k)$, $\tau^{der}_{M,\xi}(\omega) = nf(\Rightarrow_M, \xi[x_i \leftarrow s_i; 1 \leq i \leq k])$.

Theorem 4.21. K: For every $s \in T_\Sigma$ and $q \in Q^{(n+1)}$ with $n \geq 0$, the equality $\tau^{der}_{M,q}(s) = \tau^{ind}_{M,q}(s)$ holds.

L: For every $k, n \geq 0$, $\xi \in RHS(Q, \Delta, k, n)$, and $\omega = (s_1, \ldots, s_k) \in (T_\Sigma)^k$, the equality $\tau^{der}_{M,\xi}(\omega) = \tau^{ind}_{M,\xi}(\omega)$ holds.

Proof. We prove the two statements by simultaneous induction.

Proof of IB: Let $\xi \in RHS(Q, \Delta, 0, n)$ and $\omega \in (T_\Sigma)^0 = \{()\}$. Then statement L is proved by structural induction on ξ.

(i) ξ cannot have the form $q(x_i, \xi_1, \ldots, \xi_l)$.

(ii) If ξ has the form $\delta(\xi_1, \ldots, \xi_l)$, then the statement follows immediately by applying the induction hypothesis of this structural induction.

(iii) If ξ has the form $y_j \in Y_n$, then $\tau^{der}_{M,\xi}(\omega) = nf(\Rightarrow_M, \xi[\,]) = nf(\Rightarrow_M, y_j)$ $= y_j = \tau^{ind}_{M,\xi}(\omega)$.

Proof of IS1: Let $s \in T_\Sigma$ and $q \in Q^{(n+1)}$ with $n \geq 0$. Then there are $k \geq 0$, $\sigma \in \Sigma^{(k)}$, and $s_1, \ldots, s_k \in T_\Sigma$ such that $s = \sigma(s_1, \ldots, s_k)$. We represent $(s_1, \ldots, s_k)$ by ω. Then, $\tau^{der}_{M,q}(s) = nf(\Rightarrow_M, q(s, y_1, \ldots, y_n)) = nf(\Rightarrow_M, q(\sigma(s_1, \ldots, s_k), y_1, \ldots, y_n)) = nf(\Rightarrow_M, rhs(q, \sigma)[x_1 \leftarrow s_1, \ldots, x_k \leftarrow s_k]) = \tau^{der}_{M,\xi}(\omega)$ with $\xi = rhs(q, \sigma)$. By induction hypothesis on L and by Def. 4.18, we obtain $\tau^{der}_{M,\xi}(\omega) = \tau^{ind}_{M,\xi}(\omega) = \tau^{ind}_{M,q}(\sigma(s_1, \ldots, s_k))$. Hence, $\tau^{der}_{M,q}(s) = \tau^{ind}_{M,q}(s)$.

Proof of IS2: Let $\xi \in RHS(Q, \Delta, k, n)$, $k, n \geq 0$, and $\omega = (s_1, \ldots, s_k) \in (T_\Sigma)^k$. The proof is by structural induction on ξ where we represent $[x_1 \leftarrow s_1, \ldots, x_k \leftarrow s_k]$ by $[...]$.

(i) Let $\xi = p(x_i, \xi_1, \ldots, \xi_l)$ for some p, x_1, and $\xi_1, \ldots, \xi_l$. Then

$$
\begin{aligned}
& \tau^{der}_{M,p(x_i,\xi_1,\ldots,\xi_l)}(\omega) \\
= \; & nf(\Rightarrow_M, p(x_i, \xi_1, \ldots, \xi_l)[...]) \\
= \; & nf(\Rightarrow_M, p(s_i, \xi_1[...], \ldots, \xi_l[...])) \\
= \; & nf(\Rightarrow_M, p(s_i, nf(\Rightarrow_M, \xi_1[...]), \ldots, nf(\Rightarrow_M, \xi_l[...]))) \\
& \text{(by Lemma 4.14)} \\
= \; & nf(\Rightarrow_M, p(s_i, y_1, \ldots, y_l))[y_1 \leftarrow nf(\Rightarrow_M, \xi_1[...]), \ldots \\
& \ldots, y_l \leftarrow nf(\Rightarrow_M, \xi_l[...])] \\
= \; & \tau^{der}_{M,p}(s_i)[y_1 \leftarrow \tau^{der}_{M,\xi_1}(\omega), \ldots, y_l \leftarrow \tau^{der}_{M,\xi_l}(\omega)] \\
& \text{(by definition of } \tau^{der}_{M,p} \text{ and } \tau^{der}_{M,\xi}) \\
= \; & \tau^{ind}_{M,p}(s_i)[y_1 \leftarrow \tau^{ind}_{M,\xi_1}(\omega), \ldots, y_l \leftarrow \tau^{ind}_{M,\xi_l}(\omega)] \\
& \text{(by induction hypothesis of this structural induction and by} \\
& \text{induction hypothesis on } K) \\
= \; & \tau^{ind}_{M,p(x_i,\xi_1,\ldots,\xi_l)}(\omega).
\end{aligned}
$$

(ii) and (iii) are proved in the same way as in the proof of IB. □

Theorem 4.21 states the equivalence of the two approaches to defining the tree transformation induced by a macro tree transducer M: one is based on the derivation relation of M and the other is based on inductively defined functions. This equivalence means that we can omit the superscripts *der* and *ind* from now on. However, in the proofs of the subsequent lemmas and theorems we will use the approach based on inductively defined functions.

4.4 Height Property

In this section we show that the height of the output tree of a macro tree transducer can be exponentially bounded in the height of the input tree (compare this with the situation for top-down tree transducers in Lemma 3.27).

Lemma 4.22. Let $M = (Q, \Sigma, \Delta, q_0, R, E)$ be a macro tree transducer. There is a constant $c > 0$ such that, for every $s \in T_\Sigma$, the approximation $height(\tau_M(s)) \leq c^{height(s)}$ holds.

Proof. Let c be the maximal height of the right-hand sides of rules in R. The following two statements can be proved by simultaneous induction. Since the proof is straightforward, it is left to the reader.

K: For every $s \in T_\Sigma$ and $q \in Q$, the approximation $height(\tau_{M,q}(s)) \leq c^{height(s)}$ holds.

L: For every $k, n \geq 0$, $\xi \in RHS(Q, \Delta, k, n)$, and $\omega \in (T_\Sigma)^k$, the approximation $height(\tau_{M,\xi}(\omega)) \leq height(\xi) \cdot c^{max}$ holds where $max = max\{0, height(s_1), \ldots, height(s_k)\}$. □

In fact, there is a macro tree transducer which exactly performs the exponential growth of the input tree.

Lemma 4.23. There is a macro tree transducer $M_{exp} = (Q, \Sigma, \Delta, q_0, R, E)$ with a monadic input alphabet such that, for every $s \in T_\Sigma$, the equality $height(\tau_{M_{exp}}(s)) = 2^{height(s)}$ holds.

Proof. The following macro tree transducer $M_{exp} = (Q, \Sigma, \Delta, q_0, R, E)$ meets the condition on the height of the output trees:

- $Q = \{q_0^{(2)}\}$
- $\Sigma = \Delta = \{\gamma^{(1)}, \alpha^{(0)}\}$
- R consists of two rules:
  1) $q_0(\gamma(x_1), y_1) \rightarrow q_0(x_1, q_0(x_1, y_1))$
  2) $q_0(\alpha, y_1) \rightarrow \gamma(y_1)$
- $E = (\alpha)$.

In fact, M_{exp} is very similar to the macro tree transducer of Example 4.4. However, in the first rule, the nested recursive call is applied to the same descendant as the outermost call. The height condition of output trees can be verified by a simple induction on n and thus the tree transformation induced by M_{exp} is

$$\tau_{M_{exp}} = \{(\gamma^n(\alpha), \gamma^{2^n}(\alpha)) \mid n \geq 0\}.$$

□

Clearly, the composition of macro tree transducers can perform a tower of exponentiations.

Lemma 4.24. For every $\tau : T_\Sigma \to T_\Delta$ with $\tau \in MAC^k$ and $k \geq 1$, there is a constant $c > 0$ such that, for every $s \in T_\Sigma$, the approximation $height(\tau(s)) \leq exp(k, c \cdot height(s))$ holds where, for every $k, n \geq 0$, the number $exp(k, n)$ is defined by induction on k such that $exp(0, n) = n$ and $exp(k+1, n) = 2^{exp(k,n)}$.

Proof. Since τ is an element of MAC^k, there are macro tree transducers $M_1, \ldots, M_k$ such that $\tau = \tau_{M_1} \circ \ldots \circ \tau_{M_k}$. Let c_i be the constant associated with M_i by Lemma 4.22. By applying Lemma 4.22 repeatedly, we obtain

$$height(\tau(s)) \leq c_k^{\cdot^{\cdot^{c_1^{height(s)}}}}.$$

It is easy to prove by induction on k that there is a constant $c > 0$ such that, for every $r \geq 0$, the approximation

$$c_k^{\cdot^{\cdot^{c_1^{r}}}} \leq exp(k, c \cdot r)$$

holds. □

An immediate consequence of Lemma 4.23 is that top-down tree transducers are less powerful than macro tree transducers.

Theorem 4.25. $TOP \subset MAC$.

Proof. By Note 4.3, $TOP \subseteq MAC$. Now assume that $TOP = MAC$ holds. Then there is a contradiction between the height-lemma for top-down tree transducers (Lemma 3.27) and the fact that there is a macro tree transducer M_{exp} which increases the height of the input tree in an exponential way (Lemma 4.23). □

4.5 Composition and Decomposition Results

In this section we show that the many-fold composition of the class MAC yields a proper hierarchy. Moreover, we characterize MAC as the composition of some of its particular subclasses.

Theorem 4.26. For every $k \geq 1$, $MAC^k \subset MAC^{k+1}$.

Proof. By composing $k+1$ times the macro tree transducer M_{exp} of Lemma 4.23, we obtain a tree transformation in MAC^{k+1} which cannot be in MAC^k because of Lemma 4.24. □

Now we introduce two important subclasses of macro tree transducers by imposing restrictions on the form of the rules. In these subclasses, the number of times that a subtree of the input tree can be referenced, is restricted.

Definition 4.27. Let $M = (Q, \Sigma, \Delta, q_0, R, E)$ be a macro tree transducer.

1) M is called *linear* if, for every $k \geq 0$, $\sigma \in \Sigma^{(k)}$, $1 \leq i \leq k$, and $q \in Q$, the variable x_i occurs at most once in $rhs(q, \sigma)$.

2) M is called *superlinear*, if M is linear and for every $k \geq 0$, $\sigma \in \Sigma^{(k)}$, and states $q, q' \in Q$, the set $Var_X(rhs(q, \sigma)) \cap Var_X(rhs(q', \sigma)) = \emptyset$. □

The class of tree transformations induced by linear (and superlinear) macro tree transducers is denoted by *l-MAC* (and *sl-MAC*, respectively). By definition *sl-MAC* $\subseteq$ *l-MAC*.

Note 4.28. *l-TOP* $\subseteq$ *l-MAC* and *sl-TOP* $\subseteq$ *sl-MAC*.

Let Σ be a ranked alphabet and let Δ be such that $\Delta = \{\sigma' \mid \sigma \in \Sigma\} \cup \{nil\}$. Every sequence $(t_1, \ldots, t_n)$ of Σ-trees can be written as $(\sigma(s_1, \ldots, s_m), t_2, \ldots, t_n)$ if $t_1 = \sigma(s_1, \ldots, s_m)$. Then we define the function *bin* which takes a sequence of trees over Σ as argument and delivers a tree over Δ as result:

- For every $n \geq 1$, define $bin((\sigma(s_1, \ldots, s_m), t_2, \ldots, t_n)) = \sigma'(bin((s_1, \ldots, s_m)), bin((t_2, \ldots, t_n)))$
- For $n = 0$, define $bin(()) = nil$.

Example 4.29. Consider the following macro tree transducer $M_{bin} = (Q, \Sigma, \Delta, q, R, E)$ where

- $Q = \{q^{(2)}\}$
- $\Sigma = \{\delta^{(3)}, \sigma^{(2)}, \gamma^{(1)}, \alpha^{(0)}\}$
- $\Delta = \{\delta'^{(2)}, \sigma'^{(2)}, \gamma'^{(2)}, \alpha'^{(2)}\} \cup \{nil^{(0)}\}$
- R contains the rules:
  1) $q(\delta(x_1, x_2, x_3), y_1) \to \delta'(q(x_1, q(x_2, q(x_3, nil))), y_1)$
  2) $q(\sigma(x_1, x_2), y_1) \to \sigma'(q(x_1, q(x_2, nil)), y_1)$
  3) $q(\gamma(x_1), y_1) \to \gamma'(q(x_1, nil), y_1)$
  4) $q(\alpha, y_1) \to \alpha'(nil, y_1)$
- $E = (nil)$

Clearly, M_{bin} is a superlinear macro tree transducer, because it has just one state.

It can be proved that the tree transformation induced by M is

$$\tau_{M_{bin}} = \{(s, bin((s))) \mid s \in T_\Sigma\}.$$

□

An important subclass of superlinear macro tree transducers is the class of YIELD-functions. Intuitively, they capture the second-order term substitution which is inherent in the derivation relation of macro tree transducers. The second-order term substitution allows the replacement of a symbol at any node of a tree by another tree of appropriate rank, whereas the first-order term substitution only allows such replacements at leaves. In the definition of $YIELD$ the set $Z = \{z_1, z_2, \ldots\}$ of substitution variables is used; for $r \geq 0$, let $Z_r = \{z_1, z_2, \ldots, z_r\}$.

Definition 4.30. Let Σ and Δ be ranked alphabets, $r \geq 0$ be an integer, $g : \Sigma^{(0)} \to T_\Delta(Z_r)$ be a function such that, for every $\alpha \in \Sigma^{(0)}$ and every $1 \leq j \leq r$, $g(\alpha)$ contains at most one occurrence of z_j, and let $\tilde{t} = (t_1, \ldots, t_r) \in (T_\Delta)^r$. The *YIELD-function induced by g and $\tilde{t}$*, denoted by $yield_{g,\tilde{t}}$, is the function of type $T_\Sigma \to T_\Delta$ defined by

$$yield_{g,\tilde{t}}(s) = yield_g(s)[z_1 \leftarrow t_1, \ldots, z_r \leftarrow t_r].$$

The function $yield_g : T_\Sigma \to T_\Delta(Z_r)$ is defined inductively on the structure of its argument.

(i) For every $\alpha \in \Sigma^{(0)}$, we define $yield_g(\alpha) = g(\alpha)$.
(ii) For every $\sigma \in \Sigma^{(k+1)}$ with $k \geq 0$ and $s_0, s_1, \ldots, s_k \in T_\Sigma$, we define $yield_g(\sigma(s_0, s_1, \ldots, s_k)) = yield_g(s_0)[z_1 \leftarrow yield_g(s_1), \ldots, z_k \leftarrow yield_g(s_k)]$. □

The class of all YIELD-functions is denoted by $YIELD$. The restriction that $g(\alpha)$ should be linear in Z (i.e., every z_i occurs at most once in $g(\alpha)$) becomes clear when proving $l\text{-}TOP \circ YIELD \subseteq l\text{-}MAC$ in Lemma 4.36. Also note that, if in case (ii) $k > r$, then the substitution of $yield_g(s_{r+1})$, ..., $yield_g(s_k)$ for $z_{r+1}, \ldots, z_k$, respectively, has no effect on $yield_g(s_0)$.

Example 4.31. Consider the ranked alphabets $\Sigma = \Delta = \{\delta^{(3)}, \sigma^{(2)}, \gamma^{(1)}, \alpha^{(0)}, \beta^{(0)}\}$. Let $r = 2$ and define the function $g : \Sigma^{(0)} \to T_\Delta(Z_2)$ as follows: $g(\alpha) = \sigma(z_1, z_2)$ and $g(\beta) = \delta(z_1, \alpha, z_2)$. Moreover, let $\tilde{t} = (\beta, \gamma(\alpha))$.

Now consider the input tree $s = \delta(\gamma(\beta), \alpha, \sigma(\alpha, \alpha))$. Then $yield_{g,\tilde{t}}(s) = yield_g(s)[z_1 \leftarrow \beta, z_2 \leftarrow \gamma(\alpha)]$.

$$\begin{aligned}
yield_g(s) &= yield_g(\gamma(\beta))[z_1 \leftarrow yield_g(\alpha), z_2 \leftarrow yield_g(\sigma(\alpha, \alpha))] \\
&= (yield_g(\beta)[])[z_1 \leftarrow g(\alpha), z_2 \leftarrow yield_g(\alpha)[z_1 \leftarrow yield_g(\alpha)]] \\
&= (g(\beta)[])[z_1 \leftarrow \sigma(z_1, z_2), z_2 \leftarrow \sigma(z_1, z_2)[z_1 \leftarrow \sigma(z_1, z_2)]] \\
&= (\delta(z_1, \alpha, z_2))[z_1 \leftarrow \sigma(z_1, z_2), z_2 \leftarrow \sigma(\sigma(z_1, z_2), z_2)] \\
&= \delta(\sigma(z_1, z_2), \alpha, \delta(z_1, \alpha, z_2))
\end{aligned}$$

Thus $yield_{g,\tilde{t}}(s) = \delta(\sigma(\beta, \gamma(\alpha)), \alpha, \gamma(\alpha))$. □

Indeed, every YIELD-function can be computed by a superlinear macro tree transducer.

Lemma 4.32. *$YIELD \subseteq sl\text{-}MAC$.*

Proof. Let Σ and Δ be ranked alphabets. Let $g : \Sigma^{(0)} \to T_\Delta(Z_r)$ be a function and let $\tilde{t} = (t_1, \ldots, t_r) \in (T_\Delta)^r$. We construct a superlinear macro tree transducer $M = (Q, \Sigma, \Delta, q_0, R, E)$ such that $\tau_M = yield_{g,\tilde{t}}$.

Let $Q = \{p^{(r+1)}\}$ be the set of states; hence, $q_0 = p$. For every $k \geq 0$ and $\sigma \in \Sigma^{(k)}$, the set R consists of the rules:

- If $r + 1 \leq k$, then the following rule is in R:
  $p(\sigma(x_1, \ldots, x_k), y_1, \ldots, y_r) \to$
  $\quad p(x_1, p(x_2, y_1, \ldots, y_r), \ldots, p(x_{r+1}, y_1, \ldots, y_r))$.
- If $1 \leq k < r + 1$, then the following rule is in R:
  $p(\sigma(x_1, \ldots, x_k), y_1, \ldots, y_r) \to$
  $\quad p(x_1, p(x_2, y_1, \ldots, y_r), \ldots, p(x_k, y_1, \ldots, y_r), y_k, \ldots, y_r)$.
- If $k = 0$, then the following rule is in R:

$$p(\sigma, y_1, \ldots, y_r) \to g(\sigma)[z_1 \leftarrow y_1, \ldots, z_r \leftarrow y_r].$$

- $E = (t_1, \ldots, t_r)$.

We can prove the correctness of the construction by proving the following statement by structural induction on s.

$(*)$ For every $s \in T_\Sigma$, $yield_g(s) = \tau_{M,p}(s)[y_1 \leftarrow z_1, \ldots, y_r \leftarrow z_r]$.

In the proof we will represent the substitutions $[y_1 \leftarrow z_1, \ldots, y_r \leftarrow z_r]$ and $[z_1 \leftarrow y_1, \ldots, z_r \leftarrow y_r]$ by $[y \leftarrow z]_r$ and $[z \leftarrow y]_r$, respectively.

<u>Induction base of $(*)$:</u> Let $s = \alpha \in \Sigma^{(0)}$. Then,

$yield_g(s) = yield_g(\alpha) = g(\alpha)$
$= (g(\alpha)[z \leftarrow y]_r)[y \leftarrow z]_r$
$= rhs(p, \alpha)[y \leftarrow z]_r$ (by the definition of R)
$= \tau_{M,rhs(p,\alpha)}(())[y \leftarrow z]_r$ (by Lemma 4.20)
$= \tau_{M,p}(\alpha)[y \leftarrow z]_r$
$= \tau_{M,p}(s)[y \leftarrow z]_r$

<u>Induction step of $(*)$:</u> Let $s = \sigma(s_1, \ldots, s_k)$.

<u>$r + 1 \leq k$:</u>

$yield_g(\sigma(s_1, \ldots, s_k)) =$
$= yield_g(s_1)[z_i \leftarrow yield_g(s_{i+1}); 1 \leq i \leq k-1]$
$= (\tau_{M,p}(s_1)[y \leftarrow z]_r)[z_i \leftarrow \tau_{M,p}(s_{i+1})[y \leftarrow z]_r; 1 \leq i \leq k-1]$
(by induction hypothesis on $(*)$)

$= (\tau_{M,p}(s_1)[y_j \leftarrow \tau_{M,p}(s_{j+1}); 1 \leq j \leq r])[y \leftarrow z]_r$
$= (\tau_{M,rhs(p,\sigma)}((s_1, \ldots, s_k)))[y \leftarrow z]_r$
(by definition of R and Def. 4.18)
$= \tau_{M,p}(\sigma(s_1, \ldots, s_k))[y \leftarrow z]_r.$

<u>$1 \leq k < r+1$:</u>

$yield_g(\sigma(s_1, \ldots, s_k)) =$
$= yield_g(s_1)[z_i \leftarrow yield_g(s_{i+1}); 1 \leq i \leq k-1]$
$= (\tau_{M,p}(s_1)[y \leftarrow z]_r)[z_i \leftarrow \tau_{M,p}(s_{i+1})[y \leftarrow z]_r; 1 \leq i \leq k-1]$
$= (\tau_{M,p}(s_1)[y_j \leftarrow \tau_{M,p}(s_{j+1}); 1 \leq j \leq k-1])[y \leftarrow z]_r$
$= (\tau_{M,rhs(p,\sigma)}((s_1, \ldots, s_k)))[y \leftarrow z]_r$
$= \tau_{M,p}(\sigma(s_1, \ldots, s_k))[y \leftarrow z]_r.$

Thus, for every $s \in T_\Sigma$,

$\tau_M(s) =$
$= \tau_{M,p}(s)[y_1 \leftarrow t_1, \ldots, y_r \leftarrow t_r]$
$= (\tau_{M,p}(s)[y_1 \leftarrow z_1, \ldots, y_r \leftarrow z_r])[z_1 \leftarrow t_1, \ldots, z_r \leftarrow t_r]$
$= yield_g(s)[z_1 \leftarrow t_1, \ldots, z_r \leftarrow t_r]$
$= yield_{g,\bar{t}}(s).$

□

Example 4.33. Let us reconsider the *yield*-function of Example 4.31. According to the construction of Lemma 4.32 it can be computed by the following macro tree transducer $M = (Q, \Sigma, \Delta, q_0, R, E)$ where

- $Q = \{p^{(3)}\}$ and $p = q_0$
- R contains the rules:
  1) $p(\delta(x_1, x_2, x_3), y_1, y_2) \to p(x_1, p(x_2, y_1, y_2), p(x_3, y_1, y_2))$
  2) $p(\sigma(x_1, x_2), y_1, y_2) \to p(x_1, p(x_2, y_1, y_2), y_2)$
  3) $p(\gamma(x_1), y_1, y_2) \to p(x_1, y_1, y_2)$
  4) $p(\alpha, y_1, y_2) \to \sigma(y_1, y_2)$
  5) $p(\beta, y_1, y_2) \to \delta(y_1, \alpha, y_2)$
- $E = (\beta, \gamma(\alpha))$.

Since the computation of the transformation induced by M on input tree $s = \delta(\gamma(\beta), \alpha, \sigma(\alpha, \alpha))$ is (almost) the same as the computation of $yield_g(s)$ in Example 4.31, we do not repeat it here. □

One of the main results concerning macro tree transducers is the characterization of MAC in terms of the composition of TOP and $YIELD$, i.e., $MAC = TOP \circ YIELD$ (see Theorem 4.37). This means that we have two relationships:

1. for every macro tree transducer M, there is a top-down tree transducer T, a function g, and a sequence $\tilde{t}$ (of appropriate types) such that $\tau_M = \tau_T \circ yield_{g,\tilde{t}}$ and
2. for every top-down tree transducer T and YIELD-function $yield_{g,\tilde{t}}$, there is a macro tree transducer M such that $\tau_T \circ yield_{g,\tilde{t}} = \tau_M$.

Let us briefly discuss the decomposition result (i.e., relationship 1). Roughly speaking, the main additional feature of macro tree transducers with respect to top-down tree transducers is the facility to perform second-order term substitution, i.e, replace a symbol at *any* node n of a tree by a complete tree t and substitute the subtrees of n at leaves of t, while top-down tree transducers can only perform first-order substitution, i.e., replace a symbol only at a leaf n of a tree by a tree t.

From this point of view, the decomposition of a macro tree transducer M means that every derivation can be reorganized such that two phases appear: During the first phase, merely symbolic replacement is performed while postponing the second-order substitution. Then, in the second phase, all the second-order substitutions are done together. The first phase is performed by a top-down tree transducer T. In the output trees, T inserts additional symbols which indicate the way in which the second-order substitution will take place. Then, the YIELD-function "reads" these additional symbols and carries out the second-order substitution appropriately.

Lemma 4.34.

1)	MAC	$\subseteq$	$TOP \circ YIELD$
2)	$l\text{-}MAC$	$\subseteq$	$l\text{-}TOP \circ YIELD$
3)	$sl\text{-}MAC$	$\subseteq$	$sl\text{-}TOP \circ YIELD$.

Proof. Let $M = (Q, \Sigma, \Delta, q_0, R, E)$ be a macro tree transducer with $E = (t_1, \ldots, t_r)$ and $r = rank(q_0)$. Denote $max\{n \mid Q^{(n+1)} \cup \Delta^{(n)} \neq \emptyset\}$ by mx. We construct a top-down tree transducer $T = (Q', \Sigma, \Gamma, q_0', R')$, a function $g : \Gamma^{(0)} \to T_\Delta(Z_{mx})$, and a tuple $\tilde{t} \in (T_\Delta)^{mx}$ such that $\tau_M = \tau_T \circ yield_{g,\tilde{t}}$.

- Define $Q' = \{q' \mid q \in Q\}$; for every $q \in Q$, the state q' has rank 1.
- Define Γ to be the ranked alphabet such that $\Gamma^{(0)} = \{\delta' \mid \delta \in \Delta\} \cup \{\alpha_1, \ldots, \alpha_{mx}\}$ where the α's are new symbols, and, for every k such that $Q^{(k)} \neq \emptyset$ or $\Delta^{(k-1)} \neq \emptyset$, define $\Gamma^{(k)} = \{c_k\}$ and $\Gamma^{(k)} = \emptyset$ otherwise.
- Define the function $g : \Gamma^{(0)} \to T_\Delta(Z_{mx})$ by $g(\delta') = \delta(z_1, \ldots, z_k)$ if $\delta \in \Delta^{(k)}$, and $g(\alpha_i) = z_i$ for every $i \in \{1, \ldots, mx\}$. Note that g fulfills the condition that every z_j occurs at most once in $g(\delta')$.
- If $mx = r$, then $\tilde{t} = E$, and if $r < mx$, then $\tilde{t} = (t_1, \ldots, t_r, z_{r+1}, \ldots, z_{mx})$. Note that $mx < r$ is not possible, because $q_0 \in Q^{(r+1)}$ and hence $mx \geq r$.
- For the construction of the rules of T we need the set $COMB = \{COMB_{k,n} \mid k, n \geq 0\}$ of functions $COMB_{k,n} : RHS(Q, \Delta, k, n) \to RHS(Q', \Gamma, k)$ that are defined inductively on the structure of their arguments.

(i) For every $p \in Q^{(l+1)}$ with $l \geq 0$, $1 \leq i \leq k$, and $\xi_1, \ldots, \xi_l \in RHS(Q, \Delta, k, n)$, we define $COMB_{k,n}(p(x_i, \xi_1, \ldots, \xi_l)) = c_{l+1}(p'(x_i), COMB_{k,n}(\xi_1), \ldots, COMB_{k,n}(\xi_l))$.

(ii) For every $\delta \in \Delta^{(l)}$ with $l \geq 0$ and $\xi_1, \ldots, \xi_l \in RHS(Q, \Delta, k, n)$, we define $COMB_{k,n}(\delta(\xi_1, \ldots, \xi_l)) = c_{l+1}(\delta', COMB_{k,n}(\xi_1), \ldots, COMB_{k,n}(\xi_l))$.

(iii) For every $1 \leq j \leq n$, we define $COMB_{k,n}(y_j) = \alpha_j$.

- Define
$$R' = \{q'(\sigma(x_1, \ldots, x_k)) \to COMB_{k,n}(\xi) \mid (q(\sigma(x_1, \ldots, x_k), y_1, \ldots, y_n) \to \xi) \in R\}.$$

Now assume that M is linear. Since the function $COMB_{k,n}$ does not copy variables x_i, T is also linear. Now assume that M is superlinear. Since every rule of M is transformed into exactly one rule of T, T is also superlinear.

The correctness of the construction can be shown by proving the following statements by simultaneous induction:

K: For every $n \geq 0$, $q \in Q^{(n+1)}$, and $s \in T_\Sigma$, the relation $yield_g(\tau_{T,q'}(s)) \in T_\Delta(Z_n)$ holds and the equality $\tau_{M,q}(s) = yield_g(\tau_{T,q'}(s))[z_1 \leftarrow y_1, \ldots, z_n \leftarrow y_n]$ is true.

L: For every $k, n \geq 0$, $\xi \in RHS(Q, \Delta, k, n)$, and $\omega \in (T_\Sigma)^k$, the relation $yield_g(\tau_{T,COMB_{k,n}(\xi)}(\omega)) \in T_\Delta(Z_n)$ holds and the equality $\tau_{M,\xi}(\omega) = yield_g(\tau_{T,COMB_{k,n}(\xi)}(\omega))[z_1 \leftarrow y_1, \ldots, z_n \leftarrow y_n]$ is true.

The proof is standard and hence it is left to the reader.

Now we can conclude as follows where we represent the substitutions $[z_1 \leftarrow y_1, \ldots, z_n \leftarrow y_n]$, $[y_1 \leftarrow t_1, \ldots, y_r \leftarrow t_r]$, and $[z_1 \leftarrow t_1, \ldots, z_r \leftarrow t_r]$ by $[z \leftarrow y]_n$, $[y \leftarrow t]_r$, and $[z \leftarrow t]_r$, respectively. (Recall that $r + 1 = rank(q_0)$.)

$$\begin{aligned}
&\tau_M(s) = \\
&= \tau_{M,q_0}(s)[y \leftarrow t]_r \\
&= (yield_g(\tau_{T,q_0'}(s))[z \leftarrow y]_r)[y \leftarrow t]_r \\
&= yield_g(\tau_{T,q_0'}(s))[z \leftarrow t]_r \\
&= yield_g(\tau_{T,q_0'}(s))[z \leftarrow t]_{mx} \qquad \text{because } yield_g(\tau_{T,q_0'}(s)) \in T_\Delta(Z_r) \\
&= yield_g(\tau_T(s))[z \leftarrow t]_{mx} \\
&= yield_{g,\bar{t}}(\tau_T(s)).
\end{aligned}$$
□

Example 4.35. Let us reconsider the macro tree transducer $M_{bal} = (Q, \Sigma, \Delta, q, R, E)$ of Example 4.4, where:

- $Q = \{q^{(2)}\}$
- $\Sigma = \Delta = \{\sigma^{(2)}, \alpha^{(0)}\}$
- R consists of the rules:
  1) $q(\sigma(x_1, x_2), y_1) \to q(x_2, q(x_1, y_1))$
  2) $q(\alpha, y_1) \to \sigma(y_1, y_1)$
- $E = (\alpha)$.

We apply the decomposition of Lemma 4.34 to M_{bal}. Then $mx = 2$ and we obtain the top-down tree transducer $T = (Q', \Sigma, \Gamma, q', R')$ with

- $Q' = \{q'^{(1)}\}$
- $\Gamma = \{\sigma'^{(0)}, \alpha'^{(0)}\} \cup \{\alpha_1^{(0)}\} \cup \{c_3^{(3)}, c_2^{(2)}\}$

1') $q'(\sigma(x_1, x_2)) \rightarrow c_2(q'(x_2), c_2(q'(x_1), \alpha_1))$
2') $q'(\alpha) \rightarrow c_3(\sigma', \alpha_1, \alpha_1)$

and the function $g : \Gamma^{(0)} \rightarrow T_\Delta(Z_1)$ with $g(\sigma') = \sigma(z_1, z_2)$, $g(\alpha') = \alpha$, and $g(\alpha_1) = z_1$.

Moreover, we obtain the tuple $\tilde{t} = (\alpha, z_2)$ because $r = 1 < 2 = mx$.

For the input tree $s = \sigma(\alpha, \alpha)$ we show a derivation of M_{bal} and the corresponding derivation of T in Fig. 4.3. In fact, the normal form of M_{bal}'s derivation is the image of the normal form of T's derivation under $yield_g$. □

In general, for every derivation of T there is a corresponding derivation of M, and for every call-by-value derivation of M there is corresponding derivation of T. A derivation is call-by-value, if in Def. 4.5 the φ_i's do not contain states. In fact, for the (non call-by-value) derivation

$$\begin{array}{ll} & q(\sigma(\alpha,\alpha), y_1) \\ \Rightarrow_{M_{bal}} & q(\alpha, q(\alpha, y_1)) \\ \Rightarrow_{M_{bal}} & \sigma(q(\alpha, y_1), q(\alpha, y_1)) \\ \Rightarrow_{M_{bal}} & \sigma(\sigma(y_1, y_1), q(\alpha, y_1)) \\ \Rightarrow_{M_{bal}} & \sigma(\sigma(y_1, y_1), \sigma(y_1, y_1)) \end{array}$$

of M_{bal} there is no corresponding derivation of T.

Lemma 4.36.

$$\begin{array}{llcl} 1) & TOP \circ YIELD & \subseteq & MAC \\ 2) & l\text{-}TOP \circ YIELD & \subseteq & l\text{-}MAC \\ 3) & sl\text{-}TOP \circ YIELD & \subseteq & sl\text{-}MAC. \end{array}$$

Proof. Let $T = (Q, \Sigma, \Gamma, q_0, R)$ be a top-down tree transducer. Let Δ be a ranked alphabet, $g : \Gamma^{(0)} \rightarrow T_\Delta(Z_r)$ be a function such that, for every $\alpha \in \Gamma^{(0)}$, $g(\alpha)$ is linear in Z, and $\tilde{t} = (t_1, \ldots, t_r) \in (T_\Delta)^r$ for some $r \geq 0$. We construct a macro tree transducer $M = (Q', \Sigma, \Delta, q_0', R', E')$ such that $\tau_T \circ yield_{g,\tilde{t}} = \tau_M$.

- Define $Q' = \{q' \mid q \in Q\}$ and, for every $q \in Q$, the state q' has rank $r+1$.
- Let $mx = max\{k \mid \Sigma^{(k)} \neq \emptyset\}$ and $Q(X_{mx}) = \{q(x_i) \mid q \in Q, x_i \in X_{mx}\}$. Extend g to $g' : \Gamma^{(0)} \cup Q(X_{mx}) \rightarrow T_{\Delta \cup Q'}(Z_r \cup X_{mx})$ by defining $g'(q(x_i)) = q'(x_i, z_1, \ldots, z_r)$. (Clearly, for every $\alpha \in \Gamma^{(0)}$ we have $g'(\alpha) = g(\alpha)$.)
- R' is the smallest set of rules such that, if $q(\sigma(x_1, \ldots, x_k)) \rightarrow \xi$ is in R, then the rule

$$q'(\sigma(x_1, \ldots, x_k), y_1, \ldots, y_r) \rightarrow yield_{g'}(\xi)[z_1 \leftarrow y_1, \ldots, z_r \leftarrow y_r]$$

  is in R'.

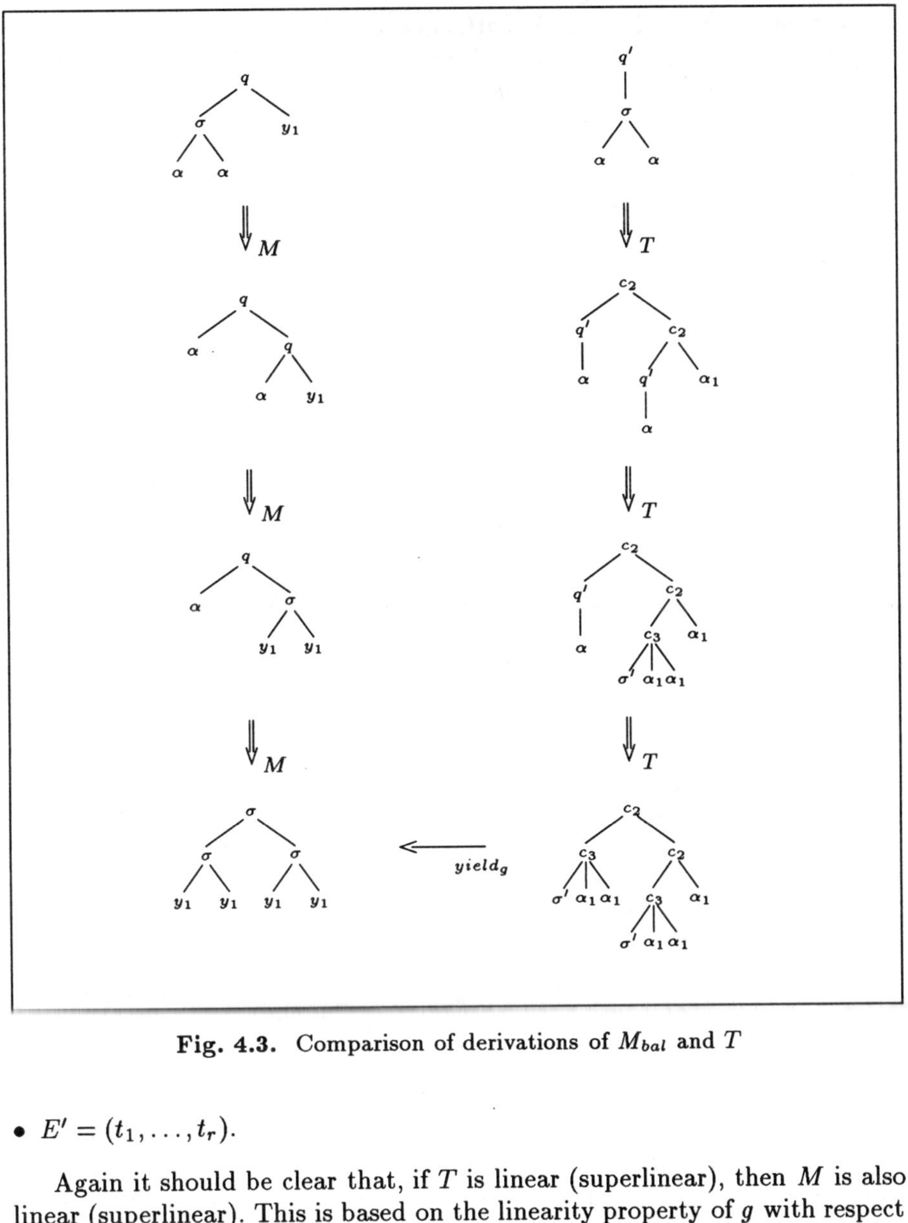

Fig. 4.3. Comparison of derivations of M_{bal} and T

- $E' = (t_1, \ldots, t_r)$.

Again it should be clear that, if T is linear (superlinear), then M is also linear (superlinear). This is based on the linearity property of g with respect to variables z_j. The correctness of the construction is verified by proving the following two statements by simultaneous induction.

K: For every $q \in Q$ and $s \in T_\Sigma$, the equality $yield_g(\tau_{T,q}(s))[z_1 \leftarrow y_1, \ldots, z_r \leftarrow y_r] = \tau_{M,q'}(s)$ holds.

L: For every $k \geq 0$, $\xi \in RHS(Q, \Gamma, k)$, and $\omega \in (T_\Sigma)^k$, the equality $yield_g(\tau_{T,\xi}(\omega))[z_1 \leftarrow y_1, \ldots, z_r \leftarrow y_r] = \tau_{M,\xi'}(\omega)$ holds where $\xi' = yield_{g'}(\xi)[z_1 \leftarrow y_1, \ldots, z_r \leftarrow y_r]$. □

From the previous two lemmas we obtain the following theorem which characterizes MAC in terms of TOP.

Theorem 4.37.

$$\begin{array}{lrcl} 1) & MAC & = & TOP \circ YIELD \\ 2) & l\text{-}MAC & = & l\text{-}TOP \circ YIELD \\ 3) & sl\text{-}MAC & = & sl\text{-}TOP \circ YIELD. \end{array}$$

As a corollary we obtain the closure of macro tree transformations with left-composition by top-down tree transformations (and also by homomorphisms).

Corollary 4.38.

$$\begin{array}{lrl} 1) & TOP \circ MAC & = MAC\ . \\ 2) & HOM \circ MAC & = MAC \end{array}$$

Proof.

$$\begin{array}{rcll} TOP \circ MAC & = & TOP \circ TOP \circ YIELD & \text{(Theorem 4.37)} \\ & = & TOP \circ YIELD & \text{(Theorem 3.39)} \\ & = & MAC & \text{(Theorem 4.37)}\ . \end{array}$$

The proof of the second equality can be derived in the same way by using $HOM \circ TOP = TOP$ (Theorem 3.39 and Lemma 2.14). □

Another consequence of the decomposition of macro tree transducers in top-down tree transducers and YIELD-functions is an analogy to the decomposition result $TOP = HOM \circ sl\text{-}TOP$ (Corollary 3.47) for MAC.

Corollary 4.39. $MAC = HOM \circ sl\text{-}MAC$.

Proof.

$$\begin{array}{rll} MAC & = TOP \circ YIELD & \text{(Theorem 4.37(1))} \\ & = HOM \circ sl\text{-}TOP \circ YIELD & \text{(Corollary 3.47)} \\ & = HOM \circ sl\text{-}MAC & \text{(Theorem 4.37(3))}. \end{array}$$

□

The composition closure of macro tree transducers from the right with top-down tree transducers or homomorphisms cannot be obtained immediately as a corollary of results known up to this point. However, in Chaps. 5 and 6 we will prove results from which these closure results follow. And since the following closure result fits here, we include it and assure the reader that there is no loop in our proof tree.

Theorem 4.40.

$$\begin{array}{ll} 1) & MAC \circ TOP = MAC \\ 2) & MAC \circ HOM = MAC. \end{array}$$

Proof.

$$\begin{aligned} MAC \circ HOM &= MAC \circ TOP && \text{(Corollary 3.32)} \\ &= HOM \circ ATT \circ TOP && \text{(Theorem 6.25)} \\ &= HOM \circ ATT && \text{(Theorem 5.47)} \\ &= MAC && \text{(Theorem 6.25).} \end{aligned}$$

□

4.6 Bibliographic Notes

In [Eng80, Eng81] Engelfriet pointed out that an archaic type of macro tree transducer had been investigated already in [Iro61]. Macro tree transducers can be considered as a combination of top-down tree transducers and context-free tree grammars [Rou70a] (see also [ES78]).

A formal model called the macro tree transducer was introduced in [Eng80]. Independently, primitive recursive program schemes were introduced in [CF82]. Actually, such schemes can be considered as many-sorted macro tree transducers. Example 4.4 is taken from [Eng80, EV86].

In [Eng81] composition, decomposition, and height results were sketched; they were investigated more thoroughly in [EV85] for the nondeterministic case, the partial case, and the total deterministic case. For instance, the composition $MAC \circ TOP = MAC$ (Theorem 4.40) has been proved in a direct way thus generalizing the construction of the composition of top-down tree transducers (Theorem 3.39). Moreover, it has been proved in [EV85] that macro tree transducers are closed under regular look-ahead.

In [EV88] macro tree transducers have been generalized to high-level tree transducers in the sense that the values of parameters are allowed to be applicative terms of a higher functional level rather than trees.

Pushdown machines for macro tree transducers have been suggested in [Eng81] and worked out in [EV86] by using the general concept of context-free grammar with storage [Eng86]. In [EV88] such pushdown machines were presented for high-level tree transducers (see also [Vog86]).

In [FHVV93] macro tree transducers were generalized to operators on classes of tree functions in the sense that some of the output symbols could be interpreted in a semantic domain with trees as the carrier set and tree functions as operations on the carrier set. It was proved that the class *PREC* of primitive recursive tree functions is closed under macro tree transformations. In particular, if the output symbols are interpreted by macro tree transformations, then so called modular tree transducers are obtained [EV91]. They characterize the class *PREC*.

In [EV94] it was proved that tree-generating top-down tree-to-graph transducers have the same transformational power as macro tree transducers. Top-down tree-to-graph transducers are top-down tree transducers in which the right-hand sides of rules are hypergraphs rather than trees. Also tree-generating bottom-up tree-to-graph transducers have the same transformational power as macro tree transducers [EV96].

In [MV95a, MV95b] a transformation strategy was defined with the aim to transform a given macro tree transducer into a tree transducer which computes output trees more time efficiently (i.e., with less derivation steps) than the original macro tree transducer. An algorithm was presented which decides whether there has been a gain in efficiency.

A pumping lemma for output tree languages of macro tree transducers has been proved in [Küh95b, Küh96].

5. Attributed Tree Transducers

In this chapter we introduce the third formal model of syntax-directed semantics: the attributed tree transducer. It is created by abstraction from attribute grammars. It can also handle context information like a macro tree transducer, and in this sense it too is more powerful than a top-down tree transducer. However, attributed tree transducers treat context information in an explicit way rather than implicitly as macro tree transducers do.

The explicit possibility of handling context information gives rise to a new phenomenon: the derivation relation of an attributed tree transducer can be circular and hence, it is not terminating. Fortunately, this property is decidable although at a high price: the decision algorithm has an inherent exponential time complexity. But there are large subclasses of the class of noncircular attributed tree transducers which can model all problems occurring in practice and which have a polynomial time membership problem. For a noncircular attributed tree transducer, the derivation relation is terminating.

The structure of this chapter is as follows.

1. Basic definitions
2. Induced tree transformation
3. Characterization of attributed tree transformation
4. Height and subtree properties
5. Composition and decomposition results
6. Bibliographic notes

5.1 Basic Definitions

Before starting the formal investigation, let us recall some features of top-down tree transducers. Figure 5.1(a) shows a sentential form of some top-down tree transducer T. Clearly, the trees $\sigma(s_1, s_2)$ and $\gamma(s_3)$ are subtrees of some original input tree s with which the derivation of T began. In Fig. 5.1(b) the same sentential form is represented in a different way: the input subtrees are replaced by their occurrences w and v, respectively, in the complete input tree s. Recall from Sect. 2.7 that occurrences are sequences of natural numbers.

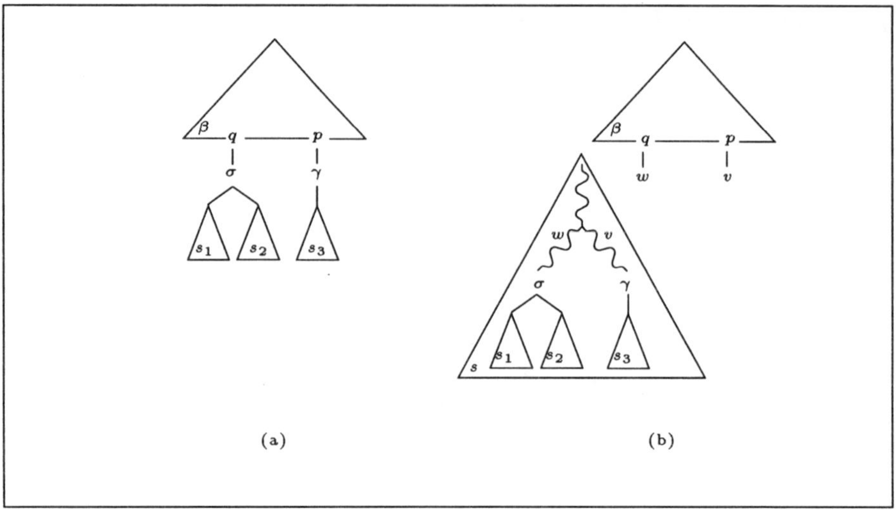

Fig. 5.1. Two representations of the same sentential form

Now it is clear that, during the derivation of T, these occurrences only grow with respect to their lengths until they have reached the leaves of s. Then, from a technical point of view, attributed tree transducers are obtained from top-down tree transducers by also allowing these occurrences of s to shrink during the derivation. Thus an arbitrary tree walk can be performed on s rather than just a recursive descent (as it is done by top-down tree transducers and by macro tree transducers). In fact, the possibility of programming an arbitrary tree walk on the input tree opens up the possibility of a circular derivation. Intuitively, the possibility of letting occurrences shrink means that, if a subtree, e.g., $\sigma(s_1, s_2)$ at occurrence w of s is considered, the computation of a meaning of $\sigma(s_1, s_2)$ may depend on the context of $\sigma(s_1, s_2)$.

In attributed tree transducers, meaning names are called *synthesized attributes*, because they synthesize the translation of subtrees. Whenever the process of increasing the length of an occurrence of s is changed to decreasing such a length, synthesized attributes call *inherited attributes.* That is, the latter attributes allow information to be passed down (or inherited) information from the context of this subtree. In order to avoid confusion, the set of synthesized attributes and that of inherited attributes should be disjoint.

Now we start the formal details by defining first (as usual) the set of right-hand sides of rules of attributed tree transducers. Since, in derivations of an attributed tree transducer, occurrences of the input tree may occur and these occurrences are manipulated (i.e., they lengthen or shrink), it is reasonable to expect a path variable in the rules of an attributed tree transducer. We introduce a path variable π which ranges over the set of occurrences of the given input tree.

Definition 5.1. Let A_s and A_i be unary ranked alphabets with $A_s \cap A_i = \emptyset$, Δ be a ranked alphabet with $\Delta \cap (A_s \cup A_i) = \emptyset$, and $k \geq 0$ be an integer. The set $RHS(A_s, A_i, \Delta, k)$ of *right-hand sides over* A_s, A_i, Δ, *and* k is the smallest subset RHS of $T_{A_s \cup A_i \cup \Delta}(\{\pi, \pi 1, ..., \pi k\})$ satisfying the following conditions.

(i) For every $a \in A_s$ and $1 \leq i \leq k$, the term $a(\pi i)$ is in RHS.
(ii) For every $b \in A_i$, the term $b(\pi)$ is in RHS.
(iii) For every $\delta \in \Delta^{(l)}$ with $l \geq 0$ and $\xi_1, \ldots, \xi_l \in RHS$, the term $\delta(\xi_1, \ldots, \xi_l)$ is in RHS. □

The symbols $\pi, \pi 1, \ldots, \pi k$ should be considered as strings, i.e., as elements of $(\{\pi\} \cup \mathbb{N})^*$, where π is a symbol. We call π a path variable because during the computation of an attributed tree transducer it will be replaced by actual occurrences of the input tree.

Note that $RHS(A_s, A_i, \Delta, 0) = T_\Delta(A_i(\{\pi\}))$ (for the definition of $A_i(\{\pi\})$ see Sect. 2.8).

Definition 5.2. An *attributed tree transducer* is a tuple $A = (Att, \Sigma, \Delta, a_0, R, E)$ satisfying the following conditions.

- Att is a unary ranked alphabet, called the set of *attributes*; Att is partitioned into the disjoint sets Att_{syn} and Att_{inh}, of which the elements are called *synthesized attributes* and *inherited attributes*, respectively.
- Σ and Δ are ranked alphabets with $Att \cap (\Sigma \cup \Delta) = \emptyset$, called the *input alphabet* and the *output alphabet*, respectively.
- $a_0 \in Att_{syn}$ is a designated attribute, called the *initial attribute*.
- R is a set $\bigcup_{\sigma \in \Sigma} R_\sigma$ of finite sets R_σ of rules; each R_σ satisfies the following two conditions.

(1) For every $a \in Att_{syn}$ and $\sigma \in \Sigma^{(k)}$ with $k \geq 0$, the set R_σ contains exactly one rule of the form

$$a(\pi) \to \xi$$

where $\xi \in RHS(Att_{syn}, Att_{inh}, \Delta, k)$.

(2) For every $b \in Att_{inh}$, $\sigma \in \Sigma^{(k)}$ with $k \geq 1$, and $1 \leq i \leq k$, the set R_σ contains exactly one rule of the form

$$b(\pi i) \to \xi$$

where $\xi \in RHS(Att_{syn}, Att_{inh}, \Delta, k)$.

- $E : Att_{inh} \to T_\Delta$ is a mapping, called the *environment*. □

Note 5.3. 1) An attributed tree transducer which does not have inherited attributes is very close to a top-down tree transducer: the synthesized attributes correspond to the states, a rule $a(\pi) \to \xi$ in R_σ with $\sigma \in \Sigma^{(k)}$ and $k \geq 0$ corresponds to the rule $a(\sigma(x_1, \ldots, x_k)) \to \xi'$, where ξ' is obtained from ξ

by replacing every $a'(\pi i)$ by $a'(x_i)$ in ξ, and the graph of the environment is empty.

2) Vice versa, for every top-down tree transducer it is straightforward to construct a corresponding attributed tree transducer with no inherited attributes. □

For an attributed tree transducer $A = (Att, \Sigma, \Delta, a_0, R, E)$ we observe the following notions and notations:

- For every $\sigma \in \Sigma^{(k)}$ with $k \geq 0$, the *set of inside attribute occurrences of* σ, denoted by $in(\sigma)$, is the set $\{a(\pi) \mid a \in Att_{syn}\} \cup \{b(\pi j) \mid b \in Att_{inh}, 1 \leq j \leq k\}$. The *set of outside attribute occurrences of* σ, denoted by $out(\sigma)$, is the set $\{b(\pi) \mid b \in Att_{inh}\} \cup \{a(\pi j) \mid a \in Att_{syn}, 1 \leq j \leq k\}$. The *set of attribute occurrences of* σ, denoted by $att(\sigma)$, is the set $in(\sigma) \cup out(\sigma)$. Note that, for every inside attribute occurrence $c(\pi\eta) \in in(\sigma)$, there is exactly one rule in R_σ with $c(\pi\eta)$ as a left-hand side. Moreover, attribute occurrences appearing in the right-hand sides of rules in R_σ are in $out(\sigma)$.
- The right-hand side of the rule $a(\pi) \rightarrow \xi$ in R_σ is represented by $rhs(a(\pi), \sigma)$ and the right-hand side of the rule $b(\pi i) \rightarrow \xi$ in R_σ is represented by $rhs(b(\pi i), \sigma)$. For every $\sigma \in \Sigma^{(k)}$, we denote by $RHS(\sigma)$ the set of right-hand sides of rules in R_σ, i.e., $RHS(\sigma) = \{rhs(c(\pi\eta), \sigma) \mid c(\pi\eta) \in in(\sigma)\}$.

Example 5.4. Consider the attributed tree transducer
$A_{shift} = (Att, \Sigma, \Delta, a_0, R, E)$ such that

- $Att = Att_{syn} \cup Att_{inh}$ with $Att_{syn} = \{a_0, a_1\}$ and $Att_{inh} = \{b\}$,
- $\Sigma = \Delta = \{\#^{(1)}, \sigma^{(2)}, \alpha_1^{(0)}, \alpha_2^{(0)}\}$,
- $E(b) = \alpha_1$,
- $R = R_\# \cup R_\sigma \cup R_{\alpha_1} \cup R_{\alpha_2}$,

  $R_\#$ contains the rules:
  1) $a_0(\pi) \rightarrow \#(a_0(\pi 1))$
  2) $a_1(\pi) \rightarrow \alpha_1$
  3) $b(\pi 1) \rightarrow a_1(\pi 1)$

  R_σ contains the rules:
  4) $a_0(\pi) \rightarrow a_0(\pi 2)$
  5) $a_1(\pi) \rightarrow a_1(\pi 1)$
  6) $b(\pi 1) \rightarrow \alpha_1$
  7) $b(\pi 2) \rightarrow \sigma(b(\pi), a_1(\pi 2))$

  R_{α_1} contains the rules:
  8) $a_0(\pi) \rightarrow b(\pi)$
  9) $a_1(\pi) \rightarrow \alpha_1$

  and R_{α_2} contains the rules:
  10) $a_0(\pi) \rightarrow b(\pi)$
  11) $a_1(\pi) \rightarrow \alpha_2$.

Let us consider the input symbol σ. The set $in(\sigma)$ of inside attribute occurrences of σ is the set $\{a_0(\pi), a_1(\pi), b(\pi 1), b(\pi 2)\}$, and the set $out(\sigma)$ of outside attribute occurrences of σ is the set $\{b(\pi), a_0(\pi 1), a_1(\pi 1), a_0(\pi 2), a_1(\pi 2)\}$. As another example for the use of abbreviations, $rhs(b(\pi 2), \sigma) = \sigma(b(\pi), a_1(\pi 2))$ (i.e., the right-hand side of rule 7), and $RHS(\sigma) = \{a_0(\pi 2), a_1(\pi 1), \alpha_1, \sigma(b(\pi), a_1(\pi 2))\}$.

In particular, if A_{shift} takes a right-growing comb s as input, then it produces a left-growing comb which has the same frontier as s. For instance, if $s = \#(\sigma(\alpha_1, \sigma(\alpha_1, \sigma(\alpha_2, \alpha_2))))$, then A_{shift} produces the output tree $\#(\sigma(\sigma(\sigma(\alpha_1, \alpha_1), \alpha_2), \alpha_2))$. □

5.2 Induced Tree Transformation

In the rest of this chapter, let $A = (Att, \Sigma, \Delta, a_0, R, E)$ be an arbitrary, but fixed, attributed tree transducer. We define the derivation relation of A, the notions of circularity and a circularity test, and the definition of the tree transformation induced by A.

Derivation relation

The derivation relation (Figs. 5.2 and 5.3) is parameterized by an input tree s which is considered as a global variable during the whole derivation. In the following definition, the occurrences of s are considered as nullary symbols.

Definition 5.5. Let $s \in T_\Sigma$. The *derivation relation induced by A on s* is the binary relation $\Rightarrow_{A,s}$ over the set $T_{Att \cup \Delta \cup occ(s)}$ such that for every $\varphi, \psi \in T_{Att \cup \Delta \cup occ(s)}$ we define $\varphi \Rightarrow_{A,s} \psi$, if

- there is a context $\beta \in C_{Att \cup \Delta \cup occ(s),1}$,
- there is an attribute $c \in Att$,
- there is a path $w \in occ(s)$,

such that $\varphi = \beta[c(w)]$ and exactly one of the following two conditions holds.

(1) $c \in Att_{syn}$, $label(s, w) = \sigma$, $\sigma \in \Sigma^{(k)}$, $k \geq 0$, and $\psi = \beta[\xi[\pi \leftarrow w]]$, where $\xi = rhs(c(\pi), \sigma)$.
(2) $c \in Att_{inh}$, $w = vi$ for some $v \in occ(s)$, $label(s, v) = \sigma$, $\sigma \in \Sigma^{(k)}$, $k \geq 1$, $1 \leq i \leq k$, and $\psi = \beta[\xi[\pi \leftarrow v]]$, where $\xi = rhs(c(\pi i), \sigma)$.

□

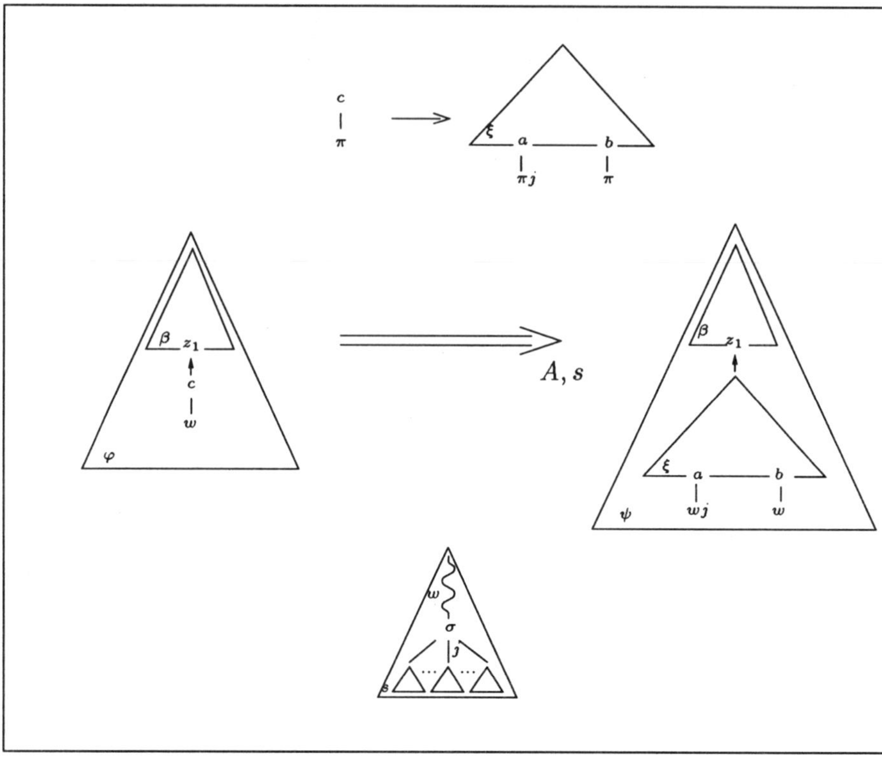

Fig. 5.2. A derivation step induced by A triggered by a synthesized attribute

Example 5.6. Let us show two typical steps of an attributed tree transducer. For this purpose, consider again the attributed tree transducer A_{shift} of Example 5.4. Let $s = \#(\sigma(\alpha_1, \sigma(\alpha_1, \alpha_2))) \in T_\Sigma$ be an input tree and let

$$\varphi = \#(a_0(122)).$$

Then there are a context $\beta = \#(z_1)$, an attribute $c = a_0$, and an occurrence $w = 122 \in occ(s)$ with $label(s, 122) = \alpha_2$, such that $\varphi = \beta[c(w)]$. Moreover, $\varphi \Rightarrow_{A_{shift},s} \psi$ using rule 10 (Fig. 5.4(a)), where

$$\begin{aligned} \psi &= \#(z_1)[rhs(a_0(\pi), \alpha_2)[\pi \leftarrow 122]] \\ &= \#(z_1)[b(\pi)[\pi \leftarrow 122]] \\ &= \#(b(122)). \end{aligned}$$

Another derivation step can be performed on ψ. Let $\beta' = \#(z_1)$, $c' = b$, $w = 122$, $v = 12$, $i = 2$ with $label(s, 12) = \sigma$. Then $\psi = \beta'[c'(w')]$. Moreover, $\psi \Rightarrow_{A_{shift},s} \psi'$ using rule 7 (Fig. 5.4(b)), where

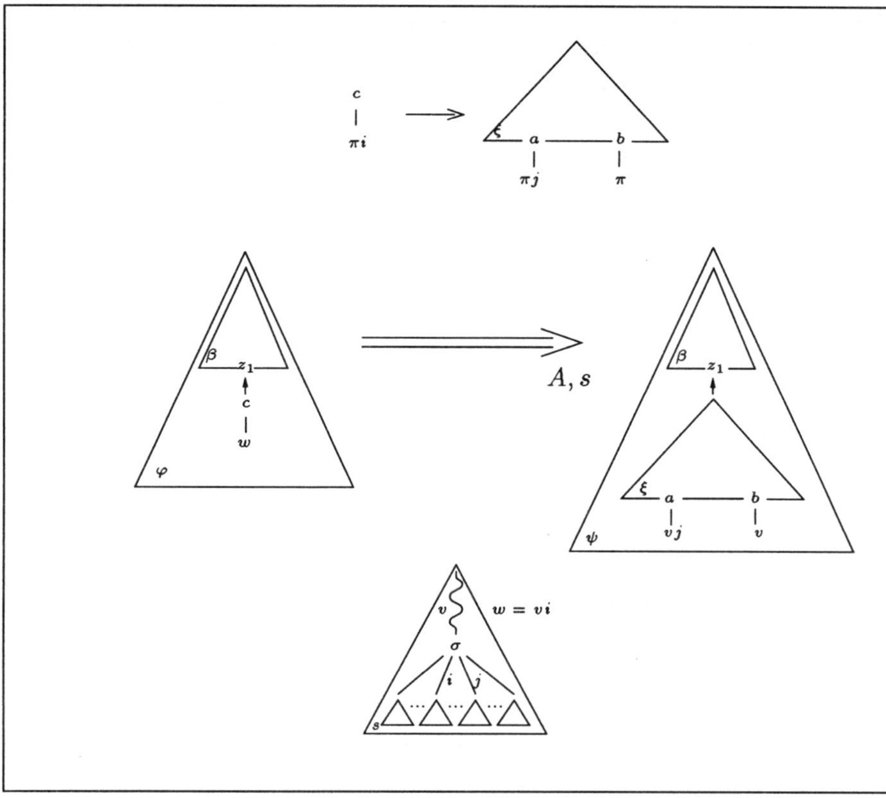

Fig. 5.3. A derivation step induced by A triggered by an inherited attribute

$$\begin{aligned}\psi' &= \#(z_1)[rhs(b(\pi 2), \sigma)[\pi \leftarrow 12]] \\ &= \#(z_1)[\sigma(b(\pi), a_1(\pi 2))[\pi \leftarrow 12]] \\ &= \#(\sigma(b(12), a_1(122))).\end{aligned}$$

□

Definition 5.7. Let $s \in T_\Sigma$. The set of *sentential forms of A and s*, denoted by $SF(A, s)$, is the smallest subset SF of $T_{Att \cup \Delta \cup occ(s)}$ for which the following conditions hold.

(i) For every $c \in Att$ and $w \in occ(s)$, the term $c(w) \in SF$.
(ii) For every $\delta \in \Delta^{(l)}$ with $l \geq 0$ and $\varphi_1, \ldots, \varphi_l \in SF$, the term $\delta(\varphi_1, \ldots, \varphi_l)$ is in SF. □

Now we state a connection between right-hand sides and sentential forms which is similar to the corresponding connection for top-down tree transducers (cf. Lemma 3.7(a)) and for macro tree transducers (cf. Lemma 4.8(a)). Since the proof is very similar to the lemmas mentioned we omit it here.

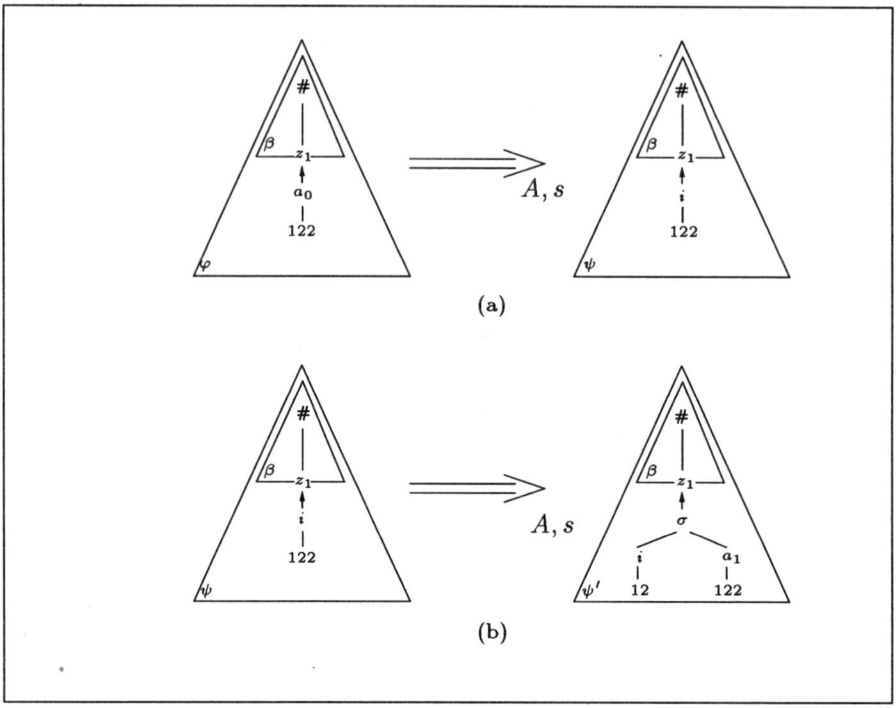

Fig. 5.4. Two particular derivation steps induced by A_{shift}

Lemma 5.8. Let $\xi \in RHS(Att_{syn}, Att_{inh}, \Delta, k)$ with $k \geq 0$, $s \in T_\Sigma$, and $w \in occ(s)$ such that for every $1 \leq i \leq k$, $wi \in occ(s)$. Then $\xi[\pi \leftarrow w] \in SF(A, s)$. □

A connection between right-hand sides and sentential forms similar to Lemma 3.7(b) for top-down tree transducers and Lemma 4.8(b) for macro tree transducers cannot be stated here. This is due to the fact that, in the right-hand side of a rule of an attributed tree transducer, the attributes refer to a node in the neighborhood of the node to which the attribute in the left-hand side refers. More precisely, in the right-hand side of a rule $a(\pi) \longrightarrow \xi$, the attributes refer to either the same node as attribute a or some descendants, and in the right-hand side of a rule $b(\pi i) \longrightarrow \xi$, the attributes refer to either some sibling of the node to which b refers, or its ancestor. However, in an arbitrary sentential form, the attributes involved need not refer to neighboring nodes of a tree – they can refer to arbitrary nodes.

We note that, in the study of top-down tree transducers and macro tree transducers, the representation of sentential forms as instances of right-hand sides (Lemmas 3.7(b) and 4.8(b), respectively) is used in the proof of termination of the corresponding derivation relation (Lemmas 3.13 and 4.11, respectively). However, for attributed tree transducers, we will prove the corresponding termination property in a different way (see Lemma 5.24).

Lemma 5.9. Let s be an input tree of A. Then $SF(A, s)$ is closed under $\Rightarrow_{A,s}$.

Proof. Let $\varphi \in SF(A,s)$ and $\psi \in T_{Att \cup \Delta}(occ(s))$ such that $\varphi \Rightarrow_{A,s} \psi$. We prove that $\psi \in SF(A,s)$ by structural induction on φ.

(i) Let $\varphi = c(w)$ for some $c \in Att$ and $w \in occ(s)$.

Case 1: Let $c \in Att_{syn}$ and let $label(s,w) = \sigma$. Then $\psi = rhs(c(\pi),\sigma)[\pi \leftarrow w]$. By Lemma 5.8, it follows that $\psi \in SF(A,s)$.

Case 2: Let $c \in Att_{inh}$ and let $w = vi$, $i \geq 1$, $label(s,v) = \sigma$. Then $\psi = rhs(c(\pi i),\sigma)[\pi \leftarrow v]$. By Lemma 5.8, it follows that $\psi \in SF(A,s)$.

(ii) Let $\varphi = \delta(\varphi_1,\ldots,\varphi_l)$ for some $\delta \in \Delta^{(l)}$, $l \geq 0$, and $\varphi_1,\ldots,\varphi_l \in SF(A,s)$. Since $\varphi \Rightarrow_{A,s} \psi$, there is an i with $1 \leq i \leq l$ and a $\psi_i \in T_{Att \cup \Delta}(occ(s))$ such that $\varphi_i \Rightarrow_{A,s} \psi_i$ and $\psi = \delta(\varphi_1,\ldots,\varphi_{i-1},\psi_i,\varphi_{i+1},\ldots,\varphi_l)$. By induction hypothesis, the term $\psi_i \in SF(A,s)$. Since $\varphi_1,\ldots,\varphi_{i-1},\varphi_{i+1},\ldots,\varphi_l \in SF(A,s)$, it follows from Def. 5.7 (ii) that $\psi \in SF(A,s)$. □

Circularity and a circularity test

Clearly, since a derivation can take an arbitrary tree walk on an input tree, it is possible that the derivation is cyclic.

Definition 5.10. 1) A is *circular*, if there is an $s \in T_\Sigma$, there is a sentential form $c(w)$ with $c \in Att$ and $w \in occ(s)$, and there is a context $\beta \in C_{Att \cup \Delta \cup occ(s),1}$ such that $c(w) \Rightarrow^+_{A,s} \beta[c(w)]$.

2) A is *noncircular*, if it is not circular. □

Example 5.11. Consider the attributed tree transducer
$A = (Att, \Sigma, \Delta, a_0, R, E)$ such that

- $Att = Att_{syn} \cup Att_{inh}$ with $Att_{syn} = \{a\}$ and $Att_{inh} = \{b\}$; $a_0 = a$,
- $\Sigma = \Delta = \{\gamma^{(1)}, \alpha^{(0)}\}$,
- $R = R_\gamma \cup R_\alpha$, where
  R_γ contains the rules
  1) $a(\pi) \rightarrow a(\pi 1)$
  2) $b(\pi 1) \rightarrow a(\pi 1)$,

  and R_α contains the rule
  3) $a(\pi) \rightarrow b(\pi)$,
- $E(b) = \alpha$.

Then there is the following infinite derivation on input tree $s = \gamma(\alpha)$:

$$a(\varepsilon) \Rightarrow_{A,s} a(1) \Rightarrow_{A,s} b(1) \Rightarrow_{A,s} a(1) \ldots$$

Hence, A is circular. □

Thus, in order to state that an attributed tree transducer is noncircular, we have to check every input tree s for cycles in the derivation relation $\Rightarrow_{A,s}$. However, there is a way to associate a so-called dependency graph with every input tree to A such that A is circular if and only if there is a cycle in the dependency graph associated with some input tree. We develop this technique.

First we define the concept of the dependency graph of an input symbol σ with respect to A. The definition of $att(\sigma)$ is as in Sect. 5.1.

Definition 5.12. Let $\sigma \in \Sigma^{(k)}$ with $k \geq 0$. The *dependency graph of* σ *for* A, denoted by $D_A(\sigma)$, is the directed graph $(att(\sigma), E(\sigma))$ with $E(\sigma) = \{(c(\pi\eta), c'(\pi\eta')) \in out(\sigma) \times in(\sigma) \mid rhs(c'(\pi\eta'), \sigma)$ contains the subtree $c(\pi\eta)\}$ as a set of arcs. □

We will use the usual graphical representation of the dependency graph $D_A(\sigma)$. It is a trapezoid with the parallel sides drawn horizontally. The symbol σ is shown on the upper line. The nodes of $D_A(\sigma)$ are placed on these parallel lines such that occurrences of inherited attributes and synthesized attributes appear to the left and right of σ, respectively. If the names of the attribute occurrences are clear from the context, then they are omitted. For example, in Fig. 5.5, the dependency graphs $D_{A_{shift}}(\#)$, $D_{A_{shift}}(\sigma)$, $D_{A_{shift}}(\alpha_1)$, and $D_{A_{shift}}(\alpha_2)$ of the attributed tree transducer A_{shift} of Example 5.4 are shown.

The local dependencies of input symbols described by the dependency graphs extend to global dependencies of attribute occurrences of input trees. Next we define the dependency graph of an input tree s for A by induction on the structure of s.

Definition 5.13. Let $s \in T_\Sigma$. The *dependency graph of* s *for* A, denoted by $D_A(s)$ is defined as follows:

(i) For every $s = \alpha \in \Sigma^{(0)}$, $D_A(s) = D_A(\sigma)$.

(ii) Let $s = \sigma(s_1, \ldots, s_k)$ for some $k \geq 1$, $\sigma \in \Sigma^{(k)}$, and $s_1, \ldots, s_k \in T_\Sigma$. Assume $D_A(s_i)$ are already defined for $1 \leq i \leq k$ and let $Att(s_i)$ and $E(s_i)$ be the set of nodes and edges of $D_A(s_i)$, respectively. Now $D_A(s)$ is the directed graph of which the set $Att(s)$ of nodes and set $E(s)$ of edges are defined as follows:
- $Att(s) = att(\sigma) \cup (\bigcup_{i=1}^{k} \{c(\pi i\eta) \mid c(\pi\eta) \in Att(s_i), 1 \leq i \leq k\})$
- $E(s) = E(\sigma) \cup (\bigcup_{i=1}^{k} \{(c(\pi i\eta), c'(\pi i\eta')) \mid (c(\pi\eta), c'(\pi\eta')) \in E(s_i), 1 \leq i \leq k\})$.

□

Roughly speaking, the dependency graph of an input tree s can be obtained by pasting together the dependency graphs of the input symbols appearing in s. In Fig. 5.6 we show the dependency graph of the input tree $\#(\sigma(\alpha_1, \sigma(\alpha_2, \sigma(\alpha_2, \alpha_2))))$.

Now let us return to the circularity problem of A. The attribute occurrences appearing in a derivation are determined by the rules applied during that derivation. Moreover, the rules induce a local dependency relation among the attribute occurrences of input symbols. The local dependencies induce a global dependency relation over the attribute occurrences of every input tree s in the way described above. Hence there is a close connection between the derivation relation $\Rightarrow_{A,s}$ and the global dependency relation shown by $D_A(s)$. Namely, any derivation by $\Rightarrow_{A,s}$ starting from an attribute occurrence (c, w) of s will "call" (or "use") another attribute occurrence (c', w') of s if and only if there is a path from (c', w') to (c, w) in $D_A(s)$.

Therefore, in order to check whether A is noncircular, we have to check, for every input tree s, whether $D_A(s)$ contains a cycle or not. A naive algorithm can certainly not do this. However there is a way to partition the set T_Σ into a finite number of equivalence classes. Every equivalence class is characterized by a so-called is-graph.

Definition 5.14. An *is-graph* (over Att_{inh} and Att_{syn}) is a directed graph (V, E) such that $V = Att_{inh} \cup Att_{syn}$ is the set of nodes and $E \subseteq Att_{inh} \times Att_{syn}$ is the set of edges. □

Thus, in an is-graph, edges may lead only from inherited attributes to synthesized attributes. Clearly, since Att is a finite set, there can only be a finite number of such is-graphs. More precisely, there are $2^{card(Att_{syn}) \cdot card(Att_{inh})}$ is-graphs. In broad terms, a single is-graph is defines the equivalence class $\{s \in T_\Sigma \mid$ there is an indirect or direct dependency between $c(\varepsilon)$ and $c'(\varepsilon)$ if and only if there is an edge between c and c' in $is\}$. We will show a lemma (Lemma 5.17) which states that it is sufficient to consider only these equivalence classes in order to test circularity. Let us now formally define the notions involved and prove the lemma on which an algorithm can be based for the decision about circularity.

First, for an attributed tree transducer A and an input tree $s \in T_\Sigma$, we define the *is-graph* of s (which is an is-graph in the sense of Def. 5.14). Intuitively, the is-graph of s shows the (global) dependencies between occurrences of inherited attributes and synthesized attributes at the root of s. In the following we also require the set of all is-graphs of A. To support the definition of the is-graph of an input tree, we define the composition of graphs and their is-graphs.

Definition 5.15. Let $\sigma \in \Sigma^{(k)}$, $k \geq 0$, and $D_A(\sigma) = (att(\sigma), E(\sigma))$ be the dependency graph of σ for A. Moreover, for every j with $1 \leq j \leq k$, let $is_j = (Att, E_j)$ be an is-graph over Att_{inh} and Att_{syn}.

1) The *composition of $D_A(\sigma)$ and $is_1, \ldots, is_k$*, denoted by $D_A(\sigma)[is_1, \ldots, is_k]$, is the directed graph $(att(\sigma), E)$ such that $E = E(\sigma) \cup \{(c(\pi j), c'(\pi j)) \mid (c, c') \in E_j, 1 \leq j \leq k\}$.

2) The *is-graph of* $D_A(\sigma)[is_1, \ldots, is_k]$, denoted by $is(D_A(\sigma)[is_1, \ldots, is_k])$, is the is-graph (Att, E) such that $E = \{(c, c') \in Att_{inh} \times Att_{syn} \mid$ there is a path in $D_A(\sigma)[is_1, \ldots, is_k]$ from $c(\pi)$ to $c'(\pi)\}$. □

We can now define the is-graph of an input tree s for A.

Definition 5.16. Let $s = \sigma(s_1, \ldots, s_k) \in T_\Sigma$.

1) The *is-graph of s for A*, denoted by $is_A(s)$, is an is-graph defined by induction on s as follows.

(i) For every $s = \alpha \in \Sigma^{(0)}$, define

$$is_A(s) = is(D_A(\alpha)[\,]).$$

(ii) For every $s = \sigma(s_1, \ldots, s_k) \in T_\Sigma$ with $k \geq 1$, define

$$is_A(s) = is(D_A(\sigma)[is_A(s_1), \ldots, is_A(s_k)]).$$

2) Define the *set of is-graphs* of A, denoted by $is\text{-}set_A$, as follows: $is\text{-}set_A := \{is_A(s) \mid s \in T_\Sigma\}$ □

Recall that $is\text{-}set_A$ is a finite set.

Now we can use the notion of the set of is-graphs to construct a circularity test for attributed tree transducers that is based on the following lemma. We formulate the lemma in such a way that it directly yields the idea of the algorithm; that is, every universal quantification is taken over a finite set. In particular, the lemma does not refer to the infinite set of input trees (as Def. 5.10 does). We will call it the circularity lemma.

Lemma 5.17. *A is circular if and only if there is a $\sigma \in \Sigma^{(k)}$ with $k \geq 1$ and there are $is_1, \ldots, is_k \in is\text{-}set_A$, such that the graph $D_A(\sigma)[is_1, \ldots, is_k]$ has a cycle.*

Proof. By our argument above, A is circular if and only if there is a tree $s \in T_\Sigma$ such that $D_A(s)$ contains a cycle.

Hence, it is sufficient to prove that there is a tree $s \in T_\Sigma$ such that $D_A(s)$ contains a cycle if and only if there is a $\sigma \in \Sigma^{(k)}$ with $k \geq 1$ and there are $is_1, \ldots, is_k \in is\text{-}set_A$ such that the graph $D_A(\sigma)[is_1, \ldots, is_k]$ has a cycle.

First we prove the 'if' clause. Assume that the graph $D_A(\sigma)[is_1, \ldots, is_k]$ has a cycle. Then, by the definition of $is\text{-}set_A$, there are trees $s_1, \ldots, s_k$, such that $is_A(s_i) = is_i$, for every $1 \leq i \leq k$. Moreover, by the definition of an is-graph, for every $1 \leq i \leq k$, there is an edge in $is_A(s_i)$ from an inherited to a synthesized attribute, if and only if there is a path in $D_A(s_i)$ from the occurrence of that inherited attribute at the root of s_i to the corresponding occurrence of that synthesized attribute. Hence, for $s = \sigma(s_1, \ldots, s_k)$, the graph $D_A(s)$ is also cyclic.

Conversely, assume that there is a tree $s \in T_\Sigma$ such that $D_A(s)$ contains a cycle. Let s' be a subtree of s such that $D_A(s')$ contains a cycle

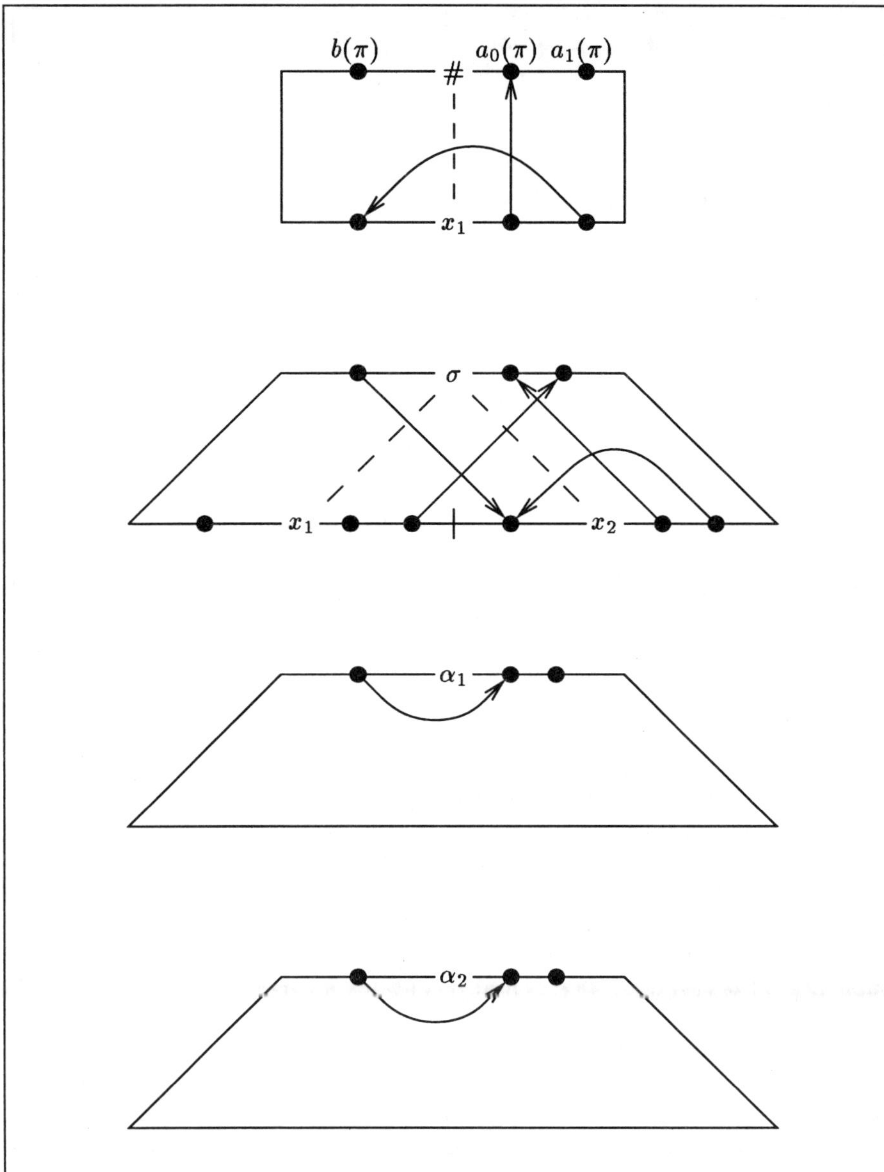

Fig. 5.5. The dependency graphs $D_{A_{shift}}(\#)$, $D_{A_{shift}}(\sigma)$, $D_{A_{shift}}(\alpha_1)$, and $D_{A_{shift}}(\alpha_2)$

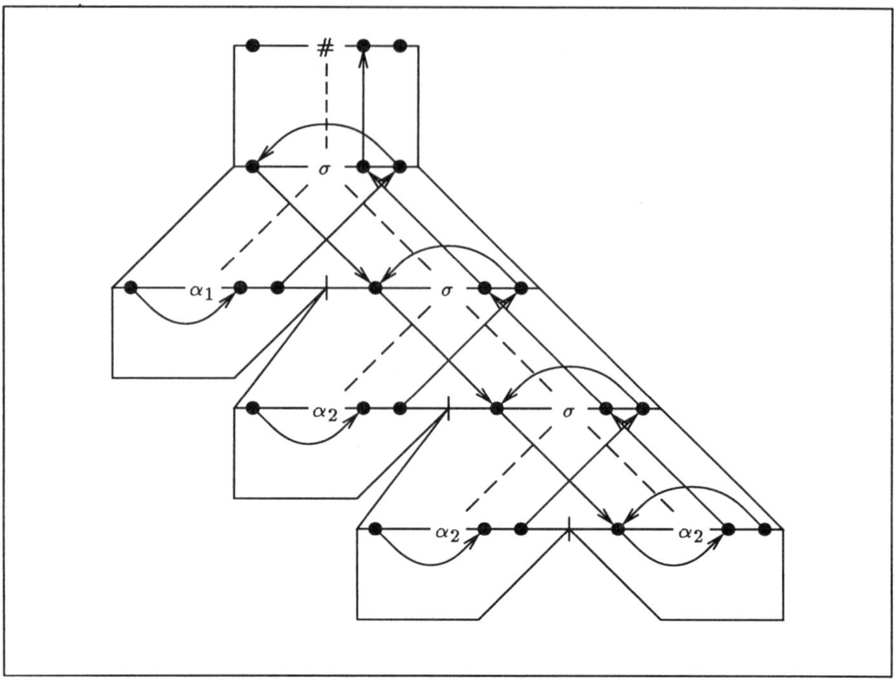

Fig. 5.6. The input tree $\#(\sigma(\alpha_1, \sigma(\alpha_2, \sigma(\alpha_2, \alpha_2))))$ with its dependency graphs

and assume that s' is minimal with respect to its size. Then s' must have the form $\sigma(s_1, \ldots, s_k)$, for some $k \geq 1$, $\sigma \in \Sigma^{(k)}$, and $s_1, \ldots, s_k \in T_\Sigma$ Since $D_A(s_1), \ldots, D_A(s_k)$ do not contain a cycle, the cycle in $D_A(s')$ should contain some edges of $D_A(\sigma)$, according to Def. 5.13. Then, since, for every $1 \leq i \leq k$, the edges of $is_A(s_i)$ represent the paths in $D_A(s_i)$ from occurrences of inherited attributes to occurrences of synthesized attributes at the root of s_i, the graph $D_A(\sigma)[is_A(s_1), \ldots, is_A(s_k)]$ also contains a cycle. □

Thus we can design an algorithm (Fig. 5.7) which constructs step by step $is\text{-}set_A$ for an attributed tree transducer A and which simultaneously checks the condition of the previous lemma. For this purpose, we use a Modula2–like notation for the presentation of the algorithm. We note that, in contrast to attribute grammars, there is no distinction between trees of different types, that is trees with different nonterminal symbols at the root. Thus, in our case it is possible to use the usual set $is\text{-}set_A$ of is-graphs for A rather than a family of sets indexed by the set of nonterminals.

Clearly, the algorithm terminates, since $is\text{-}set_A$ is a finite set. The algorithm discovers every is-graph of A, because for every $s = \sigma(s_1, \ldots, s_k) \in T_\Sigma$ with $height(s) = l$, the graph $is_A(s)$ is calculated in the l-th run through the <u>*repeat*</u>-loop out of $D_A(\sigma)$ and $is_A(s_1), \ldots, is_A(s_k)$, or it is calculated

Input: Attributed tree transducer $A = (Att, \Sigma, \Delta, a_0, R, E)$ with $Att = Att_{syn} \cup Att_{inh}$ and dependency graphs $D_A(\sigma)$ for every $\sigma \in \Sigma$.

Output: The answer "A is circular" or "A is noncircular".

Method: The algorithm needs the following variables:

is-set$_A$	set of is-graphs of A
is	new constructed is-graph
D	new constructed directed graph
change	boolean information, whether *is-set*$_A$ has changed during the last run through the *repeat*-loop

```
is-set_A := ∅;
repeat
    change := false;
    for every σ ∈ Σ^(k) do begin
        for every k–tuple (is_1, ..., is_k) with is_j ∈ is–set_A do begin
            D := D_A(σ)[is_1, ..., is_k];
            if there is a cycle in D then
                write("A is circular");
                stop
            else (* there is no cycle in D *)
                is := is(D);
                if is ∉ is-set_A then
                    is-set_A := is-set_A ∪ {is};
                    change := true
                end (* if *)
            end (* if *)
        end (* for *)
    end (* for *)
until change = false;
write("A is noncircular")
```

Fig. 5.7. Circularity test for attributed tree transducers

even earlier. Thus we conclude with the following observation concerning the correctness of the circularity test.

Observation 5.18. The algorithm shown in Fig. 5.7 with input A and dependency graphs $D_A(\sigma)$ for every $\sigma \in \Sigma$ yields the result

- "A is circular", if and only if A is circular,
- "A is noncircular", if and only if A is noncircular. □

Example 5.19. Consider an attributed tree transducer A with $Att_{inh} = \{b_1, b_2\}$ and $ATT_{syn} = \{a_1, a_2\}$. Let $\Sigma = \{\sigma^{(2)}, \alpha^{(0)}, \beta^{(0)}\}$ be the input alphabet. The dependency graphs of the input symbols are shown in Fig. 5.8 (where we have omitted the lower lines of $D_A(\alpha)$ and $D_A(\beta)$). We now show some initial steps of the algorithm in Fig. 5.7 when it is applied to A.

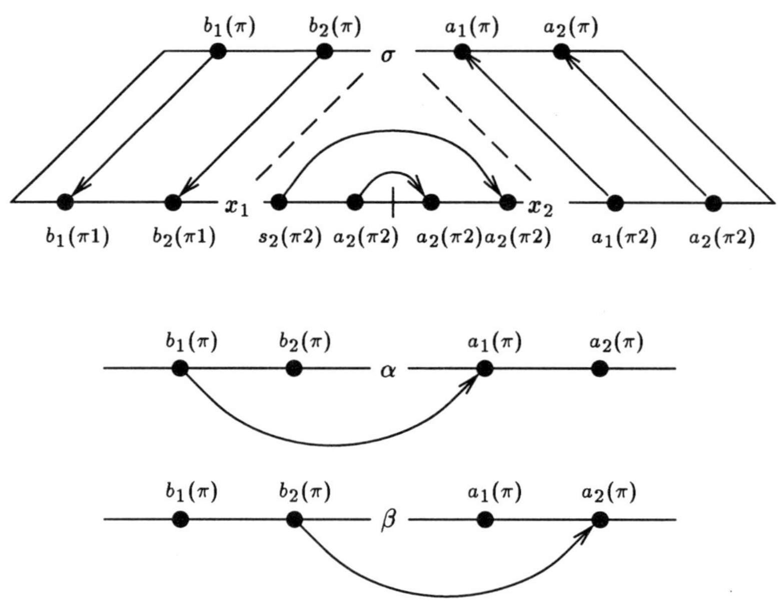

Fig. 5.8. Dependency graphs of an attributed tree transducer

$is\text{-}set_A = \emptyset$
$change = false$

<u>*for*</u> α *do*
    <u>*for*</u> () <u>*do*</u>

$is_1 = (Att, \{(b_1, a_1)\})$
$is\text{-}set_A = \{is_1\}$
$change = true$

for β do
for () *do*
$is_2 = (Att, \{(b_2, a_2)\})$
$is\text{-}set_a = \{is_1, is_2\}$
$change = true$

for σ do
for (is_1, is_1) *do*
$is_3 = (Att, \emptyset)$
$is\text{-}set_A = \{is_1, is_2, is_3\}$
$change = true$

for (is_1, is_2) *do*
$is_4 = (Att, \{(b_1, a_2)\})$
$is\text{-}set_A = \{is_1, is_2, is_3, is_4\}$
$change = true$

for (is_2, is_1) *do*
$is_5 = (Att, \{(b_2, a_1)\})$
$is\text{-}set_A = \{is_1, is_2, is_3, is_4, is_5\}$
$change = true$

for (is_2, is_2) *do*
$is_4 = (Att, \emptyset)$

$change = false$

for α do
for () *do*
$is_7 = (Att, \{(b_1, a_1)\}) = is_1 \in is\text{-}set_A$

for β do
for () *do*
$is_8 = (Att, \{(b_2, a_2)\}) = is_2 \in is\text{-}set_A$

for σ do
for (is_1, is_1) *do*
$is_9 = (Att, \emptyset) = is_3 \in is\text{-}set_A$

...

There are now in total 25 possible tuples (is, is') with $is, is' \in is\text{-}set_A$. No combination yields a new element of $is\text{-}set_A$, hence $change = false$, the *repeat*-loop terminates, and the algorithm outputs that A is noncircular. □

Definition of the tree transformation

In the rest of this chapter, let $A = (Att, \Sigma, \Delta, a_0, R, E)$ be an arbitrary, but fixed, *noncircular* attributed tree transducer.

Then we can prove that, for every input tree s, $\Rightarrow_{A,s}$ is confluent and terminating.

Lemma 5.20. For every input tree s of A, the derivation relation $\Rightarrow_{A,s}$ is locally confluent on $SF(A, s)$.

Proof. The proof of this lemma is very similar to the proof of the same property of the derivation relations induced by top-down tree transducers (cf. Lemma 3.9), and hence we omit it here. □

Now we show that, for every input tree s, the derivation relation $\Rightarrow_{A,s}$ is terminating on $SF(A, s)$. First we make some observations about infinite derivations and we coin some useful notions.

Definition 5.21. Let $s \in T_\Sigma$ and $\varphi \in SF(A, s)$.

1. The sentential form φ is *ω-reducible* if there is an infinite derivation
$$\varphi = \varphi_0 \Rightarrow_{A,s} \varphi_1 \Rightarrow_{A,s} \varphi_2 \Rightarrow_{A,s} \cdots$$
2. The sentential form φ is *minimal ω-reducible* if there is no subtree φ' of φ such that φ' is ω-reducible. □

Observation 5.22. Every ω-reducible sentential form has a minimal ω-reducible subtree. □

Observation 5.23. Every minimal ω-reducible sentential form has the form $c(w)$ for some attribute c and occurrence w. □

Recall that we are only dealing with noncircular attributed tree transducers.

Lemma 5.24. For every input tree s of A, the derivation relation $\Rightarrow_{A,s}$ is terminating on $SF(A, s)$.

Proof. We prove this fact by contradiction. Let A be a noncircular attributed tree transducer and assume that there is an $s \in T_\Sigma$ such that $\Rightarrow_{A,s}$ is *not* terminating on $SF(A, s)$. Then there is an ω-reducible sentential form $\varphi \in SF(A, s)$. By Observation 5.22, φ has a minimal ω-reducible subtree, say $\varphi_{0,0}$ and, by Observation 5.23, $\varphi_{0,0}$ has the form $c_0(w_0)$. Hence, there is an infinite derivation
$$\varphi_{0,0} \Rightarrow_{A,s} \varphi_{0,1} \Rightarrow_{A,s} \varphi_{0,2} \cdots$$
(see Fig. 5.9 where $\Rightarrow_{A,s}$ is abbreviated to $\Rightarrow$). Clearly, $\varphi_{0,1}$ is also an ω-reducible sentential form and it has a minimal ω-reducible subtree, say $\varphi_{1,0}$. There is an infinite derivation

$$\varphi_{1,0} \Rightarrow_{A,s} \varphi_{1,1} \Rightarrow_{A,s} \varphi_{1,2} \cdots$$

where again, $\varphi_{1,1}$ is ω-reducible and it has a minimal ω-reducible subtree $\varphi_{2,0}$. Obviously, this construction of minimal subtrees can be repeated and thereby we obtain two infinite sequences $(\varphi_{i,0})_{i\geq 0}$ and $(\varphi_{i,1})_{i\geq 0}$ of sentential forms such that, for every $i \geq 0$, the following three properties hold:

1. $\varphi_{i,0}$ has the form $c_i(w_i)$ for some attribute c_i and some occurrence w_i of s.
2. $\varphi_{i,0} \Rightarrow_{A,s} \varphi_{i,1}$.
3. $\varphi_{i+1,0}$ is a minimal ω-reducible subtree of $\varphi_{i,1}$, hence $\varphi_{i,1} = \beta_i[\varphi_{i+1,0}]$ for some $(Att \cup \Delta \cup occ(s), 1)$-context β_i.

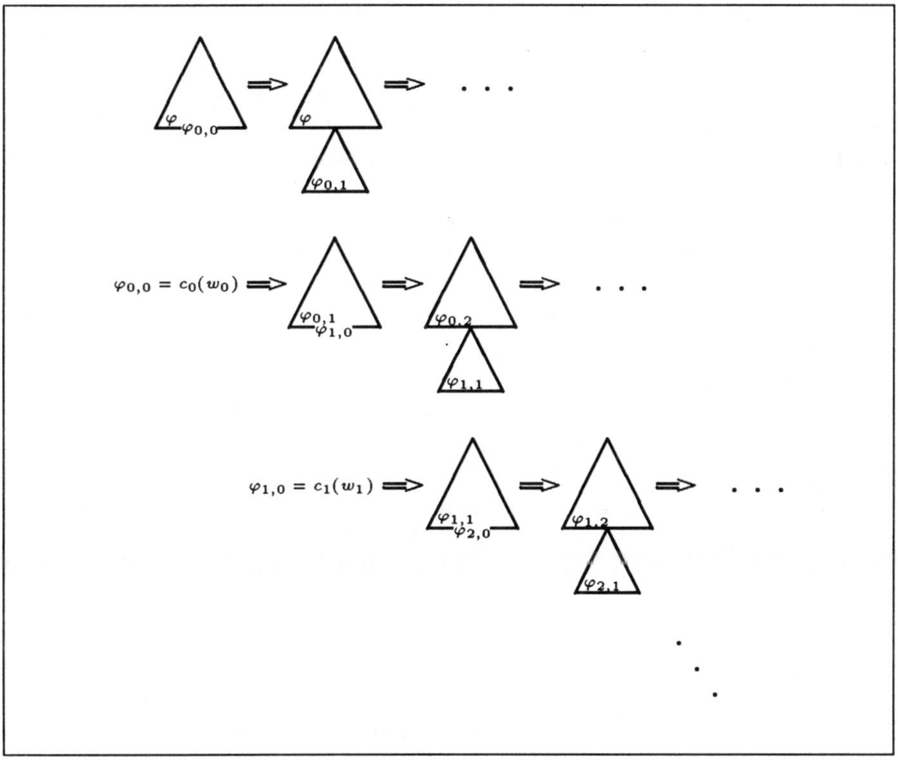

Fig. 5.9. Iterated construction of minimal ω-reducible sentential forms

This means that, for every $i, j \geq 0$ with $j > i$, there is a $(Att \cup \Delta \cup occ(s), 1)$-context $\beta_{i,j}$ such that

$$\varphi_{i,0} \Rightarrow^{+}_{A,s} \beta_{i,j}[\varphi_{j,0}]$$

where $\beta_{i,j} = \beta_i[\beta_{i+1}[\ldots[\beta_{j-1}]\ldots]]$. Since every $\varphi_{i,0}$ has the form $c_i(w_i)$ and since the set $\{c(w) \mid c \in Att, w \in occ(s)\}$ is finite, there must be indices i_0, j_0 with $j_0 > i_0$ such that $c_{i_0}(w_{i_0}) = c_{j_0}(w_{j_0})$. Hence, for $\beta = \beta_{i_0,j_0}$,

$$c_{i_0}(w_{i_0}) \Rightarrow^+_{A,s} \beta[c_{i_0}(w_{i_0})].$$

This is a contradiction to the assumption that A is noncircular. □

Lemma 5.25. For every input tree s of A, the derivation relation $\Rightarrow_{A,s}$ is confluent on $SF(A, s)$.

Proof. This follows from Lemmas 5.20, 5.24, and 2.5. □

Lemma 5.26. Let $s \in T_\Sigma$ and $\varphi \in SF(A, s)$. Then the normal form $nf(\Rightarrow_{A,s}, \varphi)$ exists.

Proof. By Lemmas 5.24 and 5.25, $\Rightarrow_{A,s}$ is terminating and confluent on $SF(A, s)$, respectively. Hence, by Corollary 2.7, the normal form $nf(\Rightarrow_{A,s}, \varphi)$ exists. □

In the following we will prove predicates concerning attributed tree transducers using a principle of simultaneous induction which is a slightly modified version of Principle 3.11. However, since the modification is very small, we will use the same name for the new principle. More precisely, we will define the notion "predicates K and L hold" to be that in Def. 3.10 except that the sets B_k now depend on an additional parameter. This parameter is a symbol σ taken from $\Sigma^{(k)}$, i.e., we consider sets $B_{k,\sigma}$ with $k \geq 0$ and $\sigma \in \Sigma^{(k)}$. Since k equals the rank of σ, it suffices to show only the parameter σ and omit the k; hence, we consider sets B_σ with $\sigma \in \Sigma^{(k)}$ for some $k \geq 0$. This additional parameter, or better: information, is motivated as follows. In the application of the principle of proof by simultaneous induction as defined in Principle 3.11, the label σ (together with some state q) determines a right-hand side ξ. Then, after having determined ξ once in some proof, it is no longer necessary to know that ξ is the right-hand side of some σ-rule. This is due to the fact that top-down tree transducers and macro tree transducers work on their input trees in a strict, recursive descent manner. In a certain sense they consume their input trees. However, for attributed tree transducers the situation is different because computations may induce an arbitrary tree walk on the input tree. To be more precise, consider the local environment of a σ-labeled node n and assume that, for some synthesized attribute a, the value of the right-hand side $\xi = rhs(a(\pi), \sigma)$ is under computation. Then, via some indirect dependencies, it may occur that the value of some inherited attribute b at some i-th descendant of n is needed. Clearly, this value is defined by some rule $b(\pi i) \to \xi'$, and this rule must be taken from the set R_σ, i.e., the same set from which the rule $a(\pi) \to \xi$ has been taken. This shows that, during the proof of properties in the environment of a σ-labeled node, it is necessary to keep the information that the node is labeled by σ.

Definition 5.27. Let Σ be a ranked alphabet. Let H be a set and, for every $a \in H$, let K_a be a predicate over T_Σ, i.e., a mapping of type $T_\Sigma \rightarrow \{true, false\}$. Moreover, for every $k \geq 0$ and $\sigma \in \Sigma^{(k)}$, let B_σ be a set and, for every $b \in B_\sigma$, let L_b be a predicate over $(T_\Sigma)^k$, i.e., a mapping of type $(T_\Sigma)^k \rightarrow \{true, false\}$. Let $K = \{K_a \mid a \in H\}$ and $L = \{L_b \mid \sigma \in \Sigma, b \in B_\sigma\}$.

1) We say that K *holds on* T_Σ, if for every $a \in H$ and $s \in T_\Sigma$, the term $K_a(s)$ has the value *true*.
2) We say that L *holds on* T_Σ, if for every $k \geq 0$, $\sigma \in \Sigma^{(k)}$, $b \in B_\sigma$, and $\omega \in (T_\Sigma)^k$, the term $L_b(\omega)$ has the value *true*. □

Principle 5.28. Let Σ, H, K_a, B_σ, and L_b be given as in Def. 5.27.
The *principle of proof by simultaneous induction* is:

If
IB: for every $\sigma \in \Sigma^{(0)}$ and $b \in B_\sigma$, $L_b(()) = true$ and
IS1: for every $k \geq 0$, $\sigma \in \Sigma^{(k)}$, $s_1, \ldots, s_k \in T_\Sigma$, if for every $b \in B_\sigma$, $L_b((s_1, \ldots, s_k)) = true$, then for every $a \in H$, $K_a(\sigma(s_1, \ldots, s_k)) = true$ and
IS2: for every $k \geq 1$ and $s_1, \ldots, s_k \in T_\Sigma$, if for every $a \in H$ and every i with $1 \leq i \leq k$, $K_a(s_i) = true$, then for every $\sigma \in \Sigma^{(k)}$, $b \in B_\sigma$, $L_b((s_1, \ldots, s_k)) = true$,
then K and L hold on T_Σ.

□

As in previous principles of simultaneous induction, the abbreviations "IB", "IS1", and "IS2" mean "induction base", "induction step 1", and "induction step 2", respectively.

In the next lemma we prove that the normal form of the sentential form $a(\varepsilon)$ with respect to some input tree s is a tree over the ranked alphabet Δ of output symbols and the set $Att_{inh}(\{\varepsilon\})$. It is clear that an occurrence $b(\varepsilon)$ with $b \in Att_{inh}$ cannot be derived further, because this would require a node above the root of the input tree.

In order to deal with the situation at a single node of a given input tree, we have to impose a linear order on the attribute occurrences. Then there is exactly one topological sorting of the set $in(\sigma)$ of inside attribute occurrences of σ. In the sorting, the condensed dependency graph $CD_A(s)$ is taken into account.

Definition 5.29. Let $s = \sigma(s_1, \ldots, s_k) \in T_\Sigma$.

The *condensed dependency graph of s for A*, denoted by $CD_A(s)$, is a directed graph over $att(\sigma)$ defined by induction on s as follows.

(i) For every $s = \alpha \in \Sigma^{(0)}$, define

$$CD_A(s) = D_A(\alpha).$$

(ii) For every $s = \sigma(s_1, \ldots, s_k) \in T_\Sigma$ with $k \geq 1$, define

$$CD_A(s) = D_A(\sigma)[is_A(s_1), \ldots, is_A(s_k)].$$

□

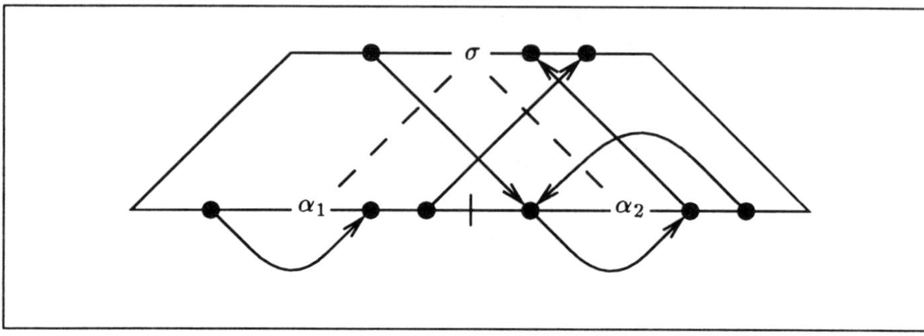

Fig. 5.10. The condensed dependency graph $D_{A_{shift}}(\sigma(\alpha_1, \alpha_2))$

Figure 5.10 shows the condensed dependency graph $D_{A_shift}(\sigma(\alpha_1, \alpha_2))$ of the input tree $\sigma(\alpha_1, \alpha_2)$ for the attributed tree transducer A_{shift} of Example 5.4.

Before studying the following definition, the reader is advised to consult Sect. 2.2 for the concept of topological sorting.

Definition 5.30. Let $s \in T_\Sigma$ and let $<$ be a linear order on *Att*.

The *topological sorting of inside attribute occurrences of $CD_A(s)$ with respect to* $<$, denoted by $topsort(A, <, s)$, is the string over $in(\sigma)$ which is obtained by applying the following two steps.

- First, topologically sort the vertices of $CD_A(s)$. If there is a choice in ordering two attribute occurrences $c(\pi\eta)$ and $c'(\pi\eta')$, then sort depth-first left-to-right, i.e., if $1 \leq \eta < \eta'$, then $c(\pi\eta)$ must occur before $c'(\pi\eta')$, and if $\eta \geq 1$ and $\eta' = 0$, then $c(\pi\eta)$ must occur before $c'(\pi\eta')$. If there is still a choice, then obey the linear order $<$. This yields a unique string w over $att(\sigma)$.
- Second, drop all the elements of $out(\sigma)$ from w. This yields $topsort(A, <, s)$.

□

Note that every inside attribute occurrence in $in(\sigma)$ occurs exactly once in $topsort(A, <, s)$. Thus $RHS(\sigma) = \{rhs(w(\nu), \sigma) \mid 1 \leq \nu \leq length(w)\}$ where $w = topsort(A, <, s)$.

For instance, if $b < a_0 < a_1$, then the topological sorting of inside attribute occurrences of $CD_{A_{shift}}(\sigma(\alpha, \alpha))$ in Fig. 5.10 is constructed as follows. After the first step we obtain the sequence $b(\pi)$, $b(\pi 1)$, $a_0(\pi 1)$, $a_1(\pi 1)$, $a_1(\pi 2)$, $b(\pi 2)$, $a_0(\pi 2)$, $a_0(\pi)$, $a_1(\pi)$. Note that, after having placed $b(\pi)$ in the sequence, the topological sorting presents a choice between the attribute occurrences $b(\pi 1)$ and $a_1(\pi 2)$. According to depth-first left-to-right, we place $b(\pi 1)$ first, followed by $a_0(\pi 1)$ and $a_1(\pi 1)$. After the second step we have: $b(\pi 1)$, $b(\pi 2)$, $a_0(\pi)$, $a_1(\pi)$.

Note 5.31. Due to the construction, $w = topsort(A, <, s)$ will have the following property. Assume that $s = \sigma(s_1, \ldots, s_k)$ for some $k \geq 0$, $\sigma \in \Sigma^{(k)}$, and $s_1, \ldots, s_k \in T_\Sigma$. Moreover, assume that $a(\pi)$ with $a \in Att_{syn}$ (resp. $b(\pi j)$ with $b \in Att_{inh}$) is an inside attribute occurrence of σ. Then every derivation $a(\varepsilon) \Rightarrow^*_{A,s} \xi$ (resp. $b(j) \Rightarrow^*_{A,s} \xi$) should be such that ξ may contain an inside attribute occurrence of σ only if that occurrence precedes $a(\pi)$ (resp. $b(\pi j)$) in w.

If $a(\pi) = w(1)$ (resp. $b(\pi j) = w(1)$), i.e., it is the first one in the linear order $topsort(A, <, s)$, then ξ cannot contain any inside attribute occurrence of σ.

Lemma 5.32. Let $a \in Att_{syn}$ and $s \in T_\Sigma$. Then $nf(\Rightarrow_{A,s}, a(\varepsilon)) \in T_\Delta(Att_{inh}(\{\varepsilon\}))$.

Proof. By simultaneous induction (using Principle 5.28), we prove that the following predicates K and L hold on T_Σ. Then, from property K the statement of the lemma follows.

K: For every $a \in Att_{syn}$ and $s \in T_\Sigma$, $nf(\Rightarrow_{A,s}, a(\varepsilon)) \in T_\Delta(Att_{inh}(\{\varepsilon\}))$.
L: For every $k \geq 0$ and $\sigma \in \Sigma^{(k)}$, $\xi \in RHS(\sigma)$, and $(s_1, \ldots, s_k) \in (T_\Sigma)^k$, $nf(\Rightarrow_{A,\sigma(s_1,\ldots,s_k)}, \xi[\pi \leftarrow \varepsilon]) \in T_\Delta(Att_{inh}(\{\varepsilon\}))$.

To be more precise, the metavariables H, K_a, B_σ, and L_b which occur in Principle 5.28, are instantiated as follows:

- $H = Att_{syn}$,
- for every $a \in Att_{syn}$, K_a is the predicate over T_Σ such that $K_a(s) = true$ if and only if $nf(\Rightarrow_{A,s}, a(\varepsilon)) \in T_\Delta(Att_{inh}(\{\varepsilon\}))$,
- $K = \{K_a \mid a \in Att_{syn}\}$

and, for every $k \geq 0$ and $\sigma \in \Sigma^{(k)}$,

- $B_\sigma = \{\sigma\} \times RHS(\sigma)$,

- for every $\xi \in RHS(\sigma)$, $L_{(\sigma,\xi)}$ is the predicate on $(T_\Sigma)^k$ such that

$$L_{(\sigma,\xi)}((s_1,\ldots,s_k)) = \mathit{true}$$

if and only if $nf(\Rightarrow_{A,\sigma(s_1,\ldots,s_k)}, \xi[\pi \leftarrow \varepsilon]) \in T_\Delta(Att_{inh}(\{\varepsilon\}))$,
- $L = \{L_{(\sigma,\xi)} \mid \sigma \in \Sigma, (\sigma,\xi) \in B_\sigma\}$.

Proof of IB: Let $\sigma \in \Sigma^{(0)}$ and $\xi \in RHS(\sigma)$. Then $\xi \in T_\Delta(Att_{inh}(\{\pi\}))$ and $nf(\Rightarrow_{A,\sigma}, \xi[\pi \leftarrow \varepsilon]) = \xi[\pi \leftarrow \varepsilon] \in T_\Delta(Att_{inh}(\{\varepsilon\}))$. Hence, $L_{(\sigma,\xi)}(()) = \mathit{true}$.

Proof of IS1: Let $k \geq 0$, $\sigma \in \Sigma^{(k)}$, and $s_1,\ldots,s_k \in T_\Sigma$. Assume that for every $\xi \in RHS(\sigma)$, $L_{(\sigma,\xi)}((s_1,\ldots,s_k)) = \mathit{true}$; clearly this is the induction hypothesis on L. Let $a \in Att_{syn}$. Since $a(\varepsilon) \Rightarrow_{A,\sigma(s_1,\ldots,s_k)} rhs(a(\pi),\sigma)[\pi \leftarrow \varepsilon]$, we obtain $nf(\Rightarrow_{A,\sigma(s_1,\ldots,s_k)}, a(\varepsilon)) = nf(\Rightarrow_{A,\sigma(s_1,\ldots,s_k)}, rhs(a(\pi),\sigma)[\pi \leftarrow \varepsilon])$. Hence, by induction hypothesis on L, $nf(\Rightarrow_{A,\sigma(s_1,\ldots,s_k)}, rhs(a(\pi),\sigma)[\pi \leftarrow \varepsilon]) \in T_\Delta(Att_{inh}(\{\varepsilon\}))$ which proves that $K_a(\sigma(s_1,\ldots,s_k)) = \mathit{true}$.

Proof of IS2: Let $k \geq 1$ and $s_1,\ldots,s_k \in T_\Sigma$. Assume that, for every $a \in Att_{syn}$ and every $1 \leq i \leq k$, $nf(\Rightarrow_{A,s_i}, a(\varepsilon)) \in T_\Delta(Att_{inh}(\{\varepsilon\}))$; clearly this is the induction hypothesis on K. Then we have to prove that, for every $\sigma \in \Sigma^{(k)}$ and $\xi \in RHS(\sigma)$, $nf(\Rightarrow_{A,\sigma(s_1,\ldots,s_k)}, \xi[\pi \leftarrow \varepsilon]) \in T_\Delta(Att_{inh}(\{\varepsilon\}))$.

Let $<$ be any linear order on Att; let w denote $topsort(A,<,\sigma(s_1,\ldots,s_k))$. Then we prove by (finite) mathematical induction on ν the following statement.

(∗) For every ν with $1 \leq \nu \leq length(w)$,
$nf(\Rightarrow_{A,\sigma(s_1,\ldots,s_k)}, rhs(w(\nu),\sigma)[\pi \leftarrow \varepsilon]) \in T_\Delta(Att_{inh}(\{\varepsilon\}))$.

Since $RHS(\sigma) = \{rhs(w(\nu),\sigma) \mid 1 \leq \nu \leq length(w)\}$, this also proves L.

Induction base of (∗) We prove the statement by structural induction on $rhs(w(1),\sigma)$ where we represent $sub(rhs(w(1),\sigma))$ by SUB.

(i) If $d(\pi j) \in SUB$ with $d \in Att_{syn}$ and $1 \leq j \leq k$, then by the assumption of IS2, i.e., by induction hypothesis on K, $nf(\Rightarrow_{A,s_j}, d(\varepsilon)) \in T_\Delta(Att_{inh}(\{\varepsilon\}))$. Moreover since $\nu = 1$, it follows that $nf(\Rightarrow_{A,s_j}, d(\varepsilon)) \in T_\Delta$, because otherwise some $b(\varepsilon) \in Att_{inh}(\{\varepsilon\})$ occurs in this normal form and hence $w(1)$ is not the first symbol of $topsort(A,<,\sigma(s_1,\ldots,s_k))$ which is a contradiction, see Note 5.31. Since $nf(\Rightarrow_{A,s_j}, d(\varepsilon)) \in T_\Delta$, we also have $nf(\Rightarrow_{A,\sigma(s_1,\ldots,s_k)}, d(\pi j)[\pi \leftarrow \varepsilon]) \in T_\Delta$ which proves the statement.

(ii) If $b(\pi) \in SUB$ with $b \in Att_{inh}$, then $nf(\Rightarrow_{A,\sigma(s_1,\ldots,s_k)}, b(\pi)[\pi \leftarrow \varepsilon]) = b(\varepsilon) \in T_\Delta(Att_{inh}(\{\varepsilon\}))$.

(iii) If $\delta(\xi_1,\ldots,\xi_l) \in SUB$, then $nf(\Rightarrow_{A,\sigma(s_1,\ldots,s_k)}, \delta(\xi_1,\ldots,\xi_l)[\pi \leftarrow \varepsilon]) = \delta(nf(\Rightarrow_{A,\sigma(s_1,\ldots,s_k)}, \xi_1[\pi \leftarrow \varepsilon]),\ldots,nf(\Rightarrow_{A,\sigma(s_1,\ldots,s_k)}, \xi_l[\pi \leftarrow \varepsilon]))$. Since, by induction hypothesis on this structural induction, every $nf(\Rightarrow_{A,\sigma(s_1,\ldots,s_k)}$

$, \xi_j[\pi \leftarrow \varepsilon]) \in T_\Delta(Att_{inh}(\{\varepsilon\}))$, where $1 \leq j \leq k$, we obtain $nf(\Rightarrow_{A,\sigma(s_1,\ldots,s_k)}$ $, \delta(\xi_1, \ldots, \xi_l)[\pi \leftarrow \varepsilon]) \in T_\Delta(Att_{inh}(\{\varepsilon\}))$.

Induction step of $(*)$ Assume that $(*)$ holds for every κ with $1 \leq \kappa \leq \nu$. We prove the statement by structural induction on $rhs(w(\nu+1), \sigma)$ where we represent $sub(rhs(w(\nu), \sigma))$ by SUB.

(i) If $d(\pi j) \in SUB$ with $d \in Att_{syn}$ and $1 \leq j \leq k$, then by the assumption of IS2, i.e., by induction hypothesis on K, $nf(\Rightarrow_{A,s_j}, d(\varepsilon)) \in T_\Delta(Att_{inh}(\{\varepsilon\}))$. Hence there is an $m \geq 0$ and a (Δ, m)-context β such that $nf(\Rightarrow_{A,s_j}$ $, d(\varepsilon)) = \beta[b_1(\varepsilon), \ldots, b_m(\varepsilon)]$ for some $b_1, \ldots, b_m \in Att_{inh}$. Since $(*)$ is proved by mathematical induction on the positions in $topsort(A, <, \sigma(s_1, \ldots, s_k))$, for every μ with $1 \leq \mu \leq m$, there is an index κ_μ with $1 \leq \kappa_\mu \leq \nu$ such that $w(\kappa_\mu) = b_\mu(\pi j)$. Hence, by induction hypothesis on $(*)$, we know that $nf(\Rightarrow_{A,\sigma(s_1,\ldots,s_k)}, rhs(b_\mu(\pi j), \sigma)[\pi \leftarrow \varepsilon]) \in T_\Delta(Att_{inh}(\{\varepsilon\}))$. Then the following statements are true, where we abbreviate $\Rightarrow_{A,\sigma(s_1,\ldots,s_k)}$ to $\Rightarrow$:

$$
\begin{aligned}
& nf(\Rightarrow, d(\pi j)[\pi \leftarrow \varepsilon]) = \\
&= nf(\Rightarrow, nf(\Rightarrow_{s_j}, d(\varepsilon))[b(\varepsilon) \leftarrow b(j)]) \\
&= nf(\Rightarrow, \beta[b_1(\varepsilon), \ldots, b_m(\varepsilon)][b(\varepsilon) \leftarrow b(j)]) \\
&= nf(\Rightarrow, \beta[b_1(j), \ldots, b_m(j)]) \\
&= \beta[nf(\Rightarrow, b_1(\pi j)), \ldots, nf(\Rightarrow, b_m(\pi j))] \\
&= \beta[nf(\Rightarrow, rhs(b_1(\pi j), \sigma)[\pi \leftarrow \varepsilon]), \ldots, nf(\Rightarrow, rhs(b_m(\pi j), \sigma)[\pi \leftarrow \varepsilon])] \\
&\in T_\Delta(Att_{inh}(\{\varepsilon\})).
\end{aligned}
$$

(ii) and (iii) are proved in the same way as in the induction base.

This finishes the proof of statement $(*)$ and of IS2. Since IB, IS1, and IS2 are proved, this shows that K and L hold on T_Σ. □

We can now define the tree transformation induced by A.

Definition 5.33. Let E be the environment of A.

(a) Let a be a synthesized attribute. The *tree transformation induced by A with a* is the mapping $\tau_{A,a}^{der} : T_\Sigma \to T_\Delta(Att_{inh}(\{\varepsilon\}))$ which is defined, for every $s \in T_\Sigma$, by $\tau_{A,a}^{der}(s) = nf(\Rightarrow_{A,s}, a(\varepsilon))$.

(b) The *tree transformation induced by A* is the mapping $\tau_A^{der} : T_\Sigma \to T_\Delta$ such that, for every $s \in T_\Sigma$, $\tau_A^{der}(s) = \tilde{E}(\tau_{A,a_0}^{der}(s))$ where $\tilde{E} : T_\Delta(Att_{inh}(\{\varepsilon\})) \to T_\Delta$ is defined by structural induction as follows.
   (i) For every $b \in Att_{inh}$, $\tilde{E}(b(\varepsilon)) = E(b)$.
   (ii) For every $\delta \in \Delta^{(l)}$ with $l \geq 0$ and $\varphi_1, \ldots, \varphi_l \in T_\Delta(Att_{inh}(\{\varepsilon\}))$, $\tilde{E}(\delta(\varphi_1, \ldots, \varphi_l)) = \delta(\tilde{E}(\varphi_1), \ldots, \tilde{E}(\varphi_l))$. □

We call a tree transformation τ an *attributed tree transformation*, if τ can be induced by some attributed tree transducer. The class of attributed tree transformations is denoted by ATT.

Example 5.34. Consider again the attributed tree transducer A_{shift} in Example 5.4. It can be proved by induction on n that, for every sequence $\alpha_{i_1}, \ldots, \alpha_{i_n}$, where $1 \leq i_1, \ldots, i_n \leq 2$
$\tau^{der}_{A_{shift}}(\#(\sigma(\alpha_{i_1}, \sigma(\alpha_{i_2}, \ldots \sigma(\alpha_{i_{n-1}}, \alpha_{i_n}) \ldots))) =$
$\#(\sigma(\sigma(\ldots \sigma(\alpha_{i_1}, \alpha_{i_2}) \ldots \alpha_{i_{n-1}}), \alpha_{i_n}))$.
Roughly speaking, $\tau^{der}_{A_{shift}}$ takes "right-growing" combs of Σ trees into "left-growing" combs in such a way that it does not change the order of the leaves. □

5.3 Characterization of Attributed Tree Transformations

As in Sects. 3.3 and 4.3 for top-down tree transducers and macro tree transducers, respectively, we provide here an inductive characterization of attributed tree transformations. We also have to modify slightly the principle of definition by simultaneous induction (Principle 3.20) so that it is suited to the context of attributed tree transducers (see the discussion after Lemma 5.26). However, as for the principle of proof by simultaneous induction, we use the same nomenclature as in Principle 3.20.

Definition 5.35. Let Σ be a ranked alphabet, let H and C be sets, and for every $\sigma \in \Sigma$, let B_σ be a set. Moreover, for every $a \in H$, let $f_a : T_\Sigma \to C$ be a function and, for every $k \geq 0$, $\sigma \in \Sigma^{(k)}$, and $b \in B_\sigma$, let $g_b : (T_\Sigma)^k \to C$ be a function.

Let
$$F = \{f_a : T_\Sigma \to C \,|\, a \in H\}$$
and let
$$G = \{g_b : (T_\Sigma)^k \to C \,|\, k \geq 0, \sigma \in \Sigma^{(k)}, b \in B_\sigma\}.$$
We say that F *is defined* if, for every $a \in H$ and $s \in T_\Sigma$, $f_a(s)$ is defined. Similarly, we say that G is defined if, for every $k \geq 0$, $\sigma \in \Sigma^{(k)}$, $b \in B_\sigma$, and $s_1, \ldots, s_k \in T_\Sigma$, $g_b((s_1, \ldots, s_k))$ is defined. □

Principle 5.36. Let F and G be two families of functions as in Def. 5.35. The *principle of definition by simultaneous induction* is:

If
IB: for every $\sigma \in \Sigma^{(0)}$ and $b \in B_\sigma$, the value $g_b(())$ is defined, and
IS1: for every $k \geq 0$, $\sigma \in \Sigma^{(k)}$, $s_1, \ldots, s_k \in T_\Sigma$, and $a \in H$, if for every $b \in B_\sigma$, the value $g_b((s_1, \ldots, s_k))$ is defined, then the value $f_a(\sigma(s_1, \ldots, s_k))$ is defined, and

IS2: for every $k \geq 1$, $\sigma \in \Sigma^{(k)}$, $s_1, \ldots, s_k \in T_\Sigma$, and $b \in B_\sigma$, if for every $a \in H$ and $1 \leq i \leq k$, the value $f_a(s_i)$ is defined, then the value $g_b((s_1, \ldots, s_k))$ is defined,

then F and G are defined.

□

We use this principle of simultaneous induction to define sets of mappings which characterize the derivation relation of attributed tree transducers. Recall that A denotes an arbitrary, but fixed, attributed tree transducer $(Att, \Sigma, \Delta, a_0, R, E)$.

Definition 5.37. Let $<$ be a linear order on Att. We define the set

$$F_{<,A} = \{\tau^{ind}_{<,A,a} : T_\Sigma \rightarrow T_\Delta(Att_{inh}(\{\varepsilon\})) \mid a \in Att_{syn}\}$$

of functions and the set

$$G_{<,A} = \{\tau^{ind}_{<,A,\sigma,\xi} : (T_\Sigma)^k \rightarrow T_\Delta(Att_{inh}(\{\varepsilon\})) \mid k \geq 0, \sigma \in \Sigma^{(k)}, \xi \in RHS(\sigma)\}$$

of functions by simultaneous induction according to Principle 5.36 where we have instantiated the sets as follows:

- $H = Att_{syn}$
- $C = T_\Delta(Att_{inh}(\{\varepsilon\}))$
- $B_\sigma = \{\sigma\} \times RHS(\sigma)$ for every $\sigma \in \Sigma$.

IB: For every $\sigma \in \Sigma^{(0)}$ and $\xi \in RHS(\sigma)$, $\tau^{ind}_{<,A,\sigma,\xi}(()) = \xi$.

IS1: For every $k \geq 0$, $\sigma \in \Sigma^{(k)}$, and $s_1, \ldots, s_k \in T_\Sigma$, and $a \in Att_{syn}$, we define

$$\tau^{ind}_{<,A,a}(\sigma(s_1, \ldots, s_k)) = \tau^{ind}_{<,A,\sigma,rhs(a(\pi),\sigma)}((s_1, \ldots, s_k)).$$

IS2: Let $k \geq 0$, $\sigma \in \Sigma^{(k)}$, and $s_1, \ldots, s_k \in T_\Sigma$. Let $w = topsort(A, <, s)$ with $s = \sigma(s_1, \ldots, s_k)$. By (finite) mathematical induction on ν with $1 \leq \nu \leq length(w)$, we define $\tau^{ind}_{<,A,\sigma,rhs(w(\nu),\sigma)}(\omega)$, where $\omega = (s_1, \ldots, s_k)$.

Induction base $\nu = 1$: The value $\tau^{ind}_{<,A,\sigma,rhs(w(1),\sigma)}(\omega)$ is defined by structural induction on $rhs(w(1), \sigma)$. Let SUB represent $sub(rhs(w(1), \sigma))$.

(i) Let $d(\pi i) \in SUB$ for some $d \in Att_{syn}$ and $1 \leq i \leq k$. Since $\nu = 1$, the value $\tau^{ind}_{<,A,d}(s_i) \in T_\Delta$; in particular, this value does not contain inherited attributes. Thus we can define $\tau^{ind}_{<,A,\sigma,d(\pi i)}(\omega) = \tau^{ind}_{<,A,d}(s_i)$.

(ii) Let $b(\pi) \in SUB$ for some $b \in Att_{inh}$. Then define $\tau^{ind}_{<,A,\sigma,b(\pi)}(\omega) = b(\varepsilon)$.

(iii) Let $\delta(\xi_1, \ldots, \xi_l) \in SUB$ for some $\delta \in \Delta^{(l)}$, $l \geq 0$, $\xi_1, \ldots, \xi_l \in SUB$. Then define $\tau^{ind}_{<,A,\sigma,\delta(\xi_1,\ldots,\xi_l)}(\omega) = \delta(\tau^{ind}_{<,A,\sigma,\xi_1}(\omega), \ldots, \tau^{ind}_{<,A,\sigma,\xi_l}(\omega))$.

Induction step $\nu \to \nu+1$: Assume that for every κ with $1 \leq \kappa \leq \nu$, the value $\tau^{ind}_{<,A,\sigma,rhs(w(\kappa),\sigma)}(\omega)$ is already defined. Then we define the value $\tau^{ind}_{<,A,rhs(w(\nu+1),\sigma)}(\omega)$ by structural induction on $rhs(w(\nu+1),\sigma)$. Let SUB represent $sub(rhs(w(\nu+1),\sigma))$.

(i) Let $d(\pi i) \in SUB$ for some $d \in Att_{syn}$ and $1 \leq i \leq k$. By the application of IS1, the value $\tau^{ind}_{<,A,d}(s_i)$ is already defined. Then there are an $m \geq 0$, a (Δ, m)-context β and $b_1, \ldots, b_m \in Att_{inh}$ such that $\tau^{ind}_{<,A,d}(s_i) = \beta[b_1(\varepsilon), \ldots, b_m(\varepsilon)]$. According to Note 5.31, for every μ with $1 \leq \mu \leq m$ there is an index κ_μ with $1 \leq \kappa_\mu \leq \nu$ such that $w(\kappa_\mu) = b_\mu(\pi i)$. Hence, by induction hypothesis, $\tau^{ind}_{<,A,\sigma,rhs(b_\mu(\pi i),\sigma)}(\omega)$ is already defined. Thus we define $\tau^{ind}_{<,A,\sigma,d(\pi i)}(\omega) = \tau^{ind}_{<,A,d}(s_i)[b_\mu(\varepsilon) \leftarrow \tau^{ind}_{<,A,\sigma,rhs(w(\kappa_\mu),\sigma)}; 1 \leq \mu \leq m]$.

(ii) and (iii) are the same as in the proof of the induction base. □

Note that IB is part of IS2 for $k = 0$.

Intuitively, it is clear that the functions $\tau^{ind}_{<,A,a}$ and $\tau^{ind}_{<,A,\sigma,\xi}$ do not depend on the order $<$. Instead, $<$ only influences the way in which the functions are defined. More precisely, if for some input symbol σ, there are two attribute occurrences $c(\pi\eta)$ and $c'(\pi\eta)$ which do not depend on the choice of the order $<_1$ or $<_2$ with $c <_1 c'$ and $c' <_2 c$, respectively. This is stated explicitly in the following lemma.

Lemma 5.38. Let $<_1$ and $<_2$ be two linear orders on the set of attributes of A. Then

- for every $a \in Att_{syn}$, $\tau^{ind}_{<_1,A,a} = \tau^{ind}_{<_2,A,a}$
- for every $k \geq 0$, $\sigma \in \Sigma^{(k)}$, $\xi \in RHS(\sigma)$, $\tau^{ind}_{<_1,A,\sigma,\xi} = \tau^{ind}_{<_2,A,\sigma,\xi}$. □

Thus, in the following, we will drop the linear order on the set of attributes from the denotation of the τ^{ind}-functions. Moreover, we show that the tree transformation τ^{der}_A induced by an attributed tree transducer A coincides with the inductively defined function τ^{ind}_A. To prove this equivalence, we introduce, for every $\sigma \in \Sigma^{(k)}$ with $k \geq 0$ and $\xi \in RHS(\sigma)$, the function $\tau^{der}_{A,\sigma,\xi} : (T_\Sigma)^k \to T_\Delta(Att_{inh}(\{\varepsilon\})$ such that $\tau^{der}_{A,\sigma,\xi}((s_1, \ldots, s_k)) = nf(\Rightarrow_{A,\sigma(s_1,\ldots,s_k)}, \xi[\pi \leftarrow \varepsilon])$.

Theorem 5.39. K: For every $a \in Att_{syn}$ and $s \in T_\Sigma$, $\tau^{der}_{A,a}(s) = \tau^{ind}_{A,a}(s)$.
L: For every $k \geq 0$, $\sigma \in \Sigma^{(k)}$, $\xi \in RHS(\sigma)$, and $\omega = (s_1, \ldots, s_k) \in (T_\Sigma)^k$, $\tau^{der}_{A,\sigma,\xi}(\omega) = \tau^{ind}_{A,\sigma,\xi}(\omega)$.

Proof. The proof of these statements follows the principle of simultaneous induction in Principle 5.28. Actually, the structure of this proof is the same as the structure of the proof of Lemma 5.32 and thus we omit it here. □

As in the case of top-down tree transducers, we will now drop the superscripts *ind* and *der* from the denotations of the τ^{der}- and the τ^{ind}-functions.

5.4 Height and Subtree Properties

In this section we estimate both the height of an output tree and the number of subtrees of an output tree with respect to the size of the corresponding input tree.

Lemma 5.40. There is a constant $c > 0$ such that, for every $s \in T_\Sigma$, $height(\tau_A(s)) \leq c \cdot size(s)$.

Proof. Consider an input tree $s \in T_\Sigma$ and the set $Att(s)$ of attribute instances of s, i.e., $Att(s) = \{c(w) \mid c \in Att, w \in occ(s)\}$. Since A is noncircular, the set $Att(s)$ can be sorted topologically such that, if $c(w)$ depends (directly or indirectly) on $c'(w')$, then $c'(w')$ occurs before $c(w)$ in this linear order. Let $sort = c_1(w_1), c_2(w_2), \ldots, c_n(w_n)$ be such a linear order. Note that $n = card(Att) \cdot size(s)$, where $card(Att)$ is the number of attributes.

Now it is easy to observe that the highest output tree is achieved if the right-hand sides of the defining rules of the attribute instances in *sort* are put on top of each other.

Then we can prove the following statement by mathematical induction.

For every i with $1 \leq i \leq n$, $height(nf(\Rightarrow_{A,s}, c_i(w_i))) \leq i \cdot max(height(RHS))$,

where $max(height(RHS))$ denotes the number $max\{height(\xi) \mid \xi \in RHS(\sigma), \sigma \in \Sigma\}$.

$i = 1$: Since $c_1(w_1)$ does not depend on any other attribute, the statement follows immediately.

$i \to i+1$: We have to distinguish two cases: either c_{i+1} is a synthesized attribute or c_{i+1} is an inherited attribute.

$c_{i+1} \in Att_{syn}$: Let $label(s, w_{i+1}) = \sigma$ and let $\xi = rhs(c_{i+1}(\pi), \sigma)$. Then there is an $m \geq 0$ and a (Δ, m)-context β, and there are attribute occurrences $d_1(\pi\eta_1), \ldots, d_m(\pi\eta_m) \in att(\sigma)$ such that $\xi = \beta[d_1(\pi\eta_1), \ldots, d_m(\pi\eta_m)]$. By the definition of *sort*, every $d_j(w_{i+1}\eta_j)$ occurs before $c_{i+1}(w_{i+1})$ and hence, we can assume that the statement is true for these occurrences. Then we can compute as follows:

$$\begin{aligned}
& height(nf(\Rightarrow_{A,s}, c_{i+1}(w_{i+1}))) \\
= \; & height(nf(\Rightarrow_{A,s}, \beta[d_1(w_{i+1}\eta_1), \ldots, d_m(w_{i+1}\eta_m)])) \\
\leq \; & height(\beta) + max\{height(nf(\Rightarrow_{A,s}, d_j(w_{i+1}\eta_j))) \mid 1 \leq j \leq m\} \\
\leq \; & max(height(RHS)) + max\{i \cdot max(height(RHS)) \mid 1 \leq j \leq m\} \\
= \; & (i+1) \cdot max(height(RHS)).
\end{aligned}$$

$c_{i+1} \in Att_{inh}$: If $w_{i+1} = \varepsilon$, then the statement is true trivially. Now let $w_{i+1} = vl$ for some $v \in occ(s)$ and $l \geq 1$. Let $label(s, v) = \sigma$ and let $\xi = rhs(c_{i+1}(\pi l), \sigma)$. Then there is an $m \geq 0$ and a (Δ, m)-context β, and there are attribute occurrences $d_1(\pi\eta_1), \ldots, d_m(\pi\eta_m) \in att(\sigma)$ such that $\xi = \beta[d_1(\pi\eta_1), \ldots, d_m(\pi\eta_m)]$. By the definition of *sort*, every $d_j(v\eta_j)$ occurs before $c_{i+1}(w_{i+1})$ and hence, we can assume that the statement is true for these occurrences. Then we can compute as follows:

$$\begin{aligned}
& height(nf(\Rightarrow_{A,s}, c_{i+1}(w_{i+1}))) \\
= \;& height(nf(\Rightarrow_{A,s}, \beta[d_1(v\eta_1), \ldots, d_m(v\eta_m)])) \\
\leq \;& height(\beta) + max\{height(nf(\Rightarrow_{A,s}, d_j(v\eta_j))) \mid 1 \leq j \leq m\} \\
\leq \;& max(height(RHS)) + max\{i \cdot max(height(RHS)) \mid 1 \leq j \leq m\} \\
= \;& (i+1) \cdot max(height(RHS)).
\end{aligned}$$

This proves the statement. By choosing $i = n$ and realizing that $n = card(Att) \cdot size(s)$, we obtain the following:

$$\begin{aligned}
& height(\tau_A(s)) \\
= \;& height(nf(\Rightarrow_{A,s}, a_0(\varepsilon))[b(\varepsilon) \leftarrow E(b); b \in Att_{inh}]) \\
\leq \;& height(nf(\Rightarrow_{A,s}, a_0(\varepsilon))) + max\{height(E(b)) \mid b \in Att_{inh}\} \\
\leq \;& card(Att) \cdot size(s) \cdot max(height(RHS)) + \\
& max\{height(E(b)) \mid b \in Att_{inh}\} \\
\leq \;& c \cdot size(s)
\end{aligned}$$

where $c = card(Att) \cdot max(height(RHS)) + max\{height(E(b)) \mid b \in Att_{inh}$. This proves the statement of the lemma. □

Corollary 5.41. There is a constant $d > 0$ such that, for every $s \in T_\Sigma$, $size(\tau_A(s)) \leq 2^{d \cdot size(s)}$.

Proof. By Lemma 2.13, there is a constant $e > 0$ such that $size(t) \leq e^{height(t)}$ for every $t \in T_\Delta$. Then, by Lemma 5.40, we obtain $size(\tau_A(s)) \leq e^{height(\tau_A(s))} \leq e^{c \cdot size(s)}$ where c is the constant of Lemma 5.40. Then it is easy to see that there is a constant d such that $size(\tau_A(s)) \leq 2^{d \cdot size(s)}$. □

Lemma 5.42. For every $\tau \in ATT^k$ with $k \geq 1$ there is a constant $c > 0$ such that, for every $s \in T_\Sigma$, the approximation $size(\tau(s)) \leq exp(k, c \cdot size(s))$ holds where, for every $k, n \geq 0$, the number $exp(k, n)$ is defined by induction on k such that $exp(0, n) = n$ and $exp(k+1, n) = 2^{exp(k,n)}$.

Proof. The proof is completely analogous to the proof of Lemma 4.24. □

In a similar way we can estimate the number of different subtrees of an output tree.

Lemma 5.43. There is a constant $c > 0$ such that for every $s \in T_\Sigma$, $card(sub(\tau_A(s))) \leq c \cdot size(s)$.

Proof. Consider an input tree $s \in T_\Sigma$ and again the set $Att(s)$ of attribute instances of s, i.e., $Att(s) = \{c(w) \mid c \in Att, w \in occ(s)\}$. Let $sort = c_1(w_1), c_2(w_2), \ldots, c_n(w_n)$ be the linear order of the attribute instances as in the proof of Lemma 5.40. Again let $n = card(Att) \cdot size(s)$, where $card(Att)$ is the number of attributes.

Now we prove the following statement by mathematical induction on i where $max(size(RHS))$ denotes the number $max\{size(\xi) \mid \xi \in RHS(\sigma), \sigma \in \Sigma\}$. Moreover, we represent $nf(\Rightarrow_{A,s}, \varphi)$ by $nf(\varphi)$.

For every i with $1 \leq i \leq n$, $card(\bigcup_{1 \leq j \leq i} sub(nf(c_j(w_j)))) \leq i \cdot max(size(RHS))$.

$\underline{i = 1}$: The statement follows immediately.

$\underline{i \to i+1}$: Again we have to distinguish two cases: either c_{i+1} is a synthesized attribute or c_{i+1} is an inherited attribute.

$c_{i+1} \in Att_{syn}$: Let $label(s, w_{i+1}) = \sigma$ and let $\xi = rhs(c_{i+1}(\pi), \sigma)$. Then there is an $m \geq 0$ and a (Δ, m)-context β, and there are attribute occurrences $d_1(\pi\eta_1), \ldots, d_m(\pi\eta_m) \in att(\sigma)$ such that $\xi = \beta[d_1(\pi\eta_1), \ldots, d_m(\pi\eta_m)]$. We note that, due to the definition of $sort$, every $d_j(w_{i+1}\eta_j)$ occurs before $c_{i+1}(w_{i+1})$. Then we can compute as follows:

$$\begin{aligned}
& card(\textstyle\bigcup_{1 \leq j \leq i+1} sub(nf(c_j(w_j)))) \\
= \;& card(sub(nf(c_{i+1}(w_{i+1}))) \cup \textstyle\bigcup_{1 \leq j \leq i} sub(nf(c_j(w_j)))) \\
= \;& card(sub(\varphi) \cup \textstyle\bigcup_{1 \leq j \leq i} sub(nf(c_j(w_j)))) \\
& \text{where } \varphi = \beta[nf(d_1(w_{i+1}\eta_1)), \ldots, nf(d_m(w_{i+1}\eta_m))] \\
= \;& card((sub(\varphi) - M) \cup M \cup \textstyle\bigcup_{1 \leq j \leq i} sub(nf(c_j(w_j)))) \\
& \text{where } M = \textstyle\bigcup_{1 \leq l \leq m} sub(nf(d_l(w_{i+1}\eta_l))), \text{ since } M \subseteq sub(\varphi) \\
= \;& card((sub(\varphi) - M) \cup \textstyle\bigcup_{1 \leq j \leq i} sub(nf(c_j(w_j)))) \\
& \text{since every } d_l(w_{i+1}\eta_l) \text{ occurs before } c_{i+1}(w_{i+1}), \\
& M \subseteq \textstyle\bigcup_{1 \leq j \leq i} sub(nf(c_j(w_j))) \\
\leq \;& card(sub(\varphi) - M) + card(\textstyle\bigcup_{1 \leq j \leq i} sub(nf(c_j(w_j)))) \\
\leq \;& card(sub(\beta)) + card(\textstyle\bigcup_{1 \leq j \leq i} sub(nf(c_j(w_j)))) \\
\leq \;& max(size(RHS)) + i \cdot max(size(RHS)) \\
= \;& (i+1) \cdot max(size(RHS)).
\end{aligned}$$

$c_{i+1} \in Att_{inh}$: This case is proved in a similar way to the first case.

This proves the statement. By taking $i = n$ and again realizing that $n = card(Att) \cdot size(s)$, we obtain

$$\begin{array}{ll} & card(sub(\tau_A(s))) \\ = & card(sub(nf(a_0(\varepsilon))[b(\varepsilon) \leftarrow E(b); b \in Att_{inh}])) \\ \leq & card(\bigcup_{1\leq j\leq n} sub(nf(c_j(w_j))) + \sum_{b\in Att_{inh}} size(E(b)) \\ \leq & card(Att) \cdot size(s) \cdot max(size(RHS)) + \sum_{b\in Att_{inh}} size(E(b)) \\ \leq & c \cdot size(s) \end{array}$$

where $c = card(Att) \cdot max(size(RHS)) + \sum_{b\in Att_{inh}} size(E(b))$. This proves the statement of the lemma. □

An immediate implication of the lemma above is that attributed tree transducers are more powerful than top-down tree transducers.

Theorem 5.44. $TOP \subset ATT$.

Proof. From Note 5.3 it is clear that $TOP \subseteq ATT$. Now consider the following attributed tree transducer $A = (Att, \Sigma, \Delta, a_0, R, E)$ with

- $Att = Att_{syn} \cup Att_{inh}$ with $Att_{syn} = \{a_0\}$ and $Att_{inh} = \{b\}$
- $\Sigma = \Delta = \{\sigma^{(2)}, \alpha^{(0)}\}$
- $R = R_\sigma \cup R_\alpha$ where
  R_σ contains the rules
  1) $a_0(\pi) \to \sigma(a_0(\pi 2), a_0(\pi 2))$
  2) $b(\pi 1) \to b(\pi)$
  3) $b(\pi 2) \to a_0(\pi 1)$,
  and R_α contains the rule
  4) $a_0(\pi) \to \sigma(b(\pi), b(\pi))$,
- $E(b) = \alpha$.

It is easy to observe that

$$\tau_A \supseteq \{(s_n, s_{2^n}) \mid n \geq 1\}$$

where s_n is the fully balanced tree over Σ defined in Example 3.29.

Now assume that $\tau_A \in TOP$. Then, by Lemma 3.27, the height of an output tree of τ_A is linearly bounded by the height of the corresponding input tree. However, this contradicts our observation and hence, $\tau_A \notin TOP$. □

5.5 Composition and Decomposition Results

As in the case of macro tree transducers, attributed tree transducers also form a strict composition hierarchy.

Theorem 5.45. $ATT^k \subset ATT^{k+1}$ for every $k \geq 1$.

Proof. By composing $(k+1)$-times the attributed tree transducer of the proof of Theorem 5.44, we obtain a tree transformation in ATT^{k+1} which cannot be in ATT^k, by Lemma 5.42. □

Finally we want to show that ATT is closed under right composition with TOP, i.e., $ATT \circ TOP \subseteq ATT$. The result generalizes Theorem 3.39 in which $TOP \circ TOP \subseteq TOP$ has been proved. The idea behind the construction involved in Theorem 3.39 can be used again: it is a common product construction.

Lemma 5.46. $ATT \circ TOP \subseteq ATT$.

Proof. Let $A = (Att_A, \Sigma, \Delta, a_0, R_A, E_A)$ be a noncircular attributed tree transducer and let $T = (Q, \Delta, \Omega, q, R_T)$ be a top-down tree transducer. We construct a noncircular attributed tree transducer $B = (Att_B, \Sigma, \Omega, (q, a_0), R_B, E_B)$ such that $\tau_A \circ \tau_T = \tau_B$.

First we modify T to be the top-down tree transducer T' which can transform the right-hand sides of rules in R_A. For this purpose, let $mx = max\{k \mid \Sigma^{(k)} \neq \emptyset\}$ and define the two ranked alphabets

$$Att_A(\pi, mx) = \{a(\pi i) \mid a \in (Att_A)_{syn}, 1 \leq i \leq mx\} \cup \{b(\pi) \mid b \in (Att_A)_{inh}\}$$

and

$$Q \times Att_A(\pi, mx) = \{(q, a)(\pi i) \mid q \in Q, a \in (Att_A)_{syn}, 1 \leq i \leq mx\}$$
$$\cup \{(q, b)(\pi) \mid q \in Q, b \in (Att_A)_{inh}\}$$

where every symbol has rank 0.

Now construct the top-down tree transducer $T' = (Q, \Delta \cup Att_A(\pi, mx), \Omega \cup Q \times Att_A(\pi, mx), q, R'_T)$ by defining

$$R'_T = R_T \cup \{q(a(\pi i)) \to (q, a)(\pi i) \mid q \in Q, a(\pi i) \in Att_A(\pi, mx)\}$$
$$\cup \{q(b(\pi)) \to (q, b)(\pi) \mid q \in Q, b(\pi) \in Att_A(\pi, mx)\}.$$

Note that "$a(\pi i) \in Att_A(\pi, mx)$" is shorthand for "$w \in Att_A(\pi, mx)$ and w has the form $a(\pi i)$" and similarly for "$b(\pi) \in Att_A(\pi, mx)$".

Next we construct $B = (Att_B, \Sigma, \Omega, (q, a_0), R_B, E_B)$ as follows:

- $Att_B = (Att_B)_{syn} \cup (Att_B)_{inh}$ where $(Att_B)_{syn} = Q \times (Att_A)_{syn}$ and $(Att_B)_{inh} = Q \times (Att_A)_{inh}$.

- R_B is the smallest set satisfying the following conditions. For every $\sigma \in \Sigma$, for every $q \in Q$, and
  - for every rule $a(\pi) \to \xi$ in $(R_A)_\sigma$, the rule $(q,a)(\pi) \to \tau_{T',q}(\xi)$ is in $(R_B)_\sigma$, and
  - for every rule $b(\pi i) \to \xi$ in $(R_A)_\sigma$, the rule $(q,b)(\pi i) \to \tau_{T',q}(\xi)$ is in $(R_B)_\sigma$.
- E_B is the function $E_B : (Att_B)_{inh} \to T_\Omega$ such that, for every $(q,b) \in (Att_B)_{inh}$, $E_B((q,b)) = \tau_{T,q}(E_A(b))$.

The following intermediate statement can be proved by structural induction on ξ.

For every $q \in Q$, $\xi \in RHS((Att_A)_{syn}, (Att_A)_{inh}, \Delta, k)$ with $0 \leq k \leq mx$, and set $\{t_{a,j} \in T_\Delta \mid a \in (Att_A)_{syn}, 1 \leq j \leq k\} \cup \{t_b \mid b \in (Att_A)_{inh}\}$:

$$\begin{aligned}\tau_{T,q}(\xi[a(\pi j) \leftarrow t_{a,j}; a(\pi j) \in Att_A(\pi, mx)][b(\pi) \leftarrow t_b; b(\pi) \in Att_A(\pi, mx)]) = \\ \tau_{T',q}(\xi)[(p,a)(\pi j) \leftarrow \tau_{T,p}(t_{a,j}); (p,a)(\pi j) \in Q \times Att_A(\pi, mx)] \\ [(p,b)(\pi) \leftarrow \tau_{T,p}(t_b); (p,b)(\pi) \in Q \times Att_A(\pi, mx)].\end{aligned}$$

Using this intermediate statement, the fact that the predicates K and L hold on T_Σ, can be proved by the principle of simultaneous induction.

K: For every $s \in T_\Sigma$, $a \in (Att_A)_{syn}$, $q \in Q$, and $E : (Att_A)_{inh} \to T_\Delta$, the equation

$$\begin{aligned}\tau_{T,q}(\tau_{A,a}(s)[b(\varepsilon) \leftarrow E(b); b \in (Att_A)_{inh}]) = \\ \tau_{B,(q,a)}(s)[(p,b)(\varepsilon) \leftarrow \tau_{T,p}(E(b)); p \in Q, b \in (Att_A)_{inh}]\end{aligned}$$

holds.

L: For every $k \geq 0$, $\sigma \in \Sigma^{(k)}$, $\xi \in RHS(\sigma)$, $\omega \in (T_\Sigma)^k$, $q \in Q$, and $E : (Att_A)_{inh} \to T_\Delta$, the equation

$$\begin{aligned}\tau_{T,q}(\tau_{A,\sigma,\xi}(\omega)[b(\varepsilon) \leftarrow E(b); b \in (Att_A)_{inh}]) = \\ \tau_{B,\sigma,\xi'}(\omega)[(p,b)(\varepsilon) \leftarrow \tau_{T,p}(E(b)); p \in Q, b \in (Att_A)_{inh}]\end{aligned}$$

holds, where $\xi' = \tau_{T',q}(\xi)$ □

Theorem 5.47. $ATT \circ TOP = ATT$.

Proof. The statement follows immediately from $ATT \circ TOP \subseteq ATT$ (Lemma 5.46), the fact that TOP contains the identical tree transformations, and Lemma 2.14. □

5.6 Bibliographic Notes

Attributed tree transducers were introduced in [Fül81] (see also [Bar81, Bar83]). The composition result $ATT \circ TOP = ATT$ (Theorem 5.47) is due to this paper. The proofs of Lemmas 5.40 and 5.43 are due to [Küh95a].

We have defined the concepts of a dependency graph, an extended dependency graph, and an is-graph for attributed tree transducers in a very similar way to the corresponding concepts for attribute grammars (see, e.g., [Cou84, Eng84]).The circularity test is also similar to the circularity test for attribute grammars [ASU86] (see also [KV94b]). The result concerning the intrinsic exponential circularity test is due to [JOR75], see also [Jaz81]. The strictness of the composition hierarchy of attributed tree transducers is due to [Fül81], see also [Eng81].

In [Eng84] attributed tree transducers are used as a formal model for the computation of attribute values in an attribute grammar (in program P4).

In [FHVV93] attributed tree transducers have been considered as program schemes in the sense that some of the output symbols can be interpreted in a semantic domain with the set of trees as the carrier set and tree functions as operations on the carrier set. There it is proved that the class of primitive recursive tree functions (cf. [Hup78]) is closed under the application of attributed tree transducers viewed as program schemes.

In [Gie88] attribute coupled grammars are introduced. It is shown that, under a certain condition (called syntactic single use requirement), the composition of two attribute coupled grammars is again an attribute coupled grammar.

A pumping lemma for output tree languages of attributed tree transducers has been investigated in [KV94a] (see also [Man96]).

6. Comparison of Induced Tree Transformation Classes

In the previous two chapters we studied two different formal models, i.e., macro tree transducers and attributed tree transducers, which extend top-down tree transducers in the sense that they can handle context information. In this chapter we aim to compare the power of macro tree transducers and attributed tree transducers. We collect all the inclusions which relate the classes TOP, $YIELD$, ATT, MAC, and their subclasses into an inclusion diagram. Moreover, we present the composition semigroup which is generated by the set $\{TOP, ATT, MAC\}$. Hence, the structure of this chapter is as follows.

1. Comparison of MAC and ATT
2. Inclusion diagram based on TOP, $YIELD$, ATT, and MAC
3. Composition semigroup generated by TOP, ATT, and MAC
4. Bibliographic notes

6.1 Comparison of MAC and ATT

Every attributed tree transducer A can be simulated by a macro tree transducer M: in M the synthesized attributes of A are considered as states, and the inherited attributes of A are simulated by parameters of states. Roughly speaking, for every state a and input symbol $\sigma \in \Sigma^{(k)}$, the right-hand side of the (a, σ)-rule is obtained by unfolding the dependency graph $D_A(\sigma)[is_f, \ldots, is_f]$ where is_f contains an edge between every inherited attribute and every synthesized attribute. Thus, is_f covers every possible dependency. The unfolding process is indicated in Fig. 6.1 where in (a) the output symbols occurring in the right-hand sides of the rules of A are integrated into the dependency graph. The unfolding terminates whenever an outside attribute occurrence is encountered for the second time (as, e.g., $a(\pi 1)$).

Lemma 6.1. $ATT \subseteq MAC$.

Proof. Let $A = (Att, \Sigma, \Delta, a_0, R, E)$ be an attributed tree transducer with $Att = Att_{syn} \cup Att_{inh}$. We construct a macro tree transducer M such that $\tau_A = \tau_M$. Let $(b_1, \ldots, b_r)$ be a linear order of the inherited attributes of A. Define $M = (Q, \Sigma, \Delta, a_0, R', E')$ as follows:

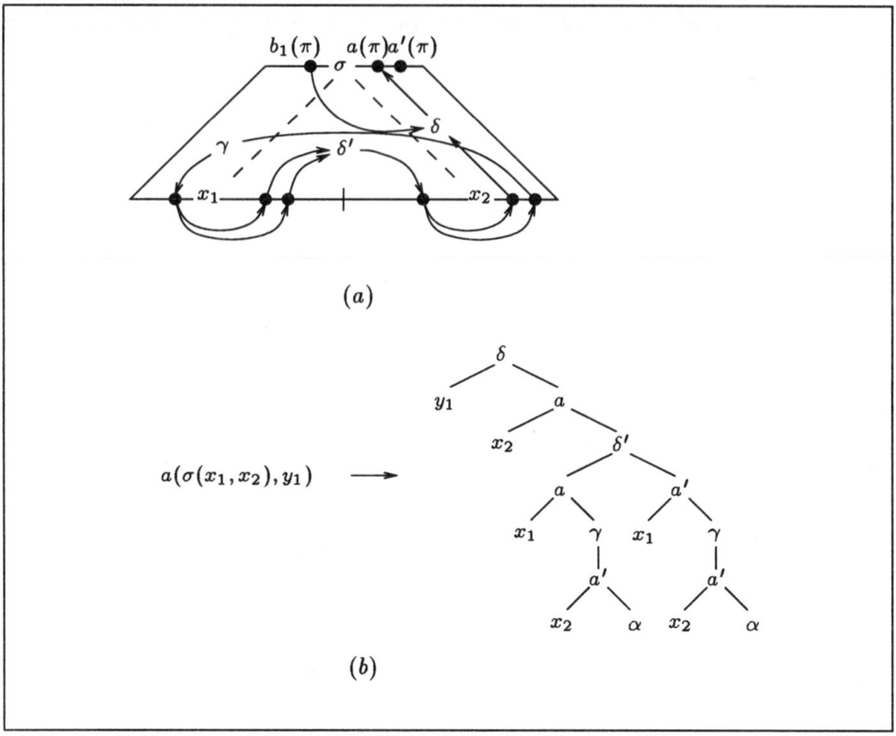

Fig. 6.1. (a) $D_A(\sigma)[is_f, \ldots, is_f]$ and (b) its unfolding

- $Q = \{a^{(r+1)} \mid a \in Att_{syn}\}$
- R' is the smallest set which contains the following rules: if $a(\pi) \to \xi$ is in R_σ for some $a \in Att_{syn}$ and $\sigma \in \Sigma^{(k)}$ with $k \geq 0$, then the rule

$$a(\sigma(x_1, \ldots, x_k), y_1, \ldots, y_r) \to \theta(\xi)$$

is in R' where $\theta(\xi)$ is obtained from ξ by performing the following substitutions:

– every $a'(\pi i)$ with $a' \in Att_{syn}$ and $1 \leq i \leq k$ is replaced by

$$a'(x_i, t\langle i, 1, \{(a', i)\}\rangle, \ldots, t\langle i, r, \{(a', i)\}\rangle)$$

– every $b_j(\pi)$ is replaced by y_j.

For every $\nu \in \{1, \ldots, k\}$, $\kappa \in \{1, \ldots, r\}$, and $P \subseteq Att_{syn} \times \{1, \ldots, k\}$, we define recursively the tree $t\langle \nu, \kappa, P\rangle$ as follows. Let $s = rhs(b_\kappa(\pi\nu), \sigma)$.

– If, for some $(c, l) \in P$, the subtree $c(\pi l)$ occurs in s, then define $t\langle \nu, \kappa, P\rangle = \alpha$ for some $\alpha \in \Delta^{(0)}$.

– Otherwise, $t\langle \nu, \kappa, P\rangle$ is obtained from s by performing the following substitutions:

- every subtree $c(\pi l)$ for $(c, l) \in (Att_{syn} \times \{1, \ldots, k\}) - P$ is replaced by

$$c(x_l, t\langle l, 1, P'\rangle, \ldots, t\langle l, r, P'\rangle)$$

where $P' = P \cup \{(c, l)\}$
- every subtree $b_j(\pi)$ for $b_j \in Att_{inh}$ is replaced by y_j.

- $E' = (E(b_1), \ldots, E(b_r))$.

This completes the construction of M. We add two remarks concerning the termination of the construction of the right-hand sides and about the role of α. During the construction of the right-hand side $\theta(\xi)$, several terms $t\langle \nu, \kappa, P\rangle$ have to be constructed. In fact, the construction proceeds in a recursive way, and in every recursion step one pair (c, i) is added to the set P. Recall that (c, i) represents the outside attribute occurrence $c(\pi i)$. However, since there is only a finite number of such outside attribute occurrences, the recursion has to stop eventually.

What about the role of $\alpha \in \Delta$? The term $t\langle \nu, \kappa, P\rangle$ is defined to be α, if, for some $(c, i) \in P$, the outside attribute occurrence $c(\pi i)$ occurs in the right-hand side $rhs(b_\kappa(\pi\nu), \sigma)$. The fact that $(c, i) \in P$ means that the construction process has visited this outside attribute occurrence $c(\pi i)$ before. However, if during the computation of A on a particular input tree s, the attribute occurrence $c(\pi i)$ is encountered twice, then A is circular and this contradicts the assumption that A is noncircular. Hence, it is reasonable to stop the construction of the $t\langle \nu, \kappa, P\rangle$ terms at this point by placing an arbitrary symbol into $\theta(\xi)$: it will not be used in the computation of M anyway.

We can prove the correctness of the construction by proving the following statements by simultaneous induction according to Principle 5.28.

K: For every $a \in Att_{syn}$ and $s \in T_\Sigma$, $\tau_{A,a}(s)[b_j(\varepsilon) \leftarrow y_j; 1 \le j \le r] = \tau_{M,a}(s)$.

L : For every $k \ge 0$, $\sigma \in \Sigma^{(k)}$, $\xi \in RHS(\sigma)$, and $\omega = (s_1, \ldots, s_k) \in (T_\Sigma)^k$, $\tau_{A,\sigma,\xi}(\omega)[b_j(\varepsilon) \leftarrow y_j; 1 \le j \le r] = \tau_{M,\theta(\xi)}(\omega)$ where $\theta(\xi)$ is defined as in the definition of R'. □

Now the question arises whether this inclusion also holds in the other direction. It does not. In fact, macro tree transducers have more transformational power than attributed tree transducers. This is due to the different capabilities of increasing the height of input trees. For attributed tree transducers, the height of an output tree is linearly bounded in the size of the input tree, whereas in Chap. 4 we presented an example of a macro tree transducer which produces output trees of which the height is exponential in the height of the corresponding input trees.

Lemma 6.2. $ATT \subset MAC$.

Proof. In Lemma 4.23, it was shown that there is a macro tree transducer M_{exp} with a monadic input alphabet such that, for every input tree s, $height(\tau_{M_{exp}}(s)) = 2^{height(s)}$. Clearly, since the input alphabet is monadic, $size(s)$ is equal to $height(s)$. Now it follows from the height-lemma of attributed tree transducers (Lemma 5.40) that there is no attributed tree transducer A such that $\tau_A = \tau_{M_{exp}}$. □

We can even define a small subclass of MAC which contains tree transformations that cannot be induced by any attributed tree transducer. This subclass is induced by so-called *basic macro tree transducers* where, in the right-hand sides of rules, no state may occur nested in the parameter position of another state.

Definition 6.3. Let $M = (Q, \Sigma, \Delta, q_0, R, E)$ be a macro tree transducer. M is *basic* if, for the right-hand side ξ of every rule in R, the following holds: if $q(x_i, \xi_1, \ldots, \xi_r)$ is a subtree of ξ with $q \in Q$, then, for every $1 \leq j \leq r$, the tree ξ_j does not contain a state. □

Let *bas-MAC* denote the class of tree transformations which are induced by basic macro tree transducers. Next we define a tree transformation τ_{path} for which we will prove that $\tau_{path} \in$ *bas-MAC* $-$ *ATT*.

Definition 6.4. Let $\Sigma = \{\gamma^{(1)}, \alpha^{(0)}\}$ and $\Delta = \{\sigma^{(2)}, 1^{(1)}, 2^{(1)}, \#^{(0)}\}$. Define the tree transformation $\tau_{path} : T_\Sigma \rightarrow T_\Delta$ such that

$$\tau_{path} = \{(\gamma^n(\alpha), apprev(s_{n+1})) \mid n \geq 0\}$$

where s_n is the fully balanced tree of height n (as defined in Example 3.29) and, $apprev(s_n)$ is the tree which is obtained from s_n by substituting, for every $w \in occ(s_n)$ such that $label(s_n, w) = \alpha$, that occurrence of α by the unary tree $w^{-1}(\#)$. Figure 6.2 illustrates the function τ_{path}. □

Theorem 6.5. *bas-MAC* $-$ *ATT* $\neq \emptyset$.

Proof. We show that $\tau_{path} \in$ *bas-MAC*. Define the macro tree transducer $M = (Q, \Sigma, \Delta, q, R, E)$ by

- $Q = \{q^{(2)}\}$,
- Σ and Δ as in Def. 6.4,
- R contains the rules:
  1) $q(\gamma(x_1), y_1) \rightarrow \sigma(q(x_1, 1(y_1)), q(x_1, 2(y_1)))$,
  2) $q(\alpha, y_1) \rightarrow y_1$.
- $E = (\#)$.

It is easy to see that $\tau_M = \tau_{path}$. Moreover, it is also easy to see that $\tau_M(\gamma^n(\alpha))$ contains 2^n different subtrees of the form $w^{-1}(\#)$. Due to the

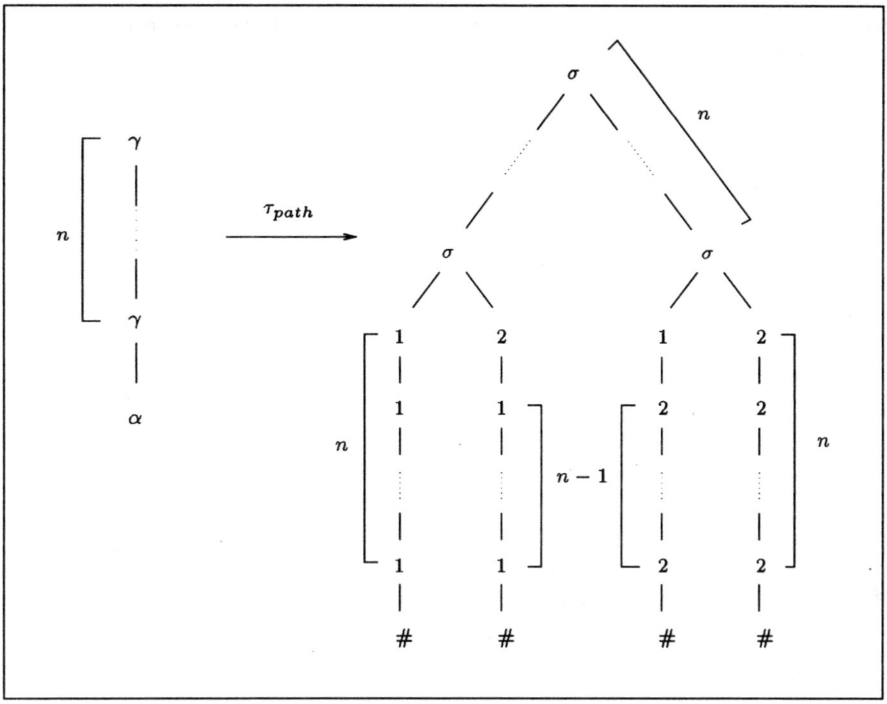

Fig. 6.2. The function τ_{path}

subtree-lemma of attributed tree transducers (Lemma 5.43), there can be no attributed tree transducer A such that $\tau_A = \tau_M$. □

How can we restrict macro tree transducers so that they have the *same* transformational power as attributed tree transducers? We do not know the answer to this question. However we can characterize a large subclass of ATT in terms of restricted macro tree transducers. The restriction amounts to considering *well-presented* macro tree transducers, and the large subclass is the class of *absolutely noncircular* attributed tree transducers.

Before defining the property well-presented, let us work through the idea behind this restriction. Consider again the basic macro tree transducer M of Theorem 6.5 which computes the tree transformation τ_{path} (Def. 6.4). Let us now try to apply the inverse of the construction in which, for some attributed tree transducer A, a macro tree transducer M has been constructed such that $\tau_A = \tau_M$ (Lemma 6.1). Clearly, it is not possible that the inverse construction will work, because otherwise $MAC \subseteq ATT$ and this contradicts Lemma 6.2. However we will investigate what makes the inverse construction fail, and this will lead to the definition of well-presented.

If we apply the inverse construction of Lemma 6.1 to M, then we obtain an attributed tree transducer A which has one synthesized attribute q and one inherited attribute y_1. Since γ is an input symbol, there must be a set R_γ of rules in A, and in R_γ there must be a rule $q(\pi) \to \xi$. By applying the inverse unfolding process to the right-hand side of the rule $q(\gamma(x_1), y_1) \to \sigma(q(x_1, 1(y_1)), q(x_1, 2(y_1)))$, it is clear that $\xi = \sigma(q(\pi 1), q(\pi 1))$. However, now there are two different terms, viz. $1(y_1)$ and $2(y_1)$, which are to be considered as the result of the unfolding process. Inverting this unfolding means that there are two rules

$$y_1(\pi 1) \to 1(y_1(\pi))$$

$$y_1(\pi 1) \to 2(y_1(\pi))$$

in R_γ which, clearly, is not allowed.

Thus, in order to use the inverse construction, the macro tree transducer should have the property that, considering the set of all σ-rules (for some input symbol σ), whenever there are two states q and p which are called on the same input subtree, then the parameter positions of q and p should contain the same terms. This is the notion of *super well-presented.*

In fact, it is possible to relax this condition a little by associating with every state q a subset $\gamma(q)$ of its parameter positions and requiring the equality of terms only for the parameter positions in the intersection $\gamma(p) \cap \gamma(q)$. This is the notion of well-presented.

Definition 6.6. Let $M = (Q, \Sigma, \Delta, q_0, R, E)$ be a macro tree transducer.

1) M is *well-presented* if there is a function $\gamma_M : Q \to \mathcal{P}(Y)$ such that for every $q, q' \in Q$, $k \geq 0$, and $\sigma \in \Sigma^{(k)}$ the following properties hold:
   (a) $\gamma_M(q) \in \mathcal{P}(Y_{rank(q)-1})$,
   (b) if $q(\sigma(x_1, \ldots, x_k), y_1, \ldots, y_n) \to \xi$ is a rule in R and y_j occurs in ξ, then $y_j \in \gamma_M(q)$, and
   (c) if $q(\sigma(x_1, \ldots, x_k), y_1, \ldots, y_n) \to \xi$ and $q'(\sigma(x_1, \ldots, x_k), y_1, \ldots, y_{n'}) \to \xi'$ are rules for σ, and $p(x_i, \xi_1, \ldots, \xi_m)$ and $p'(x_i, \xi'_1, \ldots, \xi'_{m'})$ are subtrees of ξ and ξ', respectively, then, for every $y_j \in \gamma_M(p) \cap \gamma_M(p')$, the equality $\xi_j = \xi'_j$ holds.
2) M is *super well-presented* if M is well-presented under the function γ_M such that for every $q \in Q$, the value $\gamma_M(q) = Y_{rank(q)-1}$. □

Note that, if M is super well-presented, then condition (b) is meaningless. Moreover, condition (c) requires that $\xi_1 = \xi'_1, \ldots, \xi_m = \xi'_m$ where we assume without loss of generality that $m \leq m'$. Theorem 6.5 uses a macro tree transducer which is not well-presented because it violates condition (c).

Let the class of tree transformations which are induced by (super) well-presented macro tree transducers be denoted by *wp-MAC* (and *swp-MAC*, respectively). It is clear that every super well-presented macro tree transducer is also well-presented.

Note 6.7. $swp\text{-}MAC \subseteq wp\text{-}MAC$.

Next we show a well-presented macro tree transducer which is not super well-presented.

Example 6.8. Consider a well-presented macro tree transducer M with states p and q, both of rank 3. The function γ which associates parameters to states, is defined by $\gamma(p) = \{y_1\}$ and $\gamma(q) = \{y_2\}$. Moreover, M has the rule

$$p(\sigma(x_1, x_2), y_1, y_2) \to p(x_1, q(x_1, \alpha, \beta), \gamma)$$

where α, β, and γ are nullary output symbols. Then M is well-presented, because $\gamma(p) \cap \gamma(q) = \emptyset$. However, M is not super well-presented because $p(x_1, q(x_1, \alpha, \beta), \gamma)$ and $q(x_1, \alpha, \beta)$ are two subtrees of the right-hand side of some σ-rule of M which both refer to x_1, and, e.g., $q(x_1, \alpha, \beta) \neq \alpha$. □

It is also clear that every superlinear macro tree transducer (Def. 4.27) is super well-presented.

Note 6.9. sl-$MAC \subseteq$ swp-MAC.

By transforming a well-presented macro tree transducer M into an attributed tree transducer A, we can show that A is restricted; it is ***absolutely noncircular***. Also, by transforming a super well-presented macro tree transducer M into an attributed tree transducer A, we can show that A is a restricted absolutely noncircular attributed tree transducer; it is even ***one-visit***. Finally, by transforming a basic well-presented macro tree transducer M (i.e., M is basic and well-presented) we obtain a ***nonnested*** attributed tree transducer A, where nonnested is a restriction of one-visit. Before we show the transformations, let us define the three restrictions on attributed tree transducers. The definitions of is-graph and composition of graphs were given in Defs. 5.15 and 5.16.

Definition 6.10. Let $A = (Att, \Sigma, \Delta, a_0, R, E)$ be an attributed tree transducer. A is ***absolutely noncircular*** if there is an is-graph $IS = (Att, E)$ over Att_{syn} and Att_{inh} such that, for every $\sigma \in \Sigma^{(k)}$ with $k \geq 0$, the following properties hold.

1) $D_A(\sigma)[IS, \ldots, IS]$ is acyclic where the rank $rank(\sigma)$ determines the number of occurrences of IS and
2) if (c, c') is an edge of $is(D_A(\sigma)[IS, \ldots, IS])$, then (c, c') is also an edge of IS. □

The class of tree transformations which are induced by absolutely noncircular attributed tree transducers is denoted by *anc-ATT*.

Note 6.11. anc-$ATT \subseteq ATT$.

Example 6.12. There are noncircular attributed tree transducers which are not absolutely noncircular. Figure 6.3 shows three dependency graphs of such an attributed tree transducer A. Let us assume that the inherited attributes are called b_1 and b_2 and the synthesized attributes are called a_1 and a_2. In Fig. 6.3, the attribute occurrences at every node are arranged from left to right as follows: $b_1(\pi)$, $b_2(\pi)$, $a_1(\pi)$, and $a_2(\pi)$.

Clearly, in order to fulfill condition 2 of Def. 6.10, the is-graph IS should contain edges from b_1 to a_2 (due to $D_A(\alpha)$) and from b_2 to a_1 (due to $D_A(\beta)$). But then condition 1 is violated because $D_A(\gamma)[IS]$ contains a cycle.

However, due to Lemma 5.17, A is noncircular because $is\text{-}set_A$ contains two is-graphs (one with one edge from b_1 to a_2 and the other with one edge from b_2 to a_1) and $D_A(\gamma)[is]$ does not contain a cycle for any choice of is from $is\text{-}set_A$. □

It is obvious that, in an absolutely noncircular attributed tree transducer, the is-graphs of every possible input tree are merged together resulting in an is-graph, say, $\widetilde{is}$, and it is required that the composition of $D_A(\sigma)$ (for every input symbol σ) and $\widetilde{is}$ does not contain a cycle. Then it is entirely possible (as in Example 6.12) that there is no input tree which has $\widetilde{is}$ as an is-graph.

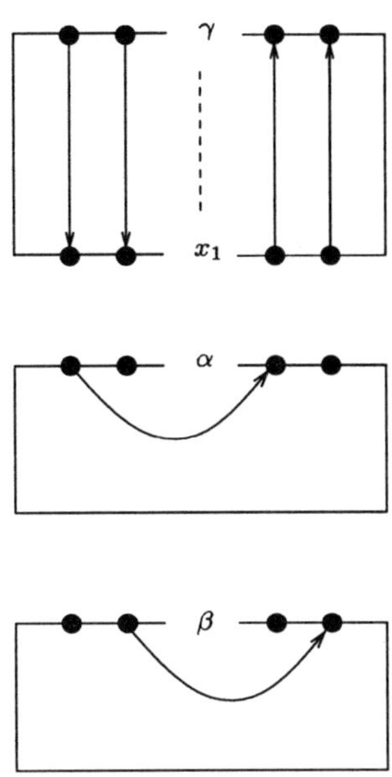

Fig. 6.3. Dependency graphs of a not absolutely noncircular attributed tree transducer

The one-visit restriction is defined by means of the notion of brother graphs.

Definition 6.13. Let $A = (Att, \Sigma, \Delta, a_0, R, E)$ be an attributed tree transducer and let $\sigma \in \Sigma^{(k)}$ for some $k \geq 0$.

The *brother graph of* σ is the directed graph $BG_\sigma = (N, E)$, where

- $N = X_k$ is the set of nodes, and
- E is the set of edges; E is the smallest subset of $X_k \times X_k$ such that for every $1 \leq i, j \leq k$, if there is a $b \in Att_{inh}$ and an $a \in Att_{syn}$ such that $a(\pi i)$ occurs in $rhs(b(\pi j), \sigma)$, then there is an arc $(x_i, x_j) \in E$.

Definition 6.14. Let $A = (Att, \Sigma, \Delta, a_0, R, E)$ be an attributed tree transducer. A is *one-visit* if, for every $\sigma \in \Sigma$, the brother graph BG_σ does not contain a cycle. □

The class of tree transformations which are induced by one-visit attributed tree transducers is denoted by $1v\text{-}ATT$.

If an attributed tree transducer A is noncircular, then a graph $D_A(\sigma)[IS, \ldots, IS]$, where $\sigma \in \Sigma^{(k)}$, $k \geq 0$, and IS is an is-graph, can contain a cycle only if the brother graph of σ is cyclic. Hence the following note is justified.

Note 6.15. $1v\text{-}ATT \subseteq anc\text{-}ATT$.

Example 6.16. There are absolutely noncircular attributed tree transducers which are not one-visit. In Fig. 6.4 we show the two dependency graphs of an attributed tree transducer A, again with inherited attributes b_1 and b_2 and synthesized attributes a_1 and a_2, and with the same arrangement as in Fig. 6.3. It is easy to check that the brother graph B_σ contains a cycle due to the dependencies $((a_1, \pi 1), (b_2, \pi 2))$ and $((a_2, \pi 2), (b_2, \pi 1))$. However, A is absolutely noncircular: consider the is-graph IS which contains edges from b_1 to a_2 and from b_2 to a_1. Both conditions of Def. 6.10 are fulfilled for this IS. □

Finally, we introduce the notion of a nonnested attributed tree transducer. Briefly, an attributed tree transducer is nonnested if there is no inherited attribute which depends on a synthesized attribute.

Definition 6.17. Let $A = (Att, \Sigma, \Delta, a_0, R, E)$ be an attributed tree transducer. A is *nonnested* if, for every $k \geq 1$, $1 \leq i \leq k$, $b \in Att_{inh}$, and $\sigma \in \Sigma^{(k)}$, the right-hand side $rhs(b(\pi i), \sigma)$ does not contain any attribute occurrence $a(\pi j)$ with $a \in Att_{syn}$ and $1 \leq j \leq k$. □

The class of tree transformations which is induced by nonnested attributed tree transducers is denoted by $nn\text{-}ATT$. It is clear that every nonnested attributed tree transducer is also one-visit because all its brother graphs are empty.

Note 6.18. $nn\text{-}ATT \subseteq 1v\text{-}ATT$.

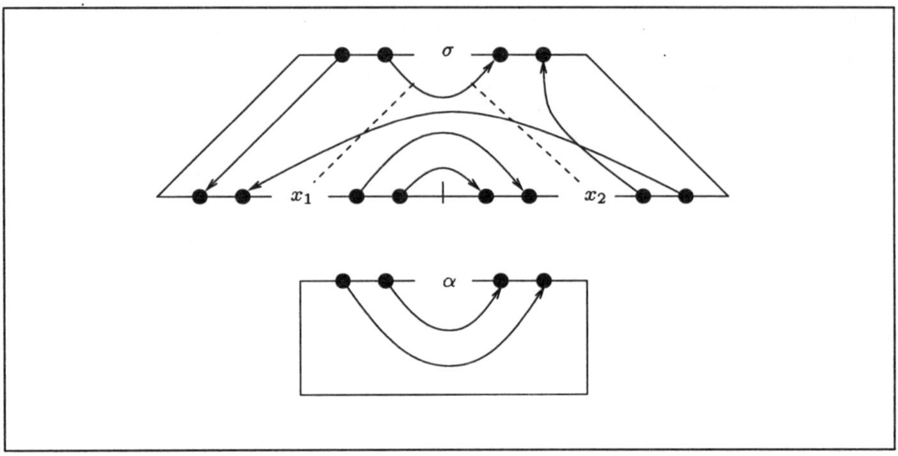

Fig. 6.4. Two dependency graphs of a not one-visit attributed tree transducer

It is also clear that every top-down tree transducer is a nonnested attributed tree transducer.

Note 6.19. $TOP \subseteq nn\text{-}ATT$.

Example 6.20. Clearly, there are also attributed tree transducers which are one-visit, but not nonnested. In Fig. 6.5 there are the two dependency graphs of a one-visit attributed tree transducer which is not nonnested. The latter (non)property is due to the rule for $b(\pi 2)$: in its right-hand side it contains $a(\pi 1)$ where a is a synthesized attribute. □

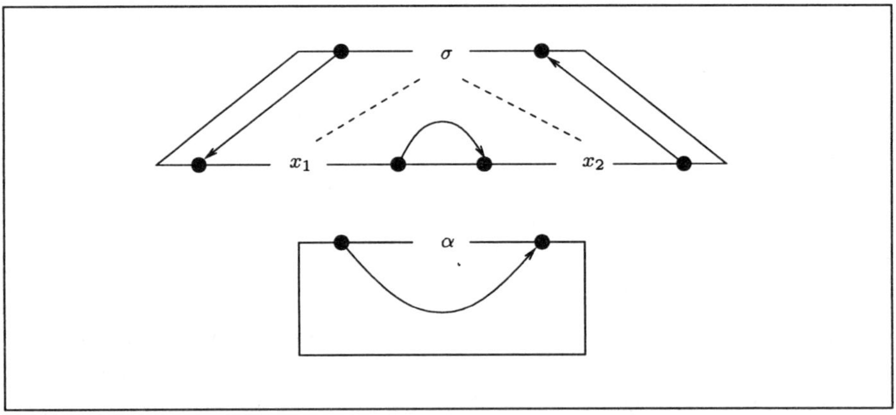

Fig. 6.5. Two dependency graphs of an attributed tree transducer which is not nonnested

Now we can prove that every well-presented macro tree transducer can be transformed into an absolutely noncircular attributed tree transducer. The same transformation is possible for super well-presented and basic well-presented macro tree transducers; this leads to one-visit attributed tree transducers and nonnested attributed tree transducers, respectively.

Lemma 6.21.
1) $wp\text{-}MAC \subseteq anc\text{-}ATT$
2) $swp\text{-}MAC \subseteq 1v\text{-}ATT$
3) $bas\text{-}wp\text{-}MAC \subseteq nn\text{-}ATT$.

Proof. We show the construction for the first statement of this lemma, and then we show that this construction leads to the desired restrictions of the second and third statements of this lemma.

Let $M = (Q, \Sigma, \Delta, q_0, R, E)$ be a well-presented macro tree transducer. Then there is a function $\gamma_M : Q \to \mathcal{P}(Y)$ which fulfills the conditions of Def. 6.6. Let α be an arbitrary element in $\Delta^{(0)}$. Denote $max\{n \mid \Sigma^{(n)} \neq \emptyset\}$ and $max\{n \mid Q^{(n+1)} \neq \emptyset\}$ by $max(M, X)$ and $max(M, Y)$, respectively. We construct an attributed tree transducer $A = (Att, \Sigma, \Delta, q_0, R', E')$ such that $\tau_A = \tau_M$. Let A be defined as follows:

- $Att = Att_{syn} \cup Att_{inh}$ where $Att_{syn} = Q$ and $Att_{inh} = \{y_j \mid 1 \leq j \leq max(M, Y)\}$
- R': First we construct the set $\bigcup_{\sigma \in \Sigma} R'_\sigma$ and second we add appropriate dummy rules to the sets of this family if necessary.
  For every $\sigma \in \Sigma$, the set R'_σ is the smallest set for which the following condition holds: if $q(\sigma(x_1, \ldots, x_k), y_1, \ldots, y_n) \to \xi$ is in R, then

  $$\{q(\pi) \to top(\xi)\} \cup DECOMPOSE(\xi) \subseteq R'_\sigma.$$

  The function $top : RHS(Q, \Delta, max(M, X), max(M, Y)) \to RHS(Att_{syn}, Att_{inh}, \Delta, max(M, X))$ is defined by structural induction on its argument.
  (i) For every $p \in Q^{(l+1)}$ with $l \geq 0$, $1 \leq i \leq max(M, X)$, and $\xi_1, \ldots, \xi_l \in RHS(Q, \Delta, max(M, X), max(M, Y))$, define $top(p(x_i, \xi_1, \ldots, \xi_l)) = p(\pi i)$.
  (ii) For every $\delta \in \Delta^{(l)}$ with $l \geq 0$ and $\xi_1, \ldots, \xi_l \in RHS(Q, \Delta, max(M, X), max(M, Y))$, define
  $top(\delta(\xi_1, \ldots, \xi_l)) = \delta(top(\xi_1), \ldots, top(\xi_l))$.
  (iii) For every $y_j \in Y_{max(M,Y)}$, define $top(y_j) = y_j(\pi)$.
  The function $DECOMPOSE : RHS(Q, \Delta, max(M, X), max(M, Y)) \to \mathcal{P}(\{l \to r \mid l \to r \text{ is a rule of an attributed tree transducer}\})$ is also defined by structural induction on its argument.
  (i) For every $p \in Q^{(l+1)}$ with $l \geq 0$, define
  $DECOMPOSE(p(x_i, \xi_1, \ldots, \xi_l)) =$
  $\{y_j(\pi i) \to top(\xi_j) \mid y_j \in \gamma_M(p)\} \cup \bigcup_{1 \leq j \leq l} DECOMPOSE(\xi_j)$.

(ii) For every $\delta \in \Delta^{(l)}$ with $l \geq 0$, define $DECOMPOSE(\delta(\xi_1, \ldots, \xi_l)) = \bigcup_{1 \leq j \leq l} DECOMPOSE(\xi_j)$.

(iii) For every $y_j \in Y_{max(M,Y)}$, define $DECOMPOSE(y_j) = \emptyset$.
Note that, due to the fact that M is well-presented, it is not possible that there are two different rules in $DECOMPOSE(p(x_i, \xi_1, \ldots, \xi_l))$ with the same left-hand side $y_j(\pi i)$. Also note that $DECOMPOSE$ terminates, because it is applied to the right-hand sides of the rules of M, i.e., finite trees. This finishes the first step in the construction of R'.
Now consider an input symbol $\sigma \in \Sigma^{(k)}$. If, for some i with $1 \leq i \leq k$ and some j with $1 \leq j \leq max(M, Y)$, there is no rule in R'_σ with left-hand side $y_j(\pi i)$, then add the dummy rule

$$y_j(\pi i) \to \alpha$$

to R'_σ.

- Let $E = (t_1, \ldots, t_r)$. Note that r is the rank of the initial state q_0 of M and that $r \leq max(M, Y)$. Then we define the function $E' : Att_{inh} \to T_\Delta$ as follows: for every $1 \leq j \leq r$, let $E'(y_j) = t_j$ and for every $r + 1 \leq j \leq max(M, Y)$, let $E'(y_j) = \alpha$.

The correctness of the construction follows from the following statements which can be proved by simultaneous induction.

K: For every $q \in Q$ and $s \in T_\Sigma$, $\tau_{M,q}(s) = \tau_{A,q}(s)[b_j(\varepsilon) \leftarrow y_j; 1 \leq j \leq max(M, Y)]$.

L: For every $q \in Q$, $k \geq 0$, $\sigma \in \Sigma^{(k)}$, $\xi \in sub(rhs(q, \sigma))$, and $\omega = (s_1, \ldots, s_k) \in (T_\Sigma)^k$, $\tau_{M,\xi}(\omega) = \tau_{A,\sigma,top(\xi)}(\omega)[b_j(\varepsilon) \leftarrow y_j; 1 \leq j \leq max(M, Y)]$.

Now we have to show that A is absolutely noncircular. Since M is well-presented, there is a function $\gamma_M : Q \to \mathcal{P}(Y)$ with the three properties given in Def. 6.6. We construct the is-graph IS with $Att_{syn} \cup Att_{inh}$ as a set of nodes and, if $y_j \in \gamma_M(q)$, there is an edge from y_j to q. Now assume that, for some $\sigma \in \Sigma^{(k)}$, the graph G defined by $G = D_A(\sigma)[IS, \ldots, IS]$ is cyclic. Then there is a sequence

$$q_1(\pi i_1), y_{j_1}(\pi i_2), \ldots q_{r-1}(\pi i_{r-1}), y_{j_{r-1}}(\pi i_r), q_r(\pi i_r)$$

of attribute occurrences such that $1 \leq i_1, \ldots, i_{r-1}, i_r \leq k$, for every pair of consecutive attribute occurrences there is a corresponding edge in G, and $q_1(\pi i_1) = q_r(\pi i_r)$ (making G cyclic).

Since $y_{j_1}(\pi i_2)$ depends on $q_1(\pi i_1)$ and since $q_1(\pi i_1) = q_r(\pi i_r)$ there must be a σ-rule $\lambda \to \rho$ in M of which the right-hand side contains the subtree $q_r(x_{i_r}, \xi_1, \ldots, \xi_l)$ for some $\xi_1, \ldots, \xi_l$. Since $(y_{j_{r-1}}(\pi i_r), q_r(\pi i_r))$ is an edge of G, there is an edge $(y_{j_{r-1}}, q_r)$ in IS, and hence, by definition of IS, $y_{j_{r-1}} \in \gamma_M(q_r)$. Since $y_{j_{r-1}}(\pi i_r)$ depends on $q_{r-1}(\pi i_{r-1})$ and since the construction will deduce the right-hand side of the rule for $y_{j_{r-1}}(\pi i_r)$ from the parameter term $\xi_{j_{r-1}}$, this parameter term has to contain the subtree $q_{r-1}(x_{i_{r-1}}, \xi'_1, \ldots, \xi'_{l'})$ for some $\xi'_1, \ldots, \xi'_{l'}$. Note that, if there is another occurrence of a subtree $q_r(x_{i_r}, \widehat{\xi_1}, \ldots, \widehat{\xi_l})$ for some $\widehat{\xi_1}, \ldots, \widehat{\xi_l}$ in some right-hand side of a σ-rule, then $\widehat{\xi_{j_{r-1}}} = \xi_{j_{r-1}}$, because M is well-presented.

Now we can repeat the same argument for $q_{r-1}(\pi i_{r-1})$, as for $q_r(\pi i_r)$. If we repeat this argument $r-1$ times, we draw the (global) observation that the j_{r-1}-th parameter term of the subtree $q_r(x_{i_r}, \xi_1, \ldots, \xi_l)$ contains the subtree $q_1(x_{i_1}, \xi''_1, \ldots, \xi''_{l''})$ for some $\xi''_1, \ldots, \xi''_{l''}$. Since $q_1(\pi i_1) = q_r(\pi i_r)$, the two terms $q_1(x_{i_1}, \xi''_1, \ldots, \xi''_l)$ and $q_r(x_{i_r}, \xi''_1, \ldots, \xi''_l)$ are syntactically equal. Moreover, by $q_1 = q_r$, we have $\gamma_M(q_1) = \gamma_M(q_r)$. Hence $y_{j_{r-1}} \in \gamma_M(q_1)$ which implies that $\xi_{j_{r-1}} = \xi''_{j_{r-1}}$. On the other hand, $\xi''_{j_{r-1}}$ is a proper subtree of $\xi_{j_{r-1}}$. Hence, there is a contradiction with the assumption that G contains a cycle.

Also we have to prove that, if (c, c') is an edge of $is(D_A(\sigma)[IS, \ldots, IS])$, then (c, c') is also an edge of IS. Let $(y_j, q) \in Att_{inh} \times Att_{syn}$ be an element of $is(D_A(\sigma)[IS, \ldots, IS])$. Hence, by Def. 5.15, there is a path in $D_A(\sigma)[IS, \ldots, IS]$ from $y_j(\pi)$ to $q(\pi)$. Considering the construction of A such a path can only exist if the right-hand side of the (q, σ)-rule contains y_j. Then, by Def. 6.6, $y_j \in \gamma_M(q)$ and thus, by the construction of IS, $(y_j, q) \in IS$ also.

Now assume that M is super well-presented. Then it is not possible that there is a right-hand side of a rule of M with a subtree of the form $p(x_i, \ldots q(x_i, \ldots) \ldots)$. Hence, by the definition of $DECOMPOSE$, A is one-visit.

Finally assume that M is basic and well-presented. Then it is immediately clear from the construction that A is nonnested. $\square$

Now we show that the inclusions of Lemma 6.21 also hold in the other direction. In fact, the construction of the following lemma is quite similar to the construction of Lemma 6.1 ($ATT \subseteq MAC$). But here the unfolding of the right-hand side is stopped by the function $\gamma_M : Q \to \mathcal{P}(Y)$ rather than by a set P.

Lemma 6.22.
1) $anc\text{-}ATT \subseteq wp\text{-}MAC$
2) $1v\text{-}ATT \subseteq swp\text{-}MAC$
3) $nn\text{-}ATT \subseteq bas\text{-}wp\text{-}MAC$.

Proof. We show the construction for the first statement of this lemma, and then we show that this construction also implies the second and third statements of this lemma.

Let $A = (Att, \Sigma, \Delta, a_0, R, E)$ be an absolutely noncircular attributed tree transducer such that $Att_{inh} = \{b_1, \ldots, b_r\}$ for some $r \geq 0$. Let $\alpha \in \Delta^{(0)}$ be an arbitrary, but fixed, output symbol.

We construct the macro tree transducer $M = (Q, \Sigma, \Delta, a_0, R', E')$ as follows. The set Q is equal to Att_{syn} and every $a \in Q$ has rank $r+1$. Next we consider a $\sigma \in \Sigma^{(k)}$ for some $k \geq 0$ and we construct the σ-rules from the elements of R_σ.

By Def. 6.10, there is an is-graph $IS = (Att, E)$ with certain properties. IS induces the function $\gamma_{IS} : Q \to \mathcal{P}(Y)$ by defining $\gamma_{IS}(a) = \{y_j \mid (b_j, a) \in E\}$ for every $a \in Q$.

Then, for every $a \in Q$, we construct the rule

$$a(\sigma(x_1, \ldots, x_k), y_1, \ldots, y_r) \to t$$

where $t = COMPOSE(rhs(a(\pi), \sigma))$. The function

$$COMPOSE : RHS(Att_{syn}, Att_{inh}, \Delta, k) \to RHS(Q, \Delta, k, r)$$

is defined inductively as follows. Let $\xi \in RHS(Att_{syn}, Att_{inh}, \Delta, k)$.

(i) If $\xi = c(\pi i)$ for some $c \in Att_{syn}$ and $1 \leq i \leq k$, then $COMPOSE(\xi) = c(x_i, t_1, \ldots, t_r)$ where, for every $1 \leq j \leq r$,
$$t_j = \begin{cases} \alpha & if \quad y_j \notin \gamma_{IS}(c) \\ COMPOSE(rhs(b_j(\pi i), \sigma)) & if \quad y_j \in \gamma_{IS}(c). \end{cases}$$
(ii) If $\xi = b_j(\pi)$ for some $b_j \in Att_{inh}$, then $COMPOSE(\xi) = y_j$.
(iii) If $\xi = \delta(\xi_1, \ldots, \xi_l)$ for $\delta \in \Delta$ and $\xi_1, \ldots, \xi_l \in RHS(Att_{syn}, Att_{inh}, \Delta, k)$, then $COMPOSE(\xi) = \delta(COMPOSE(\xi_1), \ldots, COMPOSE(\xi_l))$.

Note that the definition of t_j is acyclic, because by Def. 6.10 1) the graph $D_A(\sigma)[IS, \ldots, IS]$ has no cycle.

Moreover, let $E' = (E(b_1), \ldots, E(b_r))$.

The correctness of the construction follows from the following statements which can be proved by simultaneous induction.

K: For every $a \in Att_{syn}$ and $s \in T_\Sigma$, $\tau_{A,a}(s)[b_j(\varepsilon) \leftarrow y_j; 1 \leq j \leq r] = \tau_{M,a}(s)$.

L : For every $k \geq 0$, $\sigma \in \Sigma^{(k)}$, $\xi \in RHS(\sigma)$, and $\omega = (s_1, \ldots, s_k) \in (T_\Sigma)^k$, $\tau_{A,\sigma,\xi}(\omega)[b_j(\varepsilon) \leftarrow y_j; 1 \leq j \leq r] = \tau_{M,COMPOSE(\xi)}(\omega)$

It is easy to see why M is well-presented. For this purpose consider some (a, σ)-rule of M. By Def. 6.10 2), $is(D_A(\sigma)[IS, \ldots, IS]) \subseteq IS$, i.e., we can say that IS is closed under embedding in dependency graphs, and by definition of γ_{IS}, it follows that the elements of $\gamma_{IS}(a)$ at most appear in $rhs(a, \sigma)$. This means that Condition 1(b) of Def. 6.6 is fulfilled. Condition 1(ç) is derived from the fact that, for every inherited attribute b_j at some i-th descendant of a σ-node, there is a unique rule in R. Since $COMPOSE$ is a function, the term $COMPOSE(rhs(b_j(\pi i), \sigma))$ is also unique.

Now assume that A is one-visit. Define the function $\gamma_{full} : Q \to \mathcal{P}(Y)$ such that $\gamma_{full}(a) = \{y_1, \ldots, y_r\}$ for every $a \in Q$. Then we have to show Condition 1(c) of Def. 6.6 for γ_{full}. For this purpose consider two (not necessarily different) rules $a(\sigma(x_1, \ldots, x_k), y_1, \ldots, y_r) \to \xi$ and $a'(\sigma(x_1, \ldots, x_k), y_1, \ldots, y_r) \to \xi'$ for σ, and let $c(x_i, \xi_1, \ldots, \xi_r)$ and $c'(x_i, \xi'_1, \ldots, \xi'_r)$ be subterms of ξ and ξ', respectively. Moreover consider a parameter position j. If $c = c'$, then clearly $\xi_j = \xi'_j$. Now let $c \neq c'$. Then, if $\xi_j \neq \xi'_j$, then either ($y_j \in \gamma_{IS}(c)$ and

$y_j \notin \gamma_{IS}(c'))$ or $(y_j \notin \gamma_{IS}(c)$ and $y_j \in \gamma_{IS}(c'))$. But by the definition of γ_{full}, every y_k is in $\gamma_{full}(c)$. Hence, $\xi_j = \xi'_j$. The definition of *COMPOSE* still terminates, because A is one-visit and hence, no recursive call of *COMPOSE* on the same subtree can arise. Thus, M is super well-presented.

If A is nonnested, then, for every inherited attribute b, input symbol σ, and position i, the right-hand side $rhs(b(\pi i), \sigma)$ does not contain a synthesized attribute occurrence. Thus, *COMPOSE* does not generate a call to a state inside the parameter position of another state. Hence, M is basic. Of course, M also remains well-presented. □

From Lemmas 6.21 and 6.22 we obtain the following connection between the classes of tree transformations which are computed by macro tree transducers and attributed tree transducers.

Theorem 6.23. *1)* $wp\text{-}MAC = anc\text{-}ATT$
2) $swp\text{-}MAC = 1v\text{-}ATT$
3) $bas\text{-}wp\text{-}MAC = nn\text{-}ATT$.

By means of this theorem, we can relate the class of YIELD-functions to the class *ATT*.

Corollary 6.24. $YIELD \subseteq ATT$.

Proof.

	$YIELD$	
$\subseteq$	$sl\text{-}MAC$	(Lemma 4.32)
$\subseteq$	$swp\text{-}MAC$	(Note 6.9)
$=$	$1v\text{-}ATT$	(Theorem 6.23)
$\subseteq$	ATT	(Def. 6.14)

□

Another important consequence relates the classes of tree transformations computed by (arbitrary) macro tree transducers and by attributed tree transducers.

Theorem 6.25. *1)* $MAC = HOM \circ 1v\text{-}ATT$
2) $MAC = HOM \circ ATT$
3) $MAC = TOP \circ ATT$.

Proof.

	MAC	
$=$	$HOM \circ sl\text{-}MAC$	(Corollary 4.39)
$\subseteq$	$HOM \circ swp\text{-}MAC$	(Note 6.9)
$=$	$HOM \circ 1v\text{-}ATT$	(Theorem 6.23)
$\subseteq$	$HOM \circ ATT$	(Note 6.15 and Def. 6.14)
$\subseteq$	$TOP \circ ATT$	(Corollary 3.32)
$\subseteq$	$TOP \circ MAC$	(Lemma 6.1)
$\subseteq$	MAC	(Corollary 4.38).

□

Theorem 6.26. For every $n \geq 1$, $MAC^n = HOM \circ ATT^n$.

Proof. The proof is a simple induction on n. For $n = 1$, the statement holds by Theorem 6.25.

$$
\begin{array}{lll}
 & MAC^{n+1} & \\
= & MAC^n \circ MAC & \\
= & HOM \circ ATT^n \circ MAC & \text{(by induction hypothesis)} \\
= & HOM \circ ATT^n \circ TOP \circ YIELD & \text{(Theorem 4.37)} \\
= & HOM \circ ATT^n \circ YIELD & \text{(Theorem 5.47)} \\
\subseteq & HOM \circ ATT^{n+1} & \text{(Corollary 6.24)} \\
\subseteq & HOM \circ MAC^{n+1} & \text{(Lemma 6.1)} \\
\subseteq & MAC^{n+1} & \text{(Corollary 4.38).}
\end{array}
$$

□

6.2 Inclusion Diagram Based on *TOP*, *YIELD*, *ATT*, and *MAC*

In this section we assemble in one diagram (Fig. 6.6) most of the inclusion and incomparability results which have been proved for the classes *TOP*, *YIELD*, *ATT*, and *MAC* (and their subclasses). It is the aim of this section to prove that this diagram is an inclusion diagram in the sense of Section 2.2. The proof is performed by applying the five steps of Algorithm 2.1:

Step 1: The diagram in Fig. 6.6 is the conjectured diagram.

Step 2: The set *INCL* consists of the following 12 formal inclusions (the reader should disregard the references for the time being):

1.	*HOM*	$\subseteq$	*TOP*	(Corollary 3.32)
2.	*l-TOP*	$\subseteq$	*TOP*	(Corollary 3.32)
3.	*sl-TOP*	$\subseteq$	*l-TOP*	(Corollary 3.32)
4.	*YIELD*	$\subseteq$	*sl-MAC*	(Lemma 4.32)
5.	*sl-TOP*	$\subseteq$	*sl-MAC*	(Note 4.28)
6.	*TOP*	$\subseteq$	*nn-ATT*	(Note 6.19)
7.	*sl-MAC*	$\subseteq$	*swp-MAC*	(Note 6.9)
8.	*bas-wp-MAC*	$\subseteq$	*bas-MAC*	(Def. 6.6)
9.	*nn-ATT*	$\subseteq$	1*v-ATT*	(Note 6.18)
10.	1*v-ATT*	$\subseteq$	*ATT*	(Notes 6.15 and 6.11)
11.	*ATT*	$\subseteq$	*MAC*	(Lemma 6.1)
12.	*bas-MAC*	$\subseteq$	*MAC*	(Def. 6.3).

and the set *INEQ* consists of 38 formal inequalities the details of which are not shown here.

Step 3: Every formal inclusion is true. We have already cited in the list in Step 2 the appropriate corollaries, lemmas etc.

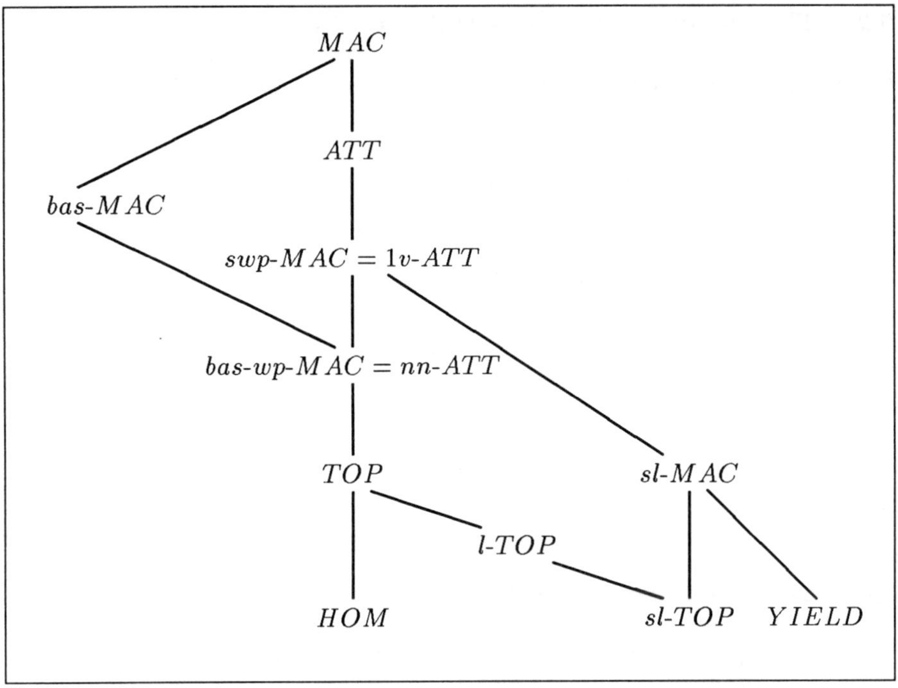

Fig. 6.6. Inclusion diagram for *TOP*, *YIELD*, *ATT*, and *MAC*

Step 4: The set *MINEQ* consists of the following 9 formal inequalities:

1. $bas\text{-}MAC - ATT \neq \emptyset$
2. $HOM - sl\text{-}MAC \neq \emptyset$
3. $l\text{-}TOP - sl\text{-}MAC \neq \emptyset$
4. $sl\text{-}TOP - HOM \neq \emptyset$
5. $sl\text{-}TOP - YIELD \neq \emptyset$
6. $YIELD - bas\text{-}MAC \neq \emptyset$
7. $ATT - 1v\text{-}ATT \neq \emptyset$
8. $nn\text{-}ATT - TOP \neq \emptyset$
9. $HOM - l\text{-}TOP \neq \emptyset$

Step 5: In the following subsections we will verify these nine conjectured inequalities so that we obtain the desired theorem:

Theorem 6.27. The diagram shown in Fig. 6.6 is an inclusion diagram in the sense of Section 2.2.

6.2.1 The inequality *bas-MAC* − *ATT* ≠ ∅

This inequality has been proved already in Theorem 6.5.

6.2.2 The inequality $HOM - sl\text{-}MAC \neq \emptyset$

In Example 3.29 we defined a tree transformation τ_{double} which translates the monadic input tree $\gamma^n(\alpha)$ into the fully balanced output tree s_{n+1}. Actually, there is also a superlinear macro tree transducer M which can compute this tree transformation. M has the following rules

$$q(\gamma(x_1), y_1) \rightarrow q(x_1, \sigma(y_1, y_1))$$

$$q(\alpha, y_1) \rightarrow y_1.$$

Thus, M uses its parameter position to accumulate the fully balanced tree s_{n+1}. In particular, the topmost γ of the input tree generates the lowest σ's of the output tree, and, vice versa, the lowest γ of the input tree generates the topmost σ of the output tree. Thus, we can say that, M reverses the order of the input symbols. Now it is easy to modify τ_{double} in such a way that the order of the input symbols is preserved in the output tree. Then, this modified tree transformation can no longer be computed by a superlinear macro tree transducer.

Definition 6.28. Let $\Sigma = \{\gamma_1^{(1)}, \gamma_2^{(1)}, \alpha^{(0)}\}$ and $\Delta = \{\sigma_1^{(2)}, \sigma_2^{(2)}, \alpha^{(0)}\}$. Let $\tau_{double-ordered} : T_\Sigma \rightarrow T_\Delta$ be the tree transformation which is computed by the homomorphism tree transducer $H = (\{q\}, \Sigma, \Delta, q, R)$ where R contains the rules

1) $q(\gamma_1(x_1)) \rightarrow \sigma_1(q(x_1), q(x_1))$
2) $q(\gamma_2(x_1)) \rightarrow \sigma_2(q(x_1), q(x_1))$
3) $q(\alpha) \rightarrow \alpha.$ □

Theorem 6.29. *$HOM - sl\text{-}MAC \neq \emptyset$.*

Proof. By Def. 6.28, $\tau_{double-ordered} \in HOM$. We will show that $\tau_{double-ordered} \notin sl\text{-}MAC$. We prove by contradiction. We assume that there is a superlinear macro tree transducer $M = (Q, \Sigma, \Delta, q_0, R, E)$ such that $\tau_M = \tau_{double-ordered}$. Assume that $E = (t_1, \ldots, t_r)$.

We proceed by making a sequence of observations.

<u>Observation 1:</u> The (q_0, γ_1)-rule contains x_1 in its right-hand side. The same holds for the (q_0, γ_2)-rule.

We prove by contradiction. We assume that, for $\xi = rhs(q_0, \gamma_1)$, $\xi \in T_\Delta(Y_r)$. Then, for every input tree $s \in T_\Sigma$, we have

$$q_0(\gamma_1(s), y_1, \ldots, y_r) \Rightarrow_M \xi,$$

and $\tau_M(\gamma_1(s)) = \xi[y_1 \leftarrow t_1, \ldots, y_r \leftarrow t_r]$. This means that $\tau_M(\gamma_1(s))$ does not depend on s, which contradicts $\tau_M = \tau_{double-ordered}$. □

Observation 2: The (q_0, γ_1)-rule in R has the form

$$q_0(\gamma_1(x_1), y_1, \ldots, y_r) \to p(x_1, u_1, \ldots, u_k),$$

for some $p \in Q^{(k+1)}$ and $u_1, \ldots, u_k \in T_\Delta(Y_r)$. The same holds for the (q_0, γ_2)-rule.

This observation can also be proved by contradiction. Assume that the root of the right-hand side of the (q_0, γ_1)-rule is an output symbol from Δ. Then, the rule should be of the form

$$q_0(\gamma_1(x_1), y_1, \ldots, y_r) \to \sigma_1(\xi_1, \xi_2),$$

where $\xi_1, \xi_2 \in RHS(Q, \Delta, 1, r)$, such that ξ_1 or ξ_2 contains x_1. The root of the right-hand side should be σ_1 because, for every tree $s \in T_\Sigma$, $\tau_M(\gamma_1(s))$ should have the form $\sigma_1(u, u)$, for a suitable tree $u \in T_\Delta$. Moreover, both ξ_1 and ξ_2 cannot contain x_1, because M is linear. Hence either ξ_1 or ξ_2 contains x_1.

Assume ξ_2 contains x_1. Then, of course, $\xi_1 \in T_\Delta(Y_r)$. Let now $s \in T_\Sigma$ be an arbitrary tree and consider the input tree $\gamma_1(s)$ to M and compute $\tau_M(\gamma_1(s))$. Then, we have

$$\begin{aligned} q_0(\gamma_1(s), y_1, \ldots, y_r) &\Rightarrow_M \sigma_1(\xi_1, \xi_2[x_1 \leftarrow s]) \\ &\Rightarrow_M^* \sigma_1(\xi_1, \bar{\xi_2}) \end{aligned}$$

such that $\bar{\xi_2} \in T_\Delta(Y_r)$ also and that

$$\tau_M(\gamma_1(s)) = \sigma_1(\xi_1[y_1 \leftarrow t_1, \ldots, y_r \leftarrow t_r], \bar{\xi_2}[y_1 \leftarrow t_1, \ldots, y_r \leftarrow t_r]).$$

However, this means that, for every input tree $s \in T_\Sigma$, the first subtree of the output tree $\tau_M(\gamma_1(s))$ is the same tree, i.e., it is $\xi_1[y_1 \leftarrow t_1, \ldots, y_r \leftarrow t_r]$. This contradicts $\tau_M = \tau_{double-ordered}$, which arises from our assumption on the shape of $rhs(q_0, \gamma_1)$. The case when ξ_1 contains x_1 also yields a contradiction, by symmetry.

With this we have proved that $rhs(q_0, \gamma_1)$ can contain x_1 only if its root is a state. Then the condition $u_1, \ldots, u_k \in T_\Delta(Y_r)$ follows from the fact that M is linear.

Since the roles of γ_1 and γ_2 are exchangeable, we have also proved the observation for γ_2. With this we have proved Observation 2. □

Observation 3: The (q_0, γ_1)-rule in R has the form

$$q_0(\gamma_1(x_1), \ldots) \to q_0(x_1, u_1, \ldots, u_r),$$

for some $u_1, \ldots, u_r \in T_\Delta(Y_r)$. The same holds for the (q_0, γ_2)-rule.

We prove by contradiction again. We assume that the (q_0, γ_1)-rule has the form

$$q_0(\gamma_1(x_1), y_1, \ldots, y_r) \to p(x_1, u_1, \ldots, u_k),$$

such that $p \neq q_0$ and $u_1, \ldots, u_k \in T_\Delta(Y_r)$. Since x_1 occurs in $rhs(q_0, \gamma_1)$ and M is superlinear, x_1 cannot occur in $rhs(p, \gamma_1)$. Hence the (p, γ_1)-rule should be of the form

$$p(\gamma_1(x_1), y_1, \ldots, y_k) \to \xi,$$

where $\xi \in T_\Delta(Y_k)$.

Consider now an input tree to M of the form $\gamma_1(\gamma_1(s))$, where $s \in T_\Sigma$ is arbitrary and compute $\tau_M(\gamma_1(\gamma_1(s)))$. We have the derivation

$$\begin{aligned} q_0(\gamma_1(\gamma_1(s)), y_1, \ldots, y_r) &\Rightarrow_M p(\gamma_1(s), u_1, \ldots, u_k) \\ &\Rightarrow_M \xi[y_1 \leftarrow u_1, \ldots, y_k \leftarrow u_k], \end{aligned}$$

hence, by letting $\bar{\xi} = \xi[y_1 \leftarrow u_1, \ldots, y_k \leftarrow u_k]$, we get $\tau_M(\gamma_1(\gamma_1(s))) = \bar{\xi}[y_1 \leftarrow t_1, \ldots, y_r \leftarrow t_r]$, regardless of s. This means that $\tau_M(\gamma_1(\gamma_1(s)))$ does not depend on s, which contradicts $\tau_M = \tau_{double-ordered}$. The contradiction arises from the assumption $p \neq q_0$.

Since the roles of γ_1 and γ_2 are exchangeable, we have proved Observation 3 for γ_2, too. □

We summarize the proof so far. The assumption $\tau_M = \tau_{double-ordered}$ implies that the set Q of (relevant) states of M is the set $Q = Q^{(r+1)} = \{q_0\}$, and that R consists of the rules

- $q_0(\gamma_1(x_1), y_1, \ldots, y_r) \to q_0(x_1, u_1, \ldots, u_r)$,
- $q_0(\gamma_2(x_1), y_1, \ldots, y_r) \to q_0(x_1, v_1, \ldots, v_r)$, and
- $q_0(\alpha, y_1, \ldots, y_r) \to t$,

where $t, u_1, \ldots, u_r, v_1, \ldots, v_r \in T_\Delta(Y_r)$.

In order to derive a contradiction, we transform M into an equivalent attributed tree transducer A. Since M is superlinear, M is also super well-presented (by Note 6.9). Hence we can apply the construction of Lemma 6.21 and obtain a one-visit attributed tree transducer A such that $\tau_M(s) = \tau_A(s)$, for every input tree s. Let E' be the environment of A. Define

$$\begin{aligned} max = max\{height(\xi) \,|\, \xi \text{ is a right-hand side of a rule of } A\} + \\ max\{height(E'(y_j)) \,|\, y_j \text{ is an inherited attribute of } A\}. \end{aligned}$$

Analyzing the set of rules of A, we find that R_{γ_1} contains the rule $q_0(\pi) \to q_0(\pi 1)$ and R_{γ_2} contains the rule $q_0(\pi) \to q_0(\pi 1)$.

Now consider an input tree s with height $max+1$ and an arbitrary derivation d of A on s of the form

$$q_0(\varepsilon) \Rightarrow^*_{A,s} \varphi$$

such that $\tau_A(s) = \varphi[y_1(\varepsilon) \leftarrow E'(y_1), \ldots, y_r(\varepsilon) \leftarrow E'(y_r)]$.

Due to the form of the rules, the derivation d starts with $height(s) - 1$ applications of the chain rule $q_0(\pi) \to q_0(\pi 1)$ without producing any output symbol. Thus, d has the form

$$q_0(\varepsilon) \Rightarrow_{A,s} q_0(1) \Rightarrow_{A,s} q_0(11) \ldots \Rightarrow_{A,s} q_0(1^{height(s)-1}) \Rightarrow^*_{A,s} \varphi.$$

Let us consider the occurrence w in s, at which A produces an output symbol for the first time during d. Due to the form of d, this output symbol is produced by an inherited attribute on the ascent of A (from the leaf to the root of s).

It is not possible that $w = \varepsilon$, i.e., that A does not produce the first output symbol until returning to the root of s again, because otherwise A would have to produce the whole output tree by means of the environment. However, every tree in the environment is too small to fulfill this requirement. It is also not possible that $w = 1$, because otherwise A would have to produce the whole output tree in its last derivation step, followed by the substitution of the environment trees. However, this is not possible either, because $height(s) = height(\tau_A(s)) > max$. Thus $w \notin \{\varepsilon, 1\}$.

Next consider the input tree s' which is obtained from s by changing the label of the root of s (i.e., if the root of s is γ_1, then the root of s' is γ_2, and conversely).

Clearly, $occ(s) = occ(s')$ and A will also produce the first output for s' at occurrence w and this output symbol will be the same as in the derivation d (since $w \notin \{\varepsilon, 1\}$). Hence, the roots of $\tau_A(s)$ and $\tau_A(s')$ will be the same. However, the roots of $\tau_{double-ordered}(s)$ and $\tau_{double-ordered}(s')$ are not equal. Since $\tau_A = \tau_{double-ordered}$, we have derived a contradiction to the assumption that $\tau_{double-ordered} \in sl\text{-}MAC$ and hence, $\tau_{double-ordered} \notin sl\text{-}MAC$. □

6.2.3 The inequality *l-TOP* − *sl-MAC* ≠ ∅

We prove the above inequality by giving a linear top-down tree transformation, called τ_{modulo}, and then proving that it cannot be induced by any superlinear macro tree transducer. The tree transformation τ_{modulo} is defined as follows.

Definition 6.30. Let $T = (Q, \Sigma, \Delta, q_{00}, R)$ be the top-down tree transducer, where

- $Q = \{q_{00}, q_{10}, q_{01}, q_{11}\}$,
- $\Sigma = \{\delta^{(1)}, \gamma^{(1)}, \alpha^{(0)}\}$, $\Delta = \{\delta_0^{(1)}, \delta_1^{(1)}, \gamma_0^{(1)}, \gamma_1^{(1)}, \alpha^{(0)}\}$,
- R consists of the following 12 rules:
  1) $q_{00}(\gamma(x_1)) \to \gamma_1(q_{10}(x_1))$,
  2) $q_{00}(\delta(x_1)) \to \delta_1(q_{01}(x_1))$,
  3) $q_{10}(\gamma(x_1)) \to \gamma_0(q_{00}(x_1))$,
  4) $q_{10}(\delta(x_1)) \to \delta_1(q_{11}(x_1))$,
  5) $q_{01}(\gamma(x_1)) \to \gamma_1(q_{11}(x_1))$,
  6) $q_{01}(\delta(x_1)) \to \delta_0(q_{00}(x_1))$,
  7) $q_{11}(\gamma(x_1)) \to \gamma_0(q_{01}(x_1))$,
  8) $q_{11}(\delta(x_1)) \to \delta_0(q_{10}(x_1))$,
  9)-12) $q_{ij}(\alpha) \to \alpha$, for every $0 \le i, j \le 1$.

By direct inspection, we see that T is linear. It is easy to see that T operates as follows: for every input tree $s \in T_\Sigma$, the output $\tau_T(s)$ can be obtained from s by substituting every occurrence of γ (and δ) by γ_0 or γ_1 (by δ_0 or δ_1) such that the i-th occurrence of γ (δ) counted from the root of s is substituted by γ_{m_i} (by δ_{m_i}), where $m_i = i \, mod \, 2$. Moreover, the leaf α of s remains unchanged. This can be done in such a way that the number of the processed γ's and δ's is encoded in the states of T. In fact, T enters state q_{ij} if and only if i (and j) is the number of the processed γ's (and δ's, respectively) modulo 2. Hence the initial state is q_{00}.

For example, compute $\tau_T(\gamma(\delta(\delta(\gamma(\alpha)))))$. We get the following derivation

$$\begin{aligned}
q_{00}(\gamma(\delta(\delta(\gamma(\alpha))))) \quad & \Rightarrow_T \quad \gamma_1(q_{10}(\delta(\delta(\gamma(\alpha))))) \\
& \Rightarrow_T \quad \gamma_1(\delta_1(q_{11}(\delta(\gamma(\alpha))))) \\
& \Rightarrow_T \quad \gamma_1(\delta_1(\delta_0(q_{10}(\gamma(\alpha))))) \\
& \Rightarrow_T \quad \gamma_1(\delta_1(\delta_0(\gamma_0(q_{00}(\alpha))))) \\
& \Rightarrow_T \quad \gamma_1(\delta_1(\delta_0(\gamma_0(\alpha)))).
\end{aligned}$$

Hence we have $\tau_T(\gamma(\delta(\delta(\gamma(\alpha))))) = \gamma_1(\delta_1(\delta_0(\gamma_0(\alpha))))$.

In the following we denote the tree transformation induced by T by τ_{modulo}. □

Theorem 6.31. *l-TOP – sl-MAC ≠ ∅.*

Proof. We prove by contradiction that τ_{modulo} cannot be induced by any superlinear macro tree transducer. Since the proof is similar to that of Theorem 6.29, the details are left to the reader. Refer to Sect. 2.7 for the method by which we denote trees over monadic ranked alphabets.

Assume that there is a superlinear macro tree transducer $M = (Q, \Sigma, \Delta, q_0, R, E)$ with $E = (t_1, \ldots, t_r)$ such that $\tau_{modulo} = \tau_M$.

We can show that R consists of the rules

- $q_0(\gamma(x_1), y_1, \ldots, y_r) \to q_0(x_1, u_1, \ldots, u_r)$,
- $q_0(\delta(x_1), y_1, \ldots, y_r) \to q_0(x_1, v_1, \ldots, v_r)$, and
- $q_0(\alpha, y_1, \ldots, y_r) \to t$,

where, for $1 \le i, j \le r$, $u_i, v_j \in T_\Delta(Y_r)$, and $t \in T_\Delta(Y_r)$.

We can prove this as follows. First let us consider the (q_0, γ)-rule. It should be clear that $rhs(q_0, \gamma)$ contains x_1. Otherwise, for every tree $s \in T_\Sigma$, the output $\tau_M(\gamma(s))$ would not depend on s: this is a contradiction, for details see the proof of Theorem 6.29. On the other hand M is linear, hence x_1 may occur at most once in $rhs(q_0, \gamma)$. So the form of the (q_0, γ)-rule should be

$$q_0(\gamma(x_1), y_1, \ldots, y_r) \to u(p(x_1, u_1, \ldots, u_k)),$$

for some $u \in (\Delta^{(1)})^*$, $k \ge 0, p \in Q^{(k+1)}$, and $u_1, \ldots, u_k \in T_\Delta(Y_r)$. Next we can show that $p = q_0$. In fact, if we assume that $p \ne q_0$, then, considering the

(p, γ)-rule, we obtain that $rhs(p, \gamma)$ cannot contain x_1. This is because, x_1 already occurs in $rhs(q_0, \gamma)$ and because M is superlinear. However, if $rhs(p, \gamma)$ does not contain x_1, then, for every tree $s \in T_\Sigma$, the output $\tau_M(\gamma\gamma(s))$ would not depend on s, which is a contradiction again. (Formally the proof follows that of Theorem 6.29.) Thus we get that $p = q_0$ and that the form of the (q_0, γ)-rule should be

$$q_0(\gamma(x_1), y_1, \ldots, y_r) \rightarrow u(q_0(x_1, u_1, \ldots, u_r)),$$

for some $u \in (\Delta^{(1)})^*$ and $u_1, \ldots, u_r \in T_\Delta(Y_r)$.

Next we can show that $u = \varepsilon$. In fact, assuming $u \neq \varepsilon$, we find that u contains at least one output symbol from Δ. Then, considering an input tree of the form $\gamma^n(\alpha)$, we obtain that $\tau_M(\gamma^n(\alpha))$ has the prefix u^n, where u contains at least one output symbol from Δ. Since, by $\tau_M = \tau_{modulo}$, $height(\tau_M(\gamma^n(\alpha))$ should also be $n + 1$, it also holds that u may contain at most one symbol. If u contains exactly one symbol, then the fact that $\tau_M(\gamma^n(\alpha))$ has the prefix u^n yields an obvious contradiction to the assumption $\tau_{modulo} = \tau_M$. Hence the only possible case is $u = \varepsilon$, which proves that the (q_0, γ)-rule has the form we stated. Since the roles of γ and δ are exchangeable, the same holds for the (q_0, δ)-rule.

In order to derive a final contradiction, we apply the technique used in the proof of Theorem 6.29. That is, we transform M into an equivalent attributed tree transducer $A = (Att, \Sigma, \Delta, a_0, R', E')$. Then A is one-visit, see the proof of Theorem 6.29, and $\tau_M(s) = \tau_A(s)$, for every input tree s. Let E' be the environment of A. Define again

$$\begin{aligned} max &= max\{height(\xi) \mid \xi \text{ is a right-hand side of a rule of } A\} \\ &+ max\{height(E'(y_j)) \mid y_j \text{ is an inherited attribute of } A\}. \end{aligned}$$

Analyzing the set of rules of A, we find that both R'_γ and R'_δ contain the rule $q_0(\pi) \rightarrow q_0(\pi 1)$.

Now consider an input tree s with height $max+1$ and an arbitrary derivation d of A on s of the form

$$q_0(\varepsilon) \Rightarrow^*_{A,s} \varphi$$

such that $\tau_A(s) = \varphi[y_1(\varepsilon) \leftarrow E'(y_1), \ldots, y_r(\varepsilon) \leftarrow E'(y_r)]$.

Due to the form of the rules, the derivation d starts with $height(s) - 1$ applications of the chain rule $q_0(\pi) \rightarrow q_0(\pi 1)$ which produce no output symbols. Thus, d has the form

$$q_0(\varepsilon) \Rightarrow_{A,s} q_0(1) \Rightarrow_{A,s} q_0(11) \ldots \Rightarrow_{A,s} q_0(1^{height(s)-1}) \Rightarrow^*_{A,s} \varphi.$$

Let us consider the occurrence w in s, at which A produces an output symbol for the first time during d. Due to the form of d, this output symbol is produced by an inherited attribute on the ascent of A (from the leaf to the root of s). As in the proof of Theorem 6.29, we can see that neither $w = \varepsilon$ nor $w = 1$ because if this were the case then A would produce an output of height not greater than max for s.

Let us change the input tree s to s' by changing the label of the root of s (i.e., if the root of s is γ, then the root of s' is δ, and conversely).

Clearly, $occ(s) = occ(s')$ and A will also produce the first output for s' at occurrence w and this output symbol will be the same as in the derivation d (since $w \notin \{\varepsilon, 1\}$). Hence, the roots of $\tau_A(s)$ and $\tau_A(s')$ will be the same. However, the roots of $\tau_{modulo}(s)$ and of $\tau_{modulo}(s')$ are not equal. Since $\tau_A = \tau_{modulo}$, we have derived a contradiction to the assumption that $\tau_{modulo} \in$ *sl-MAC*. Hence, $\tau_{modulo} \notin$ *sl-MAC*. □

6.2.4 The inequality *sl-TOP* − *HOM* ≠ ∅

This inequality has already been proved in Lemma 3.35.

6.2.5 The inequality *sl-TOP* − *YIELD* ≠ ∅

It is clear that, for every ranked alphabet Σ, the identity tree transformation $Id(T_\Sigma)$ can be computed by a superlinear top-down tree transducer (see Lemma 3.38). However, this does not hold for *YIELD*.

Lemma 6.32. There is a ranked alphabet Σ such that $Id(T_\Sigma) \notin YIELD$.

Proof. Consider the ranked alphabet $\Sigma = \{\gamma^{(1)}, \alpha^{(0)}\}$. From the definition of *YIELD* (Def. 4.30), the following property can be derived immediately for this particular monadic ranked alphabet Σ.

> For every $r \geq 0$, every function $g : \Sigma^{(0)} \to T_\Delta(Z_r)$, and every tuple $\tilde{t} \in (T_\Delta)^r$, the function $yield_{g,\tilde{t}}$ induced by g and $\tilde{t}$ has the following property:
> for every $s \in T_\Sigma$, $yield_{g,\tilde{t}}(s) = g(\alpha)[z_1 \leftarrow t_1, \ldots, z_r \leftarrow t_r]$.

Hence the range of every function $yield_{g,\tilde{t}}$ is a singleton, i.e., a set with one element. But $Id(T_\Sigma)$ has an infinite range. Thus $Id(T_\Sigma) \notin YIELD$. □

6.2.6 The inequality *YIELD* − *bas-MAC* ≠ ∅

Intuitively, the *YIELD*-functions perform a kind of nested substitution (see Def. 4.30). As shown in Lemma 4.32, every *YIELD*-function can be computed by an appropriate superlinear macro tree transducer. There it was essential that, in the right-hand sides of the rules of that tree transducer, nesting of states is allowed. Here we prove that, without this capability of nesting (i.e., if one considers basic macro tree transducers), not every *YIELD*-function can be computed.

First we define a tree transformation $\tau_{exp-leaves}$ which takes an input tree s over the ranked alphabet $\Sigma = \{\sigma^{(2)}, \alpha^{(0)}\}$ and delivers an output tree of which the height is equal to the number of leaves of s. This tree transformation can be computed by a *YIELD*-function.

Definition 6.33. Let $\Sigma = \{\sigma^{(2)}, \alpha^{(0)}\}$ and $\Delta = \{\gamma^{(1)}, \alpha^{(0)}\}$ be two ranked alphabets and let $g : \Sigma^{(0)} \to T_\Delta(Z_1)$ be the function defined by $g(\alpha) = \gamma(z_1)$. Then denote the tree transformation $yield_{g,(\alpha)} : T_\Sigma \to T_\Delta$ by $\tau_{exp-leaves}$.

Note 6.34. $\tau_{exp-leaves} \in YIELD$.

In particular, if the input tree s is a fully balanced tree over Σ, then the height of $\tau_{exp-leaves}(s)$ is $2^{height(s)-1} + 1$. Next we prove that a basic macro tree transducer can only increase the height of an input tree linearly.

Lemma 6.35. Let M be a basic macro tree transducer. There is a constant $c > 0$ such that for every $s \in T_\Sigma$,

$$height(\tau_M(s)) \leq c \cdot height(s).$$

Proof. Define $c = max\{height(\xi) \mid \xi = rhs(q, \sigma), q \in Q, \sigma \in \Sigma\}$. Next define the two statements K and L.

K: For every $s \in T_\Sigma$, $n \geq 0$, $q \in Q^{(n+1)}$, $height(\tau_{M,q}(s)) \leq c \cdot height(s)$.
L: For every $k, n \geq 0$, $\xi \in RHS(Q, \Delta, k, n)$, and $\omega \in (T_\Sigma)^k$, $height(\tau_{M,\xi}(\omega)) \leq height(\xi) + c \cdot mx$, where $mx = max\{height(s_1), \ldots, height(s_k)\}$.

We can prove K and L by using the proof by simultaneous induction, see Principle 3.11.

<u>IB:</u> Let $\xi \in RHS(Q, \Delta, 0, n)$ and $\omega = ()$. The proof is by structural induction on ξ.

(i) ξ cannot have the form $p(x_i, \xi_1, \ldots, \xi_l)$.
(ii) Let $\xi = \delta(\xi_1, \ldots, \xi_l)$ for some $\delta \in \Delta^{(l)}$ and $\xi_1, \ldots, \xi_l \in T_\Delta(Y_n)$. Assume that L is true for $\xi_1, \ldots, \xi_l$. Then we can compute as follows:

$$\begin{aligned}
& height(\tau_{M,\xi}(\omega)) \\
= \; & height(\delta(\tau_{M,\xi_1}(\omega), \ldots, \tau_{M,\xi_l}(\omega))) \\
= \; & 1 + max\{height(\tau_{M,\xi_i}(\omega)) \mid 1 \leq i \leq l\} \\
\leq \; & 1 + max\{height(\xi_i) + c \cdot max \mid 1 \leq i \leq l\} \\
& \text{(by induction hypothesis on } \xi) \\
= \; & 1 + c \cdot mx + max\{height(\xi_i) \mid 1 \leq i \leq l\} \\
= \; & height(\xi) + c \cdot mx.
\end{aligned}$$

(iii) Let $\xi = y_j$. Then we can compute:

$$\begin{aligned}
& height(\tau_{M,\xi}(\omega)) \\
= \; & height(y_j) \\
= \; & 1 \\
\leq \; & height(\xi) + c \cdot mx.
\end{aligned}$$

IS1: Let $s = \sigma(s_1, \ldots, s_k) \in T_\Sigma$, $n \geq 0$, $q \in Q^{(n+1)}$. Then we can compute:

$$\begin{array}{ll} & height(\tau_{M,q}(s)) \\ = & height(\tau_{M,rhs(q,\sigma)}((s_1, \ldots, s_k))) \\ \leq & height(rhs(q,\sigma)) + c \cdot mx \qquad \text{(by } L\text{)} \\ \leq & c + c \cdot mx \\ = & c \cdot height(s). \end{array}$$

IS2: Let $\xi \in RHS(Q, \Delta, k, n)$ be the right-hand side of the rule of a basic macro tree transducer. Let $\omega \in (T_\Sigma)^k$. The proof is by structural induction on ξ.

(i) Let $\xi = p(x_i, \xi_1, \ldots, \xi_l)$ for some $p \in Q^{(l+1)}$ with $l \geq 0$ and $\xi_1, \ldots, \xi_l \in RHS(Q, \Delta, k, n)$. Because M is basic, every $\xi_j \in T_\Delta(Y_n)$, i.e., it does not contain a state. We distinguish two cases: $l = 0$ and $l > 0$.
$l = 0$:

$$\begin{array}{ll} & height(\tau_{M,\xi}(\omega)) \\ = & height(\tau_{M,p}(s_i)) \\ \leq & c \cdot height(s_i) \qquad \text{(by K)} \\ \leq & height(\xi) + c \cdot mx. \end{array}$$

$l > 0$:

$$\begin{array}{ll} & height(\tau_{M,\xi}(\omega)) \\ = & height(\tau_{M,p}(s_i)[y_j \leftarrow \tau_{M,\xi_j}(\omega); 1 \leq j \leq l]) \\ \leq & height(\tau_{M,p}(s_i)) + max\{height(\tau_{M,\xi_j}(\omega)) \mid 1 \leq j \leq l\} \\ \leq & c \cdot height(s_i) + max\{height(\xi_j) \mid 1 \leq j \leq l\} \qquad \text{(by } K\text{)} \\ \leq & c \cdot mx + max\{height(\xi_j) \mid 1 \leq j \leq l\} \\ \leq & 1 + max\{height(\xi_j) \mid 1 \leq j \leq l\} + c \cdot mx \\ = & height(p(x_i, \xi_1, \ldots, \xi_l)) + c \cdot mx. \end{array}$$

(ii) Let $\xi = \delta(\xi_1, \ldots, \xi_l)$ for some $\delta \in \Delta^{(l)}$ and $\xi_1, \ldots, \xi_l \in T_\Delta(Y_n)$. The proof is the same as in IB.

(iii) Let $\xi = y_j$. The proof is the same as in IB. □

Lemma 6.36. *$YIELD - bas\text{-}MAC \neq \emptyset$.*

Proof. By Note 6.34, $\tau_{exp-leaves} \in YIELD$. Assume that $\tau_{exp-leaves} \in bas\text{-}MAC$. Then, by Lemma 6.35, there is a constant c such that, for every $s \in T_\Sigma$, $height(\tau_M(s)) \leq c \cdot height(s)$. But if s is a fully binary tree, then $height(\tau_{exp-leaves}(s)) = 2^{height(s)-1} + 1$ which is a contradiction. Hence, $\tau_{exp-leaves} \notin bas\text{-}MAC$. □

So far a number of statements of the form $\tau \notin C$ have been proved in this book, where τ is some tree transformation and C is some class of tree transformations. In most of the proofs the relationship between the height or size of the input tree and the corresponding output tree has been used. Here we would like to prove such a statement by means of another technique: the statement is

there is a tree transformation, called τ_{Dyck}, which is not in *bas-MAC*

and the proof technique is based on considering the paths of output trees.

First let us define the tree transformation τ_{Dyck}. It transforms trees over the input alphabet $\Sigma = \{\sigma^{(2)}, \gamma_1^{(1)}, \gamma_2^{(1)}, \alpha^{(0)}\}$ into trees over the output alphabet $\Delta = \{a^{(1)}, a'^{(1)}, b^{(1)}, b'^{(1)}, c^{(0)}\}$, thus, in particular, τ_{Dyck} computes monadic trees. Intuitively, the transformation τ_{Dyck} can be understood as follows (see Fig. 6.7 and the remark after the following definition). Consider an input tree s. Now traverse s depth-first right-to-left. Whenever the symbol γ_1 is crossed in the direction "root to frontier", then an a is produced as an output symbol; if it is crossed in the direction from "frontier to root", then an a' is produced as an output symbol. Whenever the symbol γ_2 is crossed, then, similarly, either b or b' is produced as an output symbol depending on whether the crossing direction is from "root to frontier" or from "frontier to root". Formally, we define τ_{Dyck} by means of a superlinear macro tree transducer.

Definition 6.37. Let $\Sigma = \{\sigma^{(2)}, \gamma_1^{(1)}, \gamma_2^{(1)}, \alpha^{(0)}\}$ and $\Delta = \{a^{(1)}, a'^{(1)}, b^{(1)}, b'^{(1)}, c^{(0)}\}$ be two ranked alphabets. Define the tree transformation $\tau_{Dyck} : T_\Sigma \to T_\Delta$ by $\tau_{Dyck} = \tau_M$ where $M = (Q, \Sigma, \Delta, q, R, E)$ is the superlinear macro tree transducer given by $Q = \{q^{(2)}\}$, $E = (c)$, and R contains the rules:

1) $q(\sigma(x_1, x_2), y_1) \to q(x_2, q(x_1, y_1))$
2) $q(\gamma_1(x_1), y_1) \to a(q(x_1, a'(y_1)))$
3) $q(\gamma_2(x_1), y_1) \to b(q(x_1, b'(y_1)))$
4) $q(\alpha, y_1) \to y_1$ □

Since *sl-MAC* $\subseteq$ *ATT*, the transformation τ_{Dyck} can also be specified by an attributed tree transducer. This approach is taken in Fig. 6.7 where the relation

$$\tau_{Dyck}(\gamma_1(\sigma(\gamma_2(\gamma_2(\alpha)), \gamma_1(\sigma(\gamma_2(\alpha), \alpha))))) = aabb'a'bbb'b'a'c$$

is shown.

Now we show that τ_{Dyck} is not in *bas-MAC*. The non-inclusion is due to the following reasons

- the path languages of *range*(*bas-MAC*) are linear (context-free) languages and
- the path language of *range*(τ_{Dyck}) is the Dyck language D_2 over $\{a, b\}$ and it is known that D_2 is not a linear language.

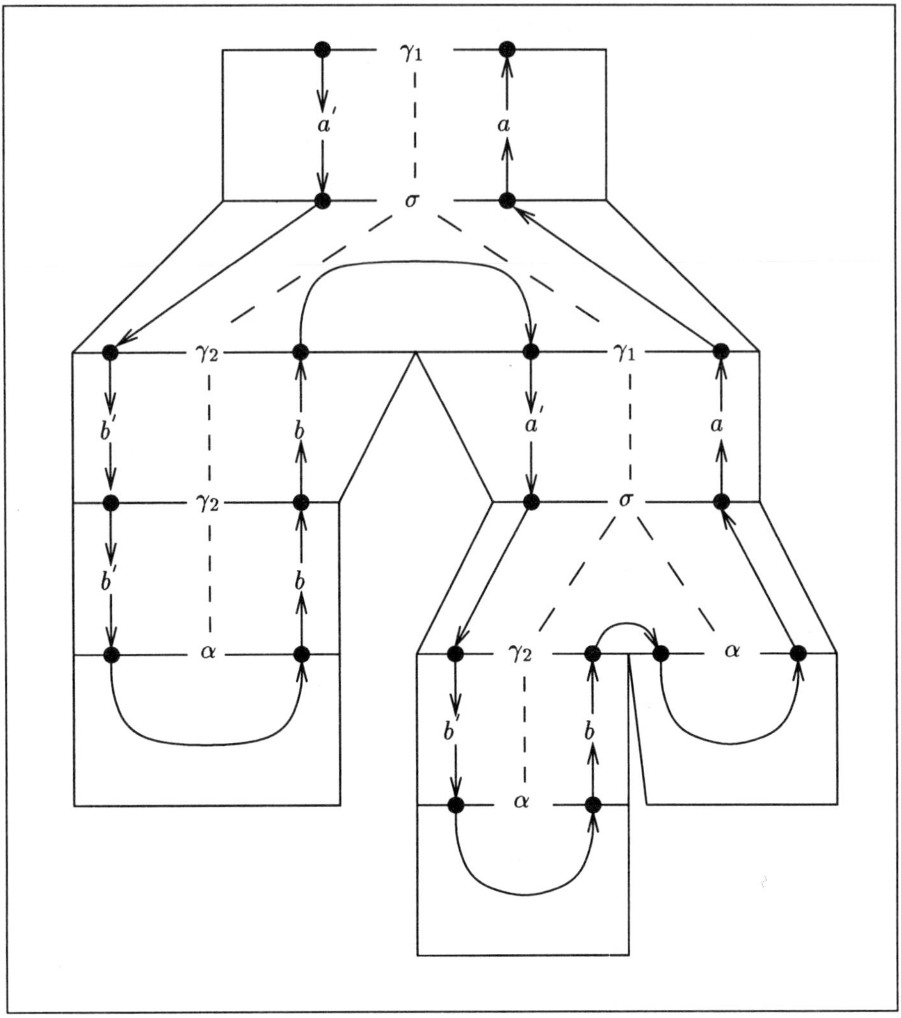

Fig. 6.7. An illustration of τ_{Dyck}

We need some terminology about path languages. Let Δ be a ranked alphabet. The *path alphabet associated with* Δ, denoted by $p\Delta$, is the finite set $\{(\delta, 0) \mid \delta \in \Delta^{(0)}\} \cup \{(\delta, i) \mid \delta \in \Delta^{(k)}$ for some $k \geq 1$ and $1 \leq i \leq k\}$. For every ranked alphabet Δ, the *path translation of* Δ is the function $path_\Delta : T_\Delta \rightarrow \mathcal{P}((p\Delta)^*)$ defined inductively as follows. (i) For $\alpha \in \Delta^{(0)}$, define $path_\Delta(\alpha) = \{(\alpha, 0)\}$. (ii) For $t = \delta(t_1, \ldots, t_k)$ with $\delta \in \Delta^{(k)}$ and $k \geq 1$, define $path_\Delta(t) = \{(\delta, i)p \mid 1 \leq i \leq k$ and $p \in path_\Delta(t_i)\}$. Let $PATH$ denote the class of all path translations $path_\Delta$, where Δ is a ranked alphabet. For a class $\mathcal{L}$ of languages, we use $PATH(\mathcal{L})$ to denote the class of all path languages $path(L)$ for $L \in \mathcal{L}$. Note that, for every $t \in T_\Delta$, $path_\Delta(t) \subseteq (p\Delta)^*$. Recall that $\mathcal{L}_{lin}$ denotes the class of linear languages.

At the same time we will also show that the path languages of ranges of top-down tree transformations are regular languages. Recall that $\mathcal{L}_{reg}$ denotes the class of regular languages.

Theorem 6.38.
1) $PATH(range(bas\text{-}MAC)) \subseteq \mathcal{L}_{lin}$.
2) $PATH(range(TOP)) \subseteq \mathcal{L}_{reg}$.

Proof. Consider a basic macro tree transducer $M = (Q, \Sigma, \Delta, q_0, R, E)$ with $rank(q_0) = r + 1$, $r \geq 0$, and $E = (t_1, \ldots, t_r)$. We construct a linear context-free grammar G such that $L(G) = path_\Delta(range(\tau_M))$.

Define $G = (N, p\Delta, P, q_{new})$ as follows.

- $N = \{q_{new}\} \cup \{[q, i] \mid q \in Q^{(n+1)}, n \geq 0, 0 \leq i \leq n\}$
- P contains the rule $q_{new} \to [q_0, 0]$ and, for every $i \in \{1, \ldots, r\}$ and $w \in path_\Delta(t_i)$, the rule $q_{new} \to [q_0, i]\ w$.
  Moreover, consider the rule $q(\sigma(x_1, \ldots, x_k), y_1, \ldots, y_n) \to \xi$ of R and consider a path $(\delta_1, \kappa_1)(\delta_2, \kappa_2) \ldots (\delta_\mu, \kappa_\mu)$ in $path_{\Delta \cup Q \cup X_k \cup Y_n}(\xi)$.
  - If, for some i with $1 \leq i < \mu$, the element δ_i is a state $p \in Q$ and $\delta_\mu \notin X_k$, then the following rule is in P:
    $$[q, \eta] \;\to\; (\delta_1, \kappa_1)(\delta_2, \kappa_2) \ldots (\delta_{i-1}, \kappa_{i-1})[p, \kappa_i - 1](\delta_{i+1}, \kappa_{i+1}) \ldots \ldots (\delta_{\mu-1}, \kappa_{\mu-1})\psi$$
    where η and ψ are determined as follows:
    - if $\delta_\mu \in \Delta$, then $\eta = 0$ and $\psi = (\delta_\mu, \kappa_\mu)$
    - if $\delta_\mu = y_j$, then $\eta = j$ and $\psi = \varepsilon$.
  - If $\delta_{\mu-1} = p \in Q$ and $\delta_\mu \in X_k$, then the rule
    $$[q, 0] \to (\delta_1, \kappa_1)(\delta_2, \kappa_2) \ldots (\delta_{\mu-2}, \kappa_{\mu-2})[p, 0]$$
    is in P.
  - If, for every i with $1 \leq i \leq \mu$, the element $\delta_i \in \Delta \cup Y$, then the rule
    $$[q, \eta] \to (\delta_1, \kappa_1)(\delta_2, \kappa_2) \ldots (\delta_{\mu-1}, \kappa_{\mu-1})\psi$$
    is in P where η and ψ are defined as in the first case.

This ends the construction of the linear grammar. If M is a top-down tree transducer, then G is clearly a regular grammar because the first of the three conditions above never holds.

Now it remains to prove the correctness of the construction. Let $mx = max\{n \mid Q^{(n+1)} \neq \emptyset\}$.

For every $q \in Q$, $i \geq 1$, and $w \in (p\Delta)^*$, we can prove the following statements.

1. $[q, i] \Rightarrow_G^* w$ iff there is an $s \in T_\Sigma$ such that $w(y_i, 0) \in path_{\Delta \cup Y_{mx}}(\tau_{M,q}(s))$.

2. $[q,0] \Rightarrow_G^* w$ iff there is an $s \in T_\Sigma$ such that $w \in path_{\Delta\cup Y_{mx}}(\tau_{M,q}(s))$. □

In fact, the classes $\mathcal{L}_{lin}$ and $PATH(range(bas\text{-}MAC))$ are very close to each other: by dropping the second component of the elements of the path alphabet, languages in $\mathcal{L}_{lin}$ can be considered as elements of $PATH(range(bas\text{-}MAC))$.

Example 6.39. Consider again the basic macro tree transducer M which was shown in Theorem 6.5. We repeat its rules:

- $q(\gamma(x_1), y_1) \to \sigma(q(x_1, 1(y_1)), q(x_1, 2(y_1)))$
- $q(\alpha, y_1) \to y_1$.

Recall that the environment E of M is the sequence $(\#)$.

According to Theorem 6.38, we construct the linear context-free grammar $G = (N, p\Delta, P, q_{new})$ as follows.

- $N = \{q_{new}\} \cup \{[q,0],[q,1]\}$
- P contains the rules
  - $q_{new} \to [q,0]$
    $q_{new} \to [q,1](\#,0)$
  - the right-hand sides of rules of M have the following paths
    1. $(\sigma,1)(q,1)(x_1,0)$
    2. $(\sigma,1)(q,2)(1,1)(y_1,0)$
    3. $(\sigma,2)(q,1)(x_1,0)$
    4. $(\sigma,2)(q,2)(2,1)(y_1,0)$
    5. $(y_1,0)$

    These paths give rise to the following rules in P:
    1. $[q,0] \to (\sigma,1)[q,0]$
    2. $[q,1] \to (\sigma,1)[q,1](1,1)$
    3. $[q,0] \to (\sigma,2)[q,0]$
    4. $[q,1] \to (\sigma,2)[q,1](2,1)$
    5. $[q,1] \to \varepsilon$.

We find that the state $[q,0]$ never occurs in a derivation $q_{new} \Rightarrow_G^* w$ for some $w \in (p\Delta)^*$, because every rule with $[q,0]$ in its left-hand side has a state in its right-hand side. □

Definition 6.40. The *Dyck language* D_2 is the formal language which is generated by the following context-free grammar $G = (\{A\}, \{a, a', b, b'\}, P, A)$ where P contains the productions:

- $A \to AA$
- $A \to aAa'$
- $A \to bAb'$
- $A \to \varepsilon$ □

Lemma 6.41. $D_2 \notin \mathcal{L}_{lin}$.

Proof. It can be proved by a pumping lemma for linear languages (see for example Theorem 6.1 of [ABB97]) that the language $L \cdot L$ with $L = \{a^n b^n \mid n \geq 0\}$ is not a linear language. Now assume that $D_2 \in \mathcal{L}_{lin}$. Let $R = \{a^k a'^l b^m b'^n \mid k, l, m, n \geq 0\}$ be a regular language and let $h : \{a, a', b, b'\}^* \to \{a, b\}^*$ be the homomorphism which maps the letters a, a', b, b' to the letters a, b, a, b, respectively. Clearly, $L \cdot L = h(D_2 \cap R)$. Since the class of linear languages is closed under intersection with regular languages and homomorphisms (see for example Section 3.1 of [MS97]), it follows that $L \cdot L \in \mathcal{L}_{lin}$ which is a contradition. □

Lemma 6.42. There is a string homomorphism h such that the Dyck language D_2 is the image of $path_\Delta(range(\tau_{Dyck}))$ under h.

Proof. Let $\Delta = \{a^{(1)}, a'^{(1)}, b^{(1)}, b'^{(1)}, c^{(0)}\}$ be the ranked alphabet of Def. 6.37. Define the string homomorphism $h : p\Delta^* \to \{a, a', b, b'\}^*$ by $h((a,1)) = a$, $h((a',1)) = a'$, $h((b,1)) = b$, $h((b',1)) = b'$, and $h((c,0)) = \varepsilon$.

Consider the macro tree transducer $M = (Q, \Sigma, \Delta, q, R, E)$ of Def. 6.37. For every $s \in T_\Sigma$, the one and only leaf of $\tau_{M,q}(s)$ is labeled by y_1. Thus, the path language $path_\Delta(range(\tau_M))$ can be generated by the context-free grammar $G = (N, p\Delta, P, q_0)$ with $N = \{q_0, q\}$ and P contains the following productions.

1) $q_0 \to q\ (c, 0)$
2) $q \to qq$
3) $q \to (a, 1)\ q\ (a', 1)$
4) $q \to (b, 1)\ q\ (b', 1)$
5) $q \to \varepsilon$

Roughly speaking, the productions of G are obtained by taking the paths of the right-hand sides of rules of M; the first production takes over the role of the environment of M in the definition of τ_M. Hence, $path_\Delta(range(\tau_M)) = L(G)$. It is clear that $h(L(G)) = D_2$. □

Theorem 6.43. $\tau_{Dyck} \notin bas\text{-}MAC$.

Proof. Assume $\tau_{Dyck} \in bas\text{-}MAC$. Then, by Theorem 6.38, $path_\Delta(range(\tau_{Dyck})) \in \mathcal{L}_{lin}$. Since $\mathcal{L}_{lin}$ is closed under homomorphisms, (according to Section 3.1 of [MS97]), it follows from Lemma 6.42 that $D_2 \in \mathcal{L}_{lin}$ which is a contradiction to Lemma 6.41. □

6.2.7 The inequality $ATT - 1v\text{-}ATT \neq \emptyset$

We prove this inequality by showing a pumping lemma for tree transformations induced by monadic one-visit attributed tree transducers and by constructing an attributed tree transducer which does not meet the statement of the pumping lemma.

Definition 6.44. An attributed tree transducer is called *monadic*, if the ranked alphabets of input symbols and output symbols are monadic.

The attributed tree transducers considered have the additional property that, for every input tree s, they always include the leaf of s in their derivations and return to the root.

Definition 6.45. Let $A = (Att, \Sigma, \Delta, a_0, R, E)$ be a monadic one-visit attributed tree transducer and let $s \in T_\Sigma$.

1) A is called *leaf touching on* s, if there is a synthesized attribute a and a $(\Delta, 1)$-context $v \in C_{\Delta,1}$ such that $a_0(\varepsilon) \Rightarrow^*_{A,s} v[a(1^{height(s)-1})]$, i.e., the derivation starting from $a_0(\varepsilon)$ calls an attribute occurrence of the leaf of s.

2) A is called *root touching on* s, if there is an inherited attribute b and a $(\Delta, 1)$-context $v \in C_{\Delta,1}$ such that $a_0(\varepsilon) \Rightarrow^*_{A,s} v[b(\varepsilon)]$, i.e., the derivation starting from $a_0(\varepsilon)$ calls another attribute occurrence of the root of s. □

In the following we use the particular ranked alphabet $\Sigma_0 = \{\gamma^{(1)}, \delta^{(1)}, \alpha^{(0)}\}$ and we consider the language $L_\delta = \{\gamma\delta^n(\alpha) \mid n \geq 0\}$.

Clearly, in order to compute an infinite output language, the attributed tree transducer should be leaf touching.

Lemma 6.46. Let A be a monadic one-visit attributed tree transducer with input alphabet Σ_0. If there is an input tree $s_0 \in L_\delta$ such that A is not leaf touching on s_0, then the set $\tau_A(L_\delta)$ is finite.

Proof. Let s_0 be an input tree such that A is not leaf touching on s_0. Due to the forms of Σ_0 and L_δ, for every tree $s \in L_\delta$ with $height(s) > height(s_0)$, there is an $m > 0$ such that s has the form $s_0[\alpha \leftarrow \delta^m(\alpha)]$. Since A is not leaf touching on s_0, A cannot distinguish between s_0 and any $s_0[\alpha \leftarrow \delta^m(\alpha)]$ with $m \geq 1$. Hence, for every $m \geq 1$, $\tau_A(s_0) = \tau_A(s_0[\alpha \leftarrow \delta^m(\alpha)])$. Since there is only a finite number of trees s in L_δ with $height(s) \leq height(s_0)$, the language $\tau_A(L_\delta)$ is finite. □

We can also transform every leaf touching attributed tree transducer into an equivalent leaf touching and root touching attributed tree transducer.

Lemma 6.47. Let A be a monadic one-visit attributed tree transducer with input alphabet Σ_0. Moreover, A has the property that, for every $s \in L_\delta$, A is leaf touching on s. Then there is a monadic one-visit attributed tree transducer A' such that $\tau_A = \tau_{A'}$ and, for every $s \in L_\delta$, A' is leaf touching and root touching on s.

Proof. Let $A = (Att, \Sigma_0, \Delta, a_0, R, E)$ be a monadic one-visit attributed tree transducer with the properties mentioned. Construct the attributed tree transducer $A' = (Att', \Sigma_0, \Delta, a_0, R', E')$ as follows:

- $Att' = Att_{syn} \cup Att'_{inh}$ where $Att'_{inh} = \{up_\beta \mid \beta \in \Delta^{(0)}\} \cup Att_{inh}$.

- E': for every $b \in Att_{inh}$, define $E'(b) = E(b)$ and for every $\beta \in \Delta^{(0)}$, define $E'(up_\beta) = \beta$.
- R' is equal to R except for the following:
  - R'_α: If there is a rule in R_α of the form $a(\pi) \to v(\beta)$ for some $a \in Att_{syn}$, $v \in (\Delta^{(1)})^*$, and $\beta \in \Delta^{(0)}$, then replace this rule by $a(\pi) \to v\, up_\beta(\pi)$.
  - R'_δ: If there is a rule in R_δ of the form $b(\pi 1) \to v(\beta)$ for some $b \in Att_{inh}$, $v \in (\Delta^{(1)})^*$, and $\beta \in \Delta^{(0)}$, then replace this rule by the rule $b(\pi 1) \to v\, up_\beta(\pi)$.
    Moreover, for every $\beta \in \Delta^{(0)}$, the rule $up_\beta(\pi 1) \to up_\beta(\pi)$ is in R'_δ.
  - R'_γ: If there is a rule in R_γ of the form $b(\pi 1) \to v(\beta)$ for some $b \in Att_{inh}$, $v \in (\Delta^{(1)})^*$, and $\beta \in \Delta^{(0)}$, then replace this rule by the rule $b(\pi 1) \to v\, up_\beta(\pi)$. Moreover, for every $\beta \in \Delta^{(0)}$, add the rule $up_\beta(\pi 1) \to up_\beta(\pi)$ to R_γ.

It is obvious that, for every input tree $s \in L_\delta$, A' is leaf-touching and root-touching on s and that $\tau_A = \tau_{A'}$. □

Now we prove the pumping lemma for tree transformations induced by monadic one-visit attributed tree transducers.

Lemma 6.48. Let A be a monadic one-visit attributed tree transducer with input alphabet Σ_0. Moreover, A has the property that, for every $s \in L_\delta$, A is leaf touching and root touching on s.

There is a constant n_0, such that, for every $n > n_0$ and input tree $\gamma\delta^n(\alpha)$, there are $\kappa_1, \kappa_2 \in (\Sigma_0^{(1)})^*$ and $\kappa_3 \in (\Sigma_0^{(1)})^*\Sigma_0^{(0)}$, and there are $v_1, v_2, v_3, v_4 \in (\Delta^{(1)})^*$ and $v_5 \in (\Delta^{(1)})^*\Delta^{(0)}$ and

- $\gamma\delta^n(\alpha) = \kappa_1\kappa_2\kappa_3$, $\tau_A(\gamma\delta^n(\alpha)) = v_1 v_2 v_3 v_4 v_5$ and $\kappa_2 \neq \varepsilon$
- for every $k \geq 0$, $\tau_A(\kappa_1\kappa_2^k\kappa_3) = v_1 {v_2}^k v_3 {v_4}^k v_5$.

Proof. Let $A = (Att, \Sigma_0, \Delta, a_0, R, E)$ be a monadic one-visit attributed tree transducer with the properties mentioned. Define $n_0 = card(Att_{syn}) \cdot card(Att_{inh})$.

Now consider the derivation of A on the input tree $s = \gamma\delta^n(\alpha)$ for some $n > n_0$. Since A is leaf touching and root touching on s, there are $a_{i_0}, \ldots, a_{i_{n+1}} \in Att_{syn}$, $u_1, \ldots, u_{n+1} \in (\Delta^{(1)})^*$, $v \in (\Delta^{(1)})^*$, $b_{j_0}, \ldots, b_{j_n} \in Att_{inh}$, and $w_1, \ldots, w_n \in (\Delta^{(1)})^*$ and there is a $b \in Att_{inh}$ and a $w \in (\Delta^{(1)})^*$ such that $a_0 = a_{i_0}$ and

$$
\begin{aligned}
a_{i_0}(\varepsilon) &\Rightarrow_{A,s} u_1\, a_{i_1}(1) \\
&\Rightarrow_{A,s} u_1\, u_2\, a_{i_2}(11) \\
&\ldots \\
&\Rightarrow_{A,s} u_1\, u_2\, \ldots u_{n+1}\, a_{i_{n+1}}(1^{n+1}) \\
&\Rightarrow_{A,s} u_1\, u_2\, \ldots u_{n+1}\, v\, b_{j_0}(1^{n+1}) \\
&\Rightarrow_{A,s} u_1\, u_2\, \ldots u_{n+1}\, v\, w_1\, b_{j_1}(1^n) \\
&\ldots \\
&\Rightarrow_{A,s} u_1\, u_2\, \ldots u_{n+1}\, v\, w_1 \ldots w_n\, b_{j_n}(1) \\
&\Rightarrow_{A,s} u_1\, u_2\, \ldots u_{n+1}\, v\, w_1 \ldots w_n\, w\, b(\varepsilon).
\end{aligned}
$$

The fact that A is one-visit is used in the form of the above derivation in the following way. There are no inherited attributes appearing in the derivation before an inherited attribute of the leaf α is involved. In fact, if there was at any point, then a synthesized attribute should appear later so that the derivation touches the leaf of s. However, in this case an inherited attribute of δ would depend on one of its synthesized attributes, contradicting the requirement that A is one-visit. Moreover, once an inherited attribute of α is reached, a synthesized attribute no longer appears in the derivation. This is because if it did, then an inherited attribute of δ would depend on one of its synthesized attributes, again contradicting the requirement that A is one-visit.

Since $n > n_0 = card(Att_{syn}) \cdot card(Att_{inh})$, there are l and l' such that $1 \leq l < l' \leq n$ and $a_{i_l} = a_{i_{l'}}$ and $b_{j_{n-l'+1}} = b_{j_{n-l+1}}$. Let l_0 and l'_0 be indices such that $a_{i_{l_0}} = a_{i_{l'_0}}$ and $b_{j_{n-l'_0+1}} = b_{j_{n-l_0+1}}$. Let these indexes be fixed for the rest of the proof.

Define $\kappa_1 = \gamma\delta^{l_0-1}$, $\kappa_2 = \delta^{l'_0-l_0}$, and $\kappa_3 = \delta^{n-l'_0+1}(\alpha)$. Then obviously, $\gamma\delta^n(\alpha) = \kappa_1\kappa_2\kappa_3$. Since $l_0 < l'_0$, $\kappa_2 \neq \varepsilon$.

Define

$$\begin{aligned}
v_1 &= u_1 \ldots u_{l_0} \\
v_2 &= u_{l_0+1} \ldots u_{l'_0} \\
v_3 &= u_{l'_0+1} \ldots u_{n+1}\, v\, w_1 \ldots w_{n-l'_0+1} \\
v_4 &= w_{n-l'_0+2} \ldots w_{n-l_0+1} \\
v_5 &= w_{n-l_0+2} \ldots w_n w\, w_E \text{ where } w_E = E(b)
\end{aligned}$$

Since $a_{i_{l_0}} = a_{i_{l'_0}}$ and $b_{j_{n-l'_0+1}} = b_{j_{n-l_0+1}}$, the piece κ_2 of the input tree s can be repeated without changing the derivation of A on the pieces κ_1 and κ_3. This means that, for every $k \geq 0$ and $s_k = \kappa_1\kappa_2^k\kappa_3$, the derivation of A on s_k looks as follows:

$$\begin{aligned}
a_0(\varepsilon) &\Rightarrow^*_{A,s_k} v_1\, a_{i_{l_0}}(1^{l_0}) \\
&\Rightarrow^*_{A,s_k} v_1\, v_2^k\, a_{i_{l'_0}}(1^{l_0+k\cdot(l'_0-l_0)}) \\
&\Rightarrow^*_{A,s_k} v_1\, v_2^k\, v_3\, b_{j_{n-l'_0+1}}(1^{l_0+k\cdot(l'_0-l_0)}) \\
&\Rightarrow^*_{A,s_k} v_1\, v_2^k\, v_3\, v_4^k\, b_{j_{n-l_0+1}}(1^{l_0}) \\
&\Rightarrow^*_{A,s_k} v_1\, v_2^k\, v_3\, v_4^k\, v_5
\end{aligned}$$

Thus, for every $k \geq 0$, $\tau_A(\kappa_1\kappa_2^k\kappa_3) = v_1\, v_2^k\, v_3\, v_4^k\, v_5$. □

Lemma 6.49. $ATT - 1v\text{-}ATT \neq \emptyset$.

Proof. Consider the following attributed tree transducer $A = (Att, \Sigma_0, \Delta, a_0, R, E)$ with Σ_0 as before, $\Delta = \{a^{(1)}, b^{(1)}, c^{(1)}, \alpha^{(0)}\}$, $Att_{syn} = \{a_0, s_a, s_c\}$, and $Att_{inh} = \{i_b\}$. The rules of A are as follows:

- $R_\gamma = \{a_0(\pi) \to s_a(\pi 1), i_b(\pi 1) \to s_c(\pi 1)\} \cup \{s_a(\pi) \to \alpha, s_c(\pi) \to \alpha\}$
- $R_\delta = \{s_a(\pi) \to a s_a(\pi 1), s_c(\pi) \to c s_c(\pi 1), i_b(\pi 1) \to b i_b(\pi)\} \cup \{a_0(\pi) \to \alpha\}$

- $R_\alpha = \{s_a(\pi) \to i_b(\pi), s_c(\pi) \to \alpha\} \cup \{a_0(\pi) \to \alpha\}$.

We note that the rules appearing in the second members of the unions defining R_γ, R_δ, and R_α are just dummy rules because they will never be used in any derivation starting from the attribute occurrence $a_0(\varepsilon)$ of an input tree.

Then certainly A is noncircular, but A is not one-visit because of the rule $i_b(\pi 1) \to s_c(\pi 1)$. The tree transformation induced by A on L_δ has the property that

$$\tau_A(\gamma\delta^n(\alpha)) = a^n b^n c^n(\alpha)$$

for every $n \geq 0$.

Now assume that $\tau_A \in 1v\text{-}ATT$. Since $\tau_A(L_\delta)$ is infinite, by Lemmas 6.46 and 6.47 we can assume that A is leaf-touching and root-touching on every input tree. Then, by Lemma 6.48 there is an n_0 with certain properties. Consider the input tree $s = \gamma\delta^n(\alpha)$ for some $n > n_0$. Then, by the pumping lemma, there are $\kappa_1, \kappa_2, \kappa_3$ and there are $v_1, v_2, v_3, v_4 \in (\Delta^{(1)})^*$ and $v_5 \in (\Delta^{(1)})^*\Delta^{(0)}$ such that $\gamma\delta^n(\alpha) = \kappa_1\kappa_2\kappa_3$, $\tau(\gamma\delta^n(\alpha)) = v_1v_2v_3v_4v_5$ and, for every $k \geq 0$, $\tau(\kappa_1\kappa_2^k\kappa_3) = v_1{v_2}^k v_3{v_4}^k v_5$.

Clearly, when the input tree s is pumped by repeating κ_2, the input tree grows (note that $\kappa_2 \neq \varepsilon$). Hence, because of $\tau_A(\gamma\delta^n(\alpha)) = a^n b^n c^n(\alpha)$, the output will also grow. And thus, v_2 or v_4 cannot be empty and they occur more often in the output of a pumped input tree than in the output of s. On the other hand, neither v_2 nor v_4 can contain more than one of the letters a, b, and c, otherwise these letters would not be in alphabetical order in the output of the pumped input tree. Then the output cannot be of the form $a^m b^m c^m(\alpha)$, because the number of no more than two letters can change in it. This is a contradiction to the assumption that $\tau_A \in 1v\text{-}ATT$ however the κ's and v's are chosen. □

6.2.8 The inequality $nn\text{-}ATT - TOP \neq \emptyset$

Here we follow the same line of proof as in Theorem 6.43. We construct a nonnested attributed tree transducer A such that the path language of $range(\tau_A)$ is a linear language. Then difference set is not empty because the path languages of ranges of top-down tree transducers are regular languages.

Definition 6.50. Let $\Sigma = \{\gamma^{(1)}, \alpha^{(0)}\}$ and $\Delta = \{\gamma_1^{(1)}, \gamma_2^{(1)}, \alpha^{(0)}\}$. Define the tree transformation $\tau_{repeat} : T_\Sigma \to T_\Delta$ by $\tau_{repeat} = \tau_A$ where $A = (Att, \Sigma, \Delta, a, R, E)$ is the nonnested attributed tree transducer given by $Att = Att_{syn} \cup Att_{inh}$ with $Att_{syn} = \{a\}$ and $Att_{inh} = \{b\}$, $E(b) = \alpha$, and R consists of the sets R_γ and R_α.

R_γ:

1) $a(\pi) \to \gamma_1(a(\pi 1))$
2) $b(\pi 1) \to \gamma_2(b(\pi))$

and R_α:

3) $a(\pi) \to b(\pi)$

It is obvious that $\tau_A = \{(\gamma^n(\alpha), \gamma_1^n \gamma_2^n(\alpha)) \mid n \geq 0\}$. □

Note 6.51. By standard arguments, $\{(\gamma_1, 1)^n(\gamma_2, 1)^n(\alpha, 0) \mid n \geq 0\} \notin \mathcal{L}_{reg}$.

Theorem 6.52. $\tau_{repeat} \notin TOP$.

Proof. Assume $\tau_{repeat} \in TOP$. Then, by Theorem 6.38 part 2, $L = path_\Delta(range(\tau_{repeat})) \in \mathcal{L}_{reg}$. However, by Note 6.51, $L \notin \mathcal{L}_{reg}$ contradicting the assumption. □

6.2.9 The inequality $HOM - l\text{-}TOP \neq \emptyset$

This inequality has already been proved in Lemma 3.33.

6.3 Composition Semigroup Generated by TOP, ATT, and MAC

The content of this section

In this section we consider again the problem of determining whether an inclusion (or equality or incomparability) relation holds between any two expressions built up by composition from a fixed set H of fundamental tree transformation classes. The set H now will be chosen as $H = \{TOP, ATT, MAC\}$.

We recall that the same problem has been considered in Sect. 3.7, where H consisted of certain top-down tree transformation classes. We also recall that in paragraphs (a)–(d) of that section we gave a general method with which such an algorithm can be constructed. We proved in Lemma 3.48 that the method is correct. Then we successfully implemented this method for the instance of H mentioned.

In this section we will apply the general method to the set

$$H = \{TOP, ATT, MAC\}$$

thereby constructing an algorithm for H.

We recall from Sect. 3.7 that, to develop the algorithm, we considered two semigroups in terms of H: the free semigroup H^+ with the operation of concatenation (denoted by $\cdot$) and the semigroup

$$[H] = \{U_1 \circ \cdots \circ U_m \mid m \geq 1, U_i \in H \text{ for every } 1 \leq i \leq m\}$$

generated by H with the operation of composition (denoted by $\circ$). We also used the homomorphism $|\,| : H^+ \to [H]$ which is the unique extension of the identity mapping over H and for which, therefore,

$$|U_1 \cdot \ldots \cdot U_m| = U_1 \circ \cdots \circ U_m,$$

for all elements $U_1, \ldots, U_m \in H$. Moreover, we denoted the kernel of the homomorphism $|\,|$ by θ.

Another application of the general method

In this subsection we apply the general method to the set $H = \{TOP, ATT, MAC\}$. In Sect. 3.7 it transpired that the implementation could be performed in five steps. Here we also split the implementation into five corresponding steps.

STEP 1. In the first step we produce a string rewrite system S, which is our candidate for presenting $[H]$. In the present case, let S consist of the seven rules:

$$\begin{array}{lllll}
1) & TOP \cdot TOP & \rightarrow & TOP \\
2) & TOP \cdot MAC & \rightarrow & MAC \\
3) & MAC \cdot TOP & \rightarrow & MAC \\
4) & TOP \cdot ATT & \rightarrow & MAC \\
5) & ATT \cdot TOP & \rightarrow & ATT \\
6) & ATT \cdot MAC & \rightarrow & ATT \cdot ATT \\
7) & MAC \cdot ATT & \rightarrow & MAC \cdot MAC
\end{array}$$

Roughly speaking, the rewrite rules do not change the class of tree transformations to which they are applied. Formally, this is expressed in the following lemma.

Lemma 6.53. $\Leftrightarrow_S^* \subseteq \theta$

Proof. First, in the same way as in the proof of Lemma 3.49, we show that elements of S are valid in $[H]$, i.e., that, for every $u \rightarrow v \in S$, the equality $|u| = |v|$ holds. Therefore we recall that the validity of the rules 1), 2), 3), 4) and 5) has already been verified in Theorem 3.39, Corollary 4.38, and Theorems 4.40, 6.25, and 5.47, respectively. With the following computation, we show that $MAC \circ ATT = MAC \circ MAC$, which establishes 7).

$$\begin{array}{lll}
 & MAC \circ ATT & \\
\subseteq & MAC \circ MAC & \text{(Lemma 6.1)} \\
\subseteq & MAC \circ TOP \circ YIELD & \text{(Lemma 4.34)} \\
\subseteq & MAC \circ YIELD & \text{(Theorem 4.40)} \\
\subseteq & MAC \circ sl\text{-}MAC & \text{(Lemma 4.32)} \\
\subseteq & MAC \circ ATT & \text{(Note 6.9 and Theorem 6.23).}
\end{array}$$

Next we prove that $ATT \circ MAC = ATT \circ ATT$, i.e., that 6) is valid.

$$\begin{array}{lll}
 & ATT \circ MAC & \\
\subseteq & ATT \circ HOM \circ sl\text{-}MAC & \text{(Corollary 4.39)} \\
\subseteq & ATT \circ TOP \circ sl\text{-}MAC & \text{(Corollary 3.32)} \\
\subseteq & ATT \circ sl\text{-}MAC & \text{(Lemma 5.46)} \\
\subseteq & ATT \circ ATT & \text{(Note 6.9 and Theorem 6.23)} \\
\subseteq & ATT \circ MAC & \text{(Lemma 6.1).}
\end{array}$$

Finally we note that the inclusion $\Leftrightarrow_S^* \subseteq \theta$ can be derived from the validity of the rules in S in the same way as in the proof of Lemma 3.49. □

STEP 2. Next we give a set REP which is our candidate for a set of representatives of the congruence classes of $\Leftrightarrow_S^*$. Let

$$REP = \{TOP\} \cup \{ATT^n \mid n \geq 1\} \cup \{MAC^n \mid n \geq 1\}.$$

STEP 3. In this step, we construct the inclusion diagram of the tree transformation classes represented by the elements of REP, i.e., that of the set $|REP| = \{|u| \mid u \in REP\}$. Our candidate for this inclusion diagram can be seen in Fig. 6.8. We observe that the guessed diagram is now rather special, in fact it is a chain. Therefore we do not have to follow the method presented in Sect. 2.2 to prove that it is the inclusion diagram of $|REP|$. Instead we observe that there are no incomparable classes and therefore it is sufficient to prove that all inclusions shown by the diagram are proper.

Lemma 6.54. Figure 6.8 shows the inclusion diagram of the set $|REP| = \{|u| \mid u \in REP\}$.

Proof. In Parts 1, 2, and 3, we show that all inclusions shown by the conjectured diagram are proper.

Part 1. Here we recall that the inclusion $TOP \subset ATT$ has already been proved in Theorem 5.44.

Part 2. We now prove that $ATT^n \subset MAC^n$, for every $n \geq 1$.

We note that this statement has already been proved in Lemma 6.2 for $n = 1$. We now prove it for an arbitrary $n \geq 1$.

In Lemma 6.1, it was shown that $ATT \subseteq MAC$. Hence $ATT^n \subseteq MAC^n$ also holds. We still need to prove that the inclusion is proper. We perform the proof by showing a tree transformation which is in MAC^n but not in ATT^n. Consider the macro tree transducer defined in Lemma 4.23, which induces the tree transformation $\tau = \{ (\gamma^k(\alpha), \gamma^{2^k}(\alpha)) \mid k \geq 0 \}$. Then the tree transformation

$$\tau_n = \{ (\gamma^k(\alpha), \gamma^{exp(n,k)}(\alpha) \mid k \geq 0 \},$$

i.e., the n-fold composition of τ is in MAC^n, for every $n \geq 1$ (recall the definition of exp in Lemma 4.24). Thus, for every $k \geq 0$, $height(\tau_n(\gamma^k(\alpha)))$ $= size(\tau_n(\gamma^k(\alpha))) = exp(n,k) + 1$.

Assume that $\tau_n \in ATT^n$, i.e., that, for every $1 \leq i \leq n$, there is an attributed tree transducer $A_i = (Att_i, \Sigma_i, \Sigma_{i+1}, a_i, R_i, E_i)$ such that $\Sigma_1 = \Sigma_{n+1} = \{\gamma^{(1)}, \alpha^{(0)}\}$ and

$$\tau_{A_1} \circ \ldots \circ \tau_{A_n} = \tau_n.$$

Now, for every $1 \leq i \leq n$, let c_i be the constant associated with the attributed tree transducer A_i by Lemma 5.40 such that, for every $s \in T_{\Sigma_i}$,

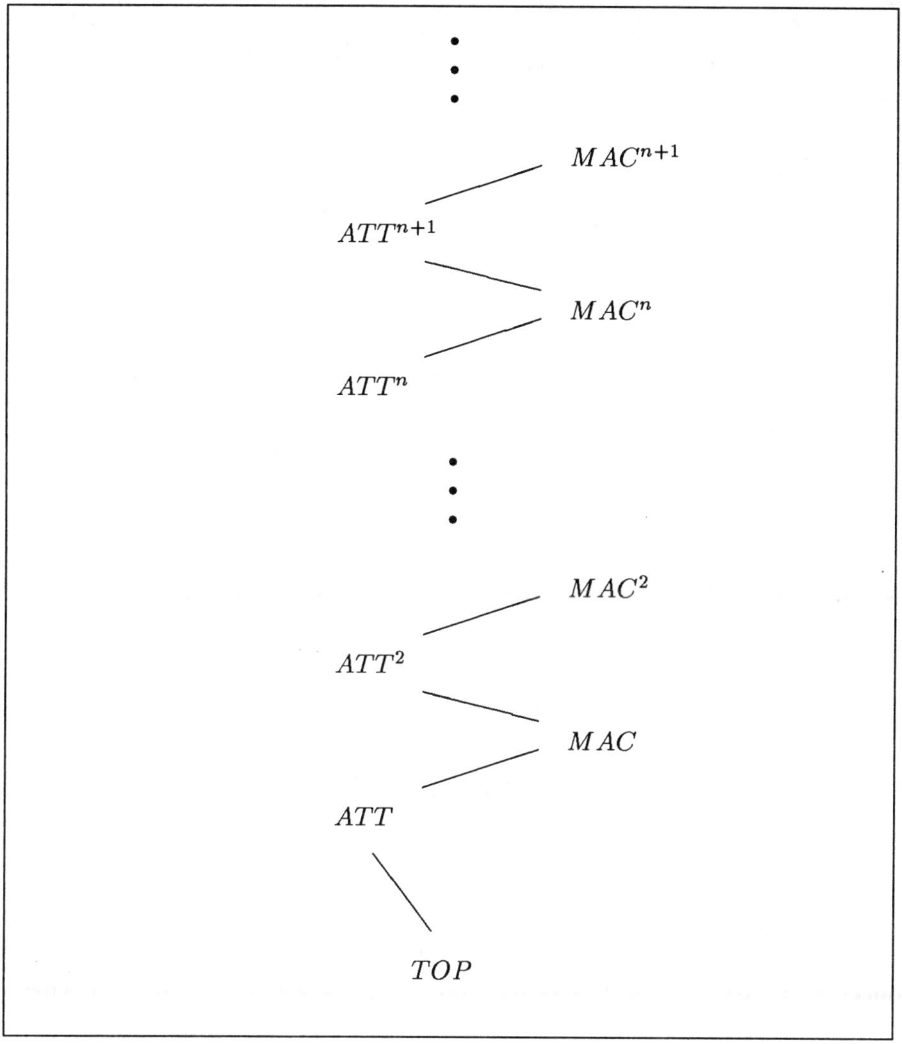

Fig. 6.8. Composition semigroup generated by $\{TOP, ATT, MAC\}$

$height(\tau_{A_i}(s)) \le c_i \cdot size(s)$. Moreover, let d_i be the constant associated with the ranked alphabet Σ_i by Lemma 2.13. Then, for every $1 \le i \le n$ and $s \in T_{\Sigma_i}$, $size(s) \le d_i^{height(s)}$ holds.

Then, for every $k \ge 0$, we can approximate $height(\tau_n(\gamma^k(\alpha)))$ by c_i's and d_i's as follows. Since $\tau_n = \tau_{A_1} \circ \ldots \circ \tau_{A_n}$, there are trees $s_1 \in T_{\Sigma_1}, \ldots, s_{n+1} \in T_{\Sigma_{n+1}}$ such that $\gamma^k(\alpha) = s_1$, $\tau_{A_1}(s_1) = s_2, \ldots, \tau_{A_n}(s_n) = s_{n+1}$ and also $\tau_n(\gamma^k(\alpha)) = s_{n+1}$. Consequently, by Lemmas 5.40 and 2.13, $height(s_{i+1}) \le c_i \cdot size(s_i)$ and $size(s_i) \le {d_i}^{height(s_i)}$, for $1 \le i \le n$. Hence, for $\tau_n(\gamma^k(\alpha))$ we get the approximation

$$height(\tau_n(\gamma^k(\alpha))) \leq c_n \cdot d_n^{c_{n-1} \cdot d_{n-1}^{\cdot^{\cdot^{\cdot^{c_2 \cdot d_2^{c_1 \cdot (k+1)}}}}}} .$$

(Note that $size(s_1) = size(\gamma^k(\alpha)) = k+1$.) Since $height(\tau_n(\gamma^k(\alpha))) = exp(n,k)+1$, we obtain

$$exp(n,k)+1 \leq c_n \cdot d_n^{c_{n-1} \cdot d_{n-1}^{\cdot^{\cdot^{\cdot^{c_2 \cdot d_2^{c_1 \cdot (k+1)}}}}}} ,$$

for every $k \geq 0$. However, this is an obvious contradiction.

Part 3. We prove that, for every $n \geq 1$, the inclusion $MAC^n \subset ATT^{n+1}$ holds.

The inclusion $MAC^n \subseteq ATT^{n+1}$ can be seen as follows. For $n = 1$, we have

$$\begin{array}{lll} & MAC & \\ = & TOP \circ YIELD & \text{(by Theorem 4.37)} \\ \subseteq & ATT \circ ATT & \text{(by Lemma 5.44 and Corollary 6.24).} \end{array}$$

Then, by induction on n we obtain

$$\begin{array}{lll} & MAC^{n+1} & \\ = & MAC^n \circ MAC & \\ \subseteq & ATT^{n+1} \circ MAC & \text{(by induction hypothesis)} \\ \subseteq & ATT^{n+1} \circ TOP \circ YIELD & \text{(by Theorem 4.37)} \\ \subseteq & ATT^{n+1} \circ YIELD & \text{(by Theorem 5.47)} \\ \subseteq & ATT^{n+2} & \text{(by Corollary 6.24).} \end{array}$$

We still have to prove that the inclusion is proper.

Therefore, recall from Example 3.29 the definition of a fully balanced binary tree over $\Sigma = \{\sigma^{(2)}, \alpha^{(0)}\}$. We will construct an attributed tree transducer $A = (Att, \Sigma, \Delta, a, R, E)$ such that, for every $k \geq 1$, $\tau_A(s_k) = s_{2^k}$ where s_k denotes the fully balanced binary tree of height k. Let Att contain the inherited attribute b and the synthesized attribute a, $\Sigma = \Delta = \{\sigma^{(2)}, \alpha^{(0)}\}$, $E(b) = \alpha$, and R contains the following sets of rules:

- R_σ:
  1) $a(\pi) \to \sigma(a(\pi 2), a(\pi 2))$
  2) $b(\pi 1) \to b(\pi)$
  3) $b(\pi 2) \to a(\pi 1)$
- R_α:
  4) $a(\pi) \to \sigma(b(\pi), b(\pi))$

Then we can prove the following statement by an easy induction on k: for every $k \geq 1$,

$$a(\varepsilon) \Rightarrow^*_{A,s_k} s_{2^k}[\alpha \leftarrow b(\varepsilon)].$$

We note that, in the induction step, we have used the fact that, for every $k', k'' \geq 1$,

$$s_{k'}[\alpha \leftarrow s_{k''}[\alpha \leftarrow b(\varepsilon)]] = s_{k'-1+k''}[\alpha \leftarrow b(\varepsilon)].$$

Now denote τ_A^{n+1} by τ.

Then it follows from the construction of A that, for every $k \geq 1$,

$$height(\tau(s_k)) = exp(n+1, k).$$

Next we prove by contradiction that τ is not an element of the class MAC^n. Thus assume that $\tau \in MAC^n$. Then by Lemma 4.24, there is a constant $c > 0$ such that, for every $s \in T_\Sigma$, the approximation

$$height(\tau(s)) \leq exp(n, c \cdot height(s))$$

holds. Hence we obtain that, for every $k \geq 1$,

$$height(\tau(s_k)) \leq exp(n, c \cdot k).$$

Combining this approximation with the calculation of $height(\tau(s_k))$ above, we obtain, for every $k \geq 1$,

$$exp(n+1, k) \leq exp(n, c \cdot k).$$

Obviously, this is a contradiction and hence, $\tau \notin MAC^n$, by which we prove our lemma. □

STEP 4. In this step we show that, from every element of H^+, we can compute the corresponding representative.

Lemma 6.55. There is an algorithm that computes, for every $w \in H^+$, a representative $u \in REP$ such that $w \Rightarrow_S^* u$.

Proof. We prove the lemma by induction on $length(w)$.

Let $length(w) = 1$, implying that $w \in H$. Then we also have $w \in REP$, by the definition of REP. Hence, for $u = w$, we have $u \in REP$ and $w \Rightarrow_S^* u$.

Now suppose that $length(w) > 1$, which means that $w = x \cdot U$, for some $x \in H^+$ and $U \in H$. Moreover, by induction hypothesis on $length(w)$, a representative $v \in REP$ can effectively be computed such that $x \Rightarrow_S^* v$. Then, regarding v, three cases are possible, each of which has three subcases concerning U. For each subcase, we compute u.

Case 1: $v = TOP$. Then $w = x \cdot U \Rightarrow_S^* v \cdot U = TOP \cdot U$.

- Subcase $U = TOP$. Then we have $u = TOP$, because $TOP \cdot U = TOP \cdot TOP \Rightarrow_S TOP$, by rule 1).
- Subcase $U = ATT$. Then $u = MAC$, because $TOP \cdot U = TOP \cdot ATT \Rightarrow_S MAC$, by rule 4).

- Subcase $U = MAC$. Then $u = MAC$, because $TOP \cdot U = TOP \cdot MAC \Rightarrow_S MAC$, by rule 2).

Case 2: $v = ATT^n$, for some $n \geq 1$. Then $w = x \cdot U \Rightarrow_S^* v \cdot U = ATT^n \cdot U$.

- Subcase $U = TOP$. Then we have $u = ATT^n$, because $ATT^n \cdot U = ATT^n \cdot TOP \Rightarrow_S ATT^n$, by rule 5).
- Subcase $U = ATT$. Then $u = ATT^{n+1}$, because $ATT^n \cdot U = ATT^n \cdot ATT = ATT^{n+1}$.
- Subcase $U = MAC$. Then $u = ATT^{n+1}$, because $ATT^n \cdot U = ATT^n \cdot MAC \Rightarrow_S ATT^{n+1}$, by rule 6).

Case 3: $v = MAC^n$, for some $n \geq 1$. Then $w = x \cdot U \Rightarrow_S^* v \cdot U = MAC^n \cdot U$

- Subcase $U = TOP$. Then we have $u = MAC^n$, because $MAC^n \cdot U = MAC^n \cdot TOP \Rightarrow_S MAC^n$, by rule 3).
- Subcase $U = ATT$. Then $u = MAC^{n+1}$, because $MAC^n \cdot U = MAC^n \cdot ATT \Rightarrow_S MAC^{n+1}$, by rule 7).
- Subcase $U = MAC$. Then $u = MAC^{n+1}$, because $MAC^n \cdot U = MAC^n \cdot MAC = MAC^{n+1}$.

We have finished the proof of our lemma. □

STEP 5. We have now constructed the desired algorithm. This is declared in the following theorem, which can be proved just as Theorem 3.54 was proved in Sect. 3.7. Therefore we omit the proof here.

Theorem 6.56. Let $H = \{TOP, ATT, MAC\}$. Then there is an algorithm that decides, given two arbitrary tree transformation classes $U_1 \circ \cdots \circ U_m$ and $V_1 \circ \cdots \circ V_n$, where $U_i, V_j \in H$, for every $1 \leq i \leq m$ and $1 \leq j \leq n$, which one of the conditions

$$\begin{array}{rrcl} \text{(i)} & U_1 \circ \cdots \circ U_m & = & V_1 \circ \cdots \circ V_n, \\ \text{(ii)} & U_1 \circ \cdots \circ U_m & \subset & V_1 \circ \cdots \circ V_n, \\ \text{(iii)} & V_1 \circ \cdots \circ V_n & \subset & U_1 \circ \cdots \circ U_m, \\ \text{(iv)} & U_1 \circ \cdots \circ U_m & \bowtie & V_1 \circ \cdots \circ V_n. \end{array}$$

holds.

An application of the algorithm

In this subsection, as an example, we apply the algorithm for two tree transformation classes. For example, consider the two tree transformation classes

$$ATT \circ TOP \circ MAC \circ ATT \text{ and } TOP \circ MAC \circ ATT \circ TOP.$$

By Lemma 3.48 we first apply the algorithm described in Lemma 6.55 to compute an element $u \in REP$ such that $ATT \cdot TOP \cdot MAC \cdot ATT \Rightarrow_S^* u$.

As in Sect. 3.7, we can construct a convenient table for applying the algorithm in Lemma 6.55. For $H = \{TOP, ATT, MAC\}$, the table looks as follows.

	TOP	ATT	MAC
TOP	TOP	MAC	MAC
ATT^n	ATT^n	ATT^{n+1}	ATT^{n+1}
MAC^n	MAC^n	MAC^{n+1}	MAC^{n+1}

We recall that the table, denoted by T, has the following properties. Its rows are labeled by elements of REP and its columns are labeled by elements of H. (In fact, REP is now an infinite set, hence the table should have an infinite number of rows. However, all elements of the form ATT^n (resp. MAC^n) of REP, where $n \geq 1$ can be handled similarly, therefore two rows are sufficient to represent the infinite number of cases.)

As we saw in Section 3.7, for $v \in REP$ and $U \in H$, the entry $T(v, U)$ at (v, U) is also an element of REP such that $v \cdot U \Rightarrow_S^* T(v, U)$.

Again, if there is a $w = U_1 \cdot U_2 \cdot \ldots \cdot U_n \in H^+$, the following small program computes $u \in REP$, for which $w \Rightarrow_S^* u$.

```
u := U_1;
for i = 2 to n do
    u := T(u, U_i)
```

Applying the program to $w = ATT \cdot TOP \cdot MAC \cdot ATT$, we obtain

$$w = ATT \cdot TOP \cdot MAC \cdot ATT \Rightarrow_S^* ATT^3,$$

which implies that

$$ATT \circ TOP \circ MAC \circ ATT = ATT^3.$$

In a similar way we obtain

$$TOP \cdot MAC \cdot ATT \cdot TOP \Rightarrow_S^* MAC^2,$$

implying that

$$TOP \circ MAC \circ ATT \circ TOP = MAC^2.$$

From the inclusion diagram in Lemma 6.54, we can see that $MAC^2 \subset ATT^3$. Hence we have also obtained

$$TOP \circ MAC \circ ATT \circ TOP \subset ATT \circ TOP \circ MAC \circ ATT.$$

Presenting [*H*] by a terminating and confluent rewrite system

In the same way as in Sect. 3.7, we can show that the string rewrite system S we have given is also terminating and confluent. Moreover, we have chosen REP in such a way that $REP = IRR(H,S)$ holds, i.e., that the irreducible elements of H^+ with respect to S are exactly the representatives we gave. This makes it possible to compute more easily, for every $w \in H^+$, the representative $u \in REP$, such that $w \Rightarrow_S^* u$.

Lemma 6.57. $REP = IRR(H,S)$

Proof. Recall that $REP = \{TOP\} \cup \{ATT^n \mid n \geq 1\} \cup \{MAC^n \mid n \geq 1\}$. The fact that $REP \subseteq IRR(H,S)$ is obvious, because none of our seven rules can be applied to TOP, ATT^n, or MAC^n, $n \geq 1$. The converse inclusion $IRR(H,S) \subseteq REP$ can easily be seen also. Let us denote the set of irreducible elements of length n with respect to S by IRR_n, for $n \geq 1$. It is sufficient to show that $IRR_n \subseteq REP$, for every $n \geq 1$. In fact, $IRR_1 = \{TOP, ATT, MAC\}$, hence we have $IRR_1 \subseteq REP$. Moreover, $IRR_n = \{ATT^n, MAC^n\}$, for every $n \geq 2$. This can be seen as follows. On the one hand, ATT^n and MAC^n are clearly irreducible. On the other hand, take a string $w \in H^+$ with $length(w) = n$ which differs both from ATT^n and MAC^n. We show that w cannot be irreducible with respect to S. We observe that w is either TOP^n or it contains at least two different symbols from H. The string TOP^n cannot be irreducible because rule 1) is applicable to it. Furthermore, if there are two different symbols X and Y in w, then there are also two different adjacent symbols X' and Y' in w , i.e., $X' \cdot Y'$ is a substring of w. However, in this case we can easily check that there is a rule among rules 2)–7) with left-hand side $X' \cdot Y'$ irrespective of X' and Y'. Then w cannot be irreducible, because this rule applies to it. Thus, $IRR_n \subseteq REP$ for $n \geq 2$ also. □

Next we show that S is terminating and confluent.

Lemma 6.58. S is terminating and confluent.

Proof. First we prove that S is terminating. We observe that rules of S have the property of not increasing length which means that, for each rule $u \to v \in S$, we have $length(u) \geq length(v)$. Hence, for every derivation $w \Rightarrow_S^+ x$, also $length(w) \geq length(x)$. This implies the equivalence between the statements

- there exists an infinite sequence $w_0 \Rightarrow_S w_1 \Rightarrow_S w_2 \ldots$ with $w_i \in H^+$ for $i = 1, 2, \ldots$ (i.e., that S is not terminating)
- there exists a string $x \in H^*$ such that $x \Rightarrow_S^+ x$.

Next we observe that if $x \Rightarrow_S^+ x$, for some $x \in H^+$, then none of the rules 1)–5) could be applied during that derivation, because they are strict length-reducing rules and thus any application of them would yield a decrease of the length of the string to which they were applied. Hence $x \Rightarrow_S^+ x$ if and only if $x \Rightarrow_S^+ x$ by applying only rules 6) and 7), which are

6) $ATT \cdot MAC \rightarrow ATT \cdot ATT$ and
7) $MAC \cdot ATT \rightarrow MAC \cdot MAC$.

Let $x = x_1 \cdot TOP \cdot x_2 \cdot TOP \cdot \ldots \cdot TOP \cdot x_m$, for some $m \geq 1$ and $x_1, \ldots, x_m \in \{ATT, MAC\}^*$. Then we obtain that $x \Rightarrow_S^+ x$ if and only if there is an i with $1 \leq i \leq m$ such that $x_i \Rightarrow_S^+ x_i$.

Therefore we can conclude that S is not terminating, if and only if there exists a string $x \in \{ATT, MAC\}^*$, such that $x \Rightarrow_S^+ x$, by applying only rules 6) and 7). (Note that the latter condition is superfluous because rules 1)–5) are not even applicable to x.) In other words, S is terminating if and only if such an x does not exist. We will prove that there is no x with this property.

To this end, we define, for every string $x \in H^*$, the weight $weight(x)$, which is a natural number. Roughly speaking, $weight(x)$ is the sum of the position numbers of the signs "$\cdot$" which are between an ATT and a MAC or between a MAC and an ATT, i.e., which are either in the situation $\ldots ATT \cdot MAC \ldots$ or in the situation $\ldots MAC \cdot ATT \ldots$ in x, where the number of the position is considered from right to left. More formally, we define $weight(x)$ by induction as follows.

(i) If $x = \varepsilon$, then $weight(x) = 0$.
(ii) If $x = MAC \cdot x'$, for some $x' \in \{ATT, MAC\}^*$, then

$$weight(x) = \begin{cases} weight(x') + length(x') & \text{if the first letter of } x' \text{ is } ATT \\ weight(x') & \text{otherwise} \end{cases}$$

(iii) Moreover, if $x = ATT \cdot x'$, for some $x' \in \{ATT, MAC\}^*$, then

$$weight(x) = \begin{cases} weight(x') + length(x') & \text{if the first letter of } x' \text{ is } MAC \\ weight(x') & \text{otherwise} \end{cases}$$

For example, $weight(\varepsilon) = 0, weight(ATT) = 0, weight(ATT \cdot ATT) = 0, weight(ATT \cdot MAC) = 1$ and $weight(ATT \cdot MAC \cdot ATT \cdot MAC) = 6$.

Finally we state that $weight$ is defined in such a way that any application of rules 6) and 7) to a string $x \in \{ATT, MAC\}^*$ reduces the weight of x, i.e., if $x \Rightarrow_S y$ by applying either 6) or 7), then we have $weight(x) > weight(y)$. This, of course, proves that $x \Rightarrow_S^+ x$ for no $x \in \{ATT, MAC\}^*$ and hence that S is terminating.

We must still verify that both 6) and 7) reduce $weight(x)$, for every $x \in \{ATT, MAC\}^*$. Assume that we have $x \Rightarrow_S y$ by applying 6). This means that there are $x', y' \in \{ATT, MAC\}^*$ such that $x = x' \cdot ATT \cdot MAC \cdot y'$ and $y = x' \cdot ATT \cdot ATT \cdot y'$. Then $weight(y) = weight(x) - 1$ if the first symbol of y' is MAC and $weight(y) = weight(x) - (length(y') + 1)$ otherwise. In either case, $weight(x) > weight(y)$. A similar reasoning can be applied for rule 7). This completes the proof that S is terminating.

We now sketch how to prove that S is confluent as well. As in Lemma 3.53, we can prove that REP is a set of representatives for $\Leftrightarrow_S^*$ which means that every congruence class of $\Leftrightarrow_S^*$ contains exactly one element of REP.

On the other hand, we have seen that $REP = IRR(H,S)$ implying that every congruence class of $\Leftrightarrow_S^*$ contains exactly one irreducible element for S. However, by Lemma 2.6, this exactly means that S is confluent.

This finishes the proof of our lemma. □

The importance of Lemmas 6.57 and 6.58 is that, when computing the representative in REP for an arbitrary string $w \in H^+$, we do not have to use the algorithm described in Lemma 6.55.

Instead, we can do the following. Given a string $w \in H^+$, we compute an irreducible element $u \in IRR(H,S)$, so that $w \Rightarrow_S^* u$. Such a u is always reachable, because S is terminating. On the other hand every derivation starting from w reaches the same irreducible element u, because S is confluent. This irreducible u will be in REP, because $IRR(H,S) = REP$.

For example, consider again the tree transformation classes

$$ATT \circ TOP \circ MAC \circ ATT \text{ and } TOP \circ MAC \circ ATT \circ TOP.$$

For $w = ATT \cdot TOP \cdot MAC \cdot ATT$, we can compute $u \in IRR(H,S)$ by

$$\begin{array}{lll} & ATT \cdot TOP \cdot MAC \cdot ATT & \\ \Rightarrow_S & ATT \cdot MAC \cdot ATT & \text{(by rule 2)} \\ \Rightarrow_S & ATT \cdot MAC \cdot MAC & \text{(by rule 7)} \\ \Rightarrow_S & ATT \cdot ATT \cdot MAC & \text{(by rule 6)} \\ \Rightarrow_S & ATT \cdot ATT \cdot ATT & \text{(by rule 6).} \end{array}$$

But also

$$\begin{array}{lll} & ATT \cdot TOP \cdot MAC \cdot ATT & \\ \Rightarrow_S & ATT \cdot MAC \cdot ATT & \text{(by rule 5)} \\ \Rightarrow_S & ATT \cdot ATT \cdot ATT & \text{(by rule 6).} \end{array}$$

Similarly, for $w = TOP \cdot MAC \cdot ATT \cdot TOP$, we have $w \Rightarrow_S^* MAC \cdot MAC$. Since

$$MAC \circ MAC \subset ATT \circ ATT \circ ATT,$$

we again obtain the inclusion

$$TOP \circ MAC \circ ATT \circ TOP \subset ATT \circ TOP \circ MAC \circ ATT.$$

6.4 Bibliographic Notes

The transformation of attribute grammars into denotational semantics has been considered in [CM79, Eng80, Mad80, May81, CF82]. The converse transformation has been investigated in [Mad80, Gan80].

The technique in the proof of the inclusion $ATT \subseteq MAC$ (Lemma 6.1) stems from [Fra82]. There, for every noncircular attribute grammar an equivalent primitive recursive program scheme with parameters has been constructed. The strictness of the inclusion $ATT \subseteq MAC$ has been proved in [Eng80]. The notions of well-presentedness and absolute noncircularity and the relationship between well-presented macro tree transducers and absolutely noncircular attributed tree transducers (Theorem 6.23) appear in [CF82].

In [EF81] the formal translation power of one-visit attribute grammars has been investigated. The notion of a brother graph has also been used there.

In [EF89] it has been proved that the transformational power of attribute grammars grows with respect to the number of visits that are allowed (see also [RS81]).

The crucial part of Theorem 6.38, which is the inclusion

$$PATH(range(bas\text{-}MAC)) \subseteq \mathcal{L}_{lin},$$

has been proved in Theorem 5.9 of [Vog87].

In [Now96] an algorithm has been proposed which generates for a given diagram the set of key lemmas. More precisely, the algorithm takes a diagram D as input, where in D the nodes are labeled by sets and the partial order is the set inclusion. The algorithm delivers a finite set M of inequalities as output, where every inequality has the form $A - B \neq \emptyset$ for two sets A and B involved in D. M has the property that, if every inequality of M can be proved, then D is a Hasse diagram.

7. Macro Attributed Tree Transducers

In Chaps. 4 and 5 we investigated two formal models of syntax-directed semantics which are able to handle context, the macro tree transducer and the attributed tree transducer, respectively. Whereas the first one handles context in an implicit way by allowing the meaning names (i.e., states) to have parameters, the second one deals with context information in an explicit way by introducing context names (i.e., inherited attributes).

In this chapter we study a formal model, called the macro attributed tree transducer, which integrates both approaches. From the point of view of a macro tree transducer, a macro attributed tree transducer is not restricted to recursive calls to states on input trees in a strictly descending way, but an arbitrary tree walk is allowed (as for attributed tree transducers). From the point of view of attributed tree transducers, a macro attributed tree transducer contains attributes, which are not restricted to denote basic values, i.e., trees, but which may denote functions over trees (as the states of macro tree transducers can do). The connection between these transducers is shown in Fig. 1.33.

Because of the possibility of arbitrary tree-walking over the input tree, macro attributed tree transducers share with attributed tree transducers the property of a possibly non-terminating derivation relation. Hence, we also have to discuss here the topics of circularity and a circularity test.

Since the growth of output trees with respect to the corresponding input trees can be deduced easily from a decomposition result of macro attributed tree transducers, we change the order of the corresponding sections appropriately. The structure of this chapter is as follows.

1. Basic definitions
2. Induced tree transformation
3. Characterization of macro attributed tree transformations
4. Composition and decomposition results
5. Height property
6. Bibliographic notes

7.1 Basic Definitions

Since a macro attributed tree transducer should be able to tree-walk on its input trees, we define this model in the same framework as the attributed tree transducer. That means that we use the notion of path variable and the separation of the input tree from the sentential form (see Fig. 5.1(b)).

Definition 7.1. Let A_s and A_i be two disjoint ranked alphabets, Δ be a ranked alphabet such that $\Delta \cap (A_s \cup A_i) = \emptyset$, and $k, n \geq 0$ be integers. The set $RHS(A_s, A_i, \Delta, k, n)$ of *right-hand sides over* A_s, A_i, Δ, k, *and* n is the smallest subset RHS of $T_{A_s \cup A_i \cup \Delta}(\{\pi, \pi 1, ..., \pi k\} \cup Y_n)$ satisfying the following conditions.

(i) For every $a \in A_s^{(l+1)}$ with $l \geq 0$, $1 \leq i \leq k$, and $\xi_1, \ldots, \xi_l \in RHS$, the term $a(\pi i, \xi_1, \ldots, \xi_l)$ is in RHS.
(ii) For every $b \in A_i^{(l+1)}$ with $l \geq 0$ and $\xi_1, \ldots, \xi_l \in RHS$, the term $b(\pi, \xi_1, \ldots, \xi_l)$ is in RHS.
(iii) For every $\delta \in \Delta^{(l)}$ with $l \geq 0$ and $\xi_1, \ldots, \xi_l \in RHS$, the term $\delta(\xi_1, \ldots, \xi_l)$ is in RHS.
(iv) $Y_n \subseteq RHS$. □

The definition of a macro attributed tree transducer follows the outline of the definition of an attributed tree transducer.

Definition 7.2. A *macro attributed tree transducer* is a tuple $N = (Att, \Sigma, \Delta, a_0, R, E)$, where

- Att is a ranked alphabet, called the set of *attributes*; every element in Att has at least rank 1; Att is partitioned into the disjoint sets Att_{syn} and Att_{inh}, of which the elements are called *synthesized attributes* and *inherited attributes*, respectively.
- Σ and Δ are ranked alphabets such that $Att \cap (\Sigma \cup \Delta) = \emptyset$, called the *input alphabet* and the *output alphabet*, respectively.
- $a_0 \in Att_{syn}^{(r+1)}$ for some $r \geq 0$ is a designated attribute, called the *initial attribute*.
- R is a set $\bigcup_{\sigma \in \Sigma} R_\sigma$ of finite sets R_σ of rules which satisfies the conditions:

(1) For every $a \in Att_{syn}^{(n+1)}$ with $n \geq 0$ and $\sigma \in \Sigma^{(k)}$ with $k \geq 0$, the set R_σ contains exactly one rule of the form

$$a(\pi, y_1, \ldots, y_n) \to \xi$$

where $\xi \in RHS(Att_{syn}, Att_{inh}, \Delta, k, n)$.

(2) For every $b \in Att_{inh}^{(n+1)}$ with $n \geq 0$, $\sigma \in \Sigma^{(k)}$ with $k \geq 1$, and $1 \leq i \leq k$, the set R_σ contains exactly one rule of the form

$$b(\pi i, y_1, \ldots, y_n) \to \xi$$

where $\xi \in RHS(Att_{syn}, Att_{inh}, \Delta, k, n)$.

- $E = (E_1, E_2)$ is the *environment*, where $E_1 = (t_1, \ldots, t_r) \in (T_\Delta)^r$ and $E_2 : Att_{inh} \to T_\Delta(Y)$ is a function such that for every $b \in Att_{inh}^{(k+1)}$ with $k \geq 0$, $E_2(b) \in T_\Delta(Y_k)$. □

Example 7.3. As example, we compute, for a given input tree s over $\{\sigma^{(2)}, \alpha^{(0)}\}$, the sum of the squares of the lengths of the paths of s by means of a macro attributed tree transducer N_{ssq}. More precisely, for a given input tree s, we compute the number

$$\sum_{w \in occ(s) \wedge label(s,w) \in \Sigma^{(0)}} (length(w))^2$$

As usual, we represent natural numbers by monadic trees over the ranked alphabet $\{\gamma^{(1)}, \alpha^{(0)}\}$.

First, we need an attribute which calculates the sum of the numbers $(length(w))^2$ for every leaf w. Let us call this synthesized attribute *sum*; it has two rules which are elements of R_σ and R_α, respectively:

- R_σ:
  - $sum(\pi, y_1) \to sum(\pi 1, sum(\pi 2, y_1))$
- R_α:
  - $sum(\pi, y_1) \to squ(\pi, y_1)$

If the synthesized attribute *sum* arrives at a leaf labeled by α, then the inherited attribute *squ* is called which calculates the square of the length of the path from this leaf to the root of the input tree. Here the equation

$$n^2 = (2n - 1) + (n - 1)^2$$

is used which is implemented by using an inherited attribute *double*; it keeps and updates the numbers $(2n - 1)$. There are the following rules for *squ* and *double* where we have omitted as usual parentheses around the unary γ:

- R_σ:
  - $squ(\pi 1, y_1) \to double(\pi, squ(\pi, y_1))$
  - $squ(\pi 2, y_1) \to double(\pi, squ(\pi, y_1))$
  - $double(\pi 1, y_1) \to \gamma\gamma double(\pi, y_1)$
  - $double(\pi 2, y_1) \to \gamma\gamma double(\pi, y_1)$

Thus, the complete macro attributed tree transducer $N_{ssq} = (Att, \Sigma, \Delta, sum, R, E)$ looks as follows.

- $Att = Att_{syn} \cup Att_{inh}$ with $Att_{syn} = \{sum^{(2)}\}$ and $Att_{inh} = \{squ^{(2)}, double^{(2)}\}$
- $\Sigma = \{\sigma^{(2)}, \alpha^{(0)}\}$
- $\Delta = \{\gamma^{(1)}, \alpha^{(0)}\}$
- R_σ:

- $sum(\pi, y_1) \to sum(\pi 1, sum(\pi 2, y_1))$
- $squ(\pi 1, y_1) \to double(\pi, squ(\pi, y_1))$
- $squ(\pi 2, y_1) \to double(\pi, squ(\pi, y_1))$
- $double(\pi 1, y_1) \to \gamma\gamma double(\pi, y_1)$
- $double(\pi 2, y_1) \to \gamma\gamma double(\pi, y_1)$

R_α:

- $sum(\pi, y_1) \to squ(\pi, y_1)$

- $E = (E_1, E_2)$ with $E_1 = (\alpha)$ and $E_2(squ) = y_1$ and $E_2(double) = \gamma(y_1)$.

After having defined the derivation relation of macro attributed tree transducers we will also show a computation of N_{ssq}. □

Note 7.4. 1) A macro attributed tree transducer in which the set *Att* of attributes is unary, is an attributed tree transducer and vice versa.

2) A macro attributed tree transducer in which there are no inherited attributes (i.e., $Att_{inh} = \emptyset$), is very close to a macro tree transducer: synthesized attributes correspond to states; a rule $a(\pi, y_1, \ldots, y_n) \to \xi$ in R_σ with $\sigma \in \Sigma^{(k)}$ corresponds to the rule $a(\sigma(x_1, \ldots, x_k), y_1, \ldots, y_n) \to \xi'$ where ξ' is obtained from ξ by replacing every subterm $a'(\pi i, \ldots)$ by $a'(x_i, \ldots)$. □

Similarly to attributed tree transducers, we fix the following notions and notations for a macro attributed tree transducer $N = (Att, \Sigma, \Delta, a_0, R, E)$:

- In the rules of R, the variable π is called *path variable.*
- The notions of *set of inside attribute occurrences of* σ, *set of outside attribute occurrences of* σ, and *set of attribute occurrences of* σ are defined in the same way as for attributed tree transducers. In particular, the fact that an attribute can have an arbitrary rank greater than 1 is ignored.
- The right-hand side of the rule $a(\pi, y_1, \ldots, y_n) \to \xi$ in R_σ is abbreviated by $rhs(a(\pi), \sigma)$ and the right-hand side of the rule $b(\pi i, y_1, \ldots, y_n) \to \xi$ in R_σ is abbreviated by $rhs(b(\pi i), \sigma)$. For $\sigma \in \Sigma^{(k)}$, we denote by $RHS(\sigma)$ the set of right-hand sides of rules in R_σ, i.e., $RHS(\sigma) = \{rhs(c(\pi\eta), \sigma) \mid c(\pi\eta) \in in(\sigma)\}$.

7.2 Induced Tree Transformation

For the rest of this chapter, let $N = (Att, \Sigma, \Delta, a_0, R, E)$ be an arbitrary, but fixed, macro attributed tree transducer. Next we define the derivation relation of a macro attributed tree transducer N on an input tree s (see Figs. 7.1 and 7.2). As in Def. 5.5, the occurrences of s are considered as nullary symbols.

Definition 7.5. Let $s \in T_\Sigma$. The *derivation relation induced by* N *on* s is the binary relation $\Rightarrow_{N,s}$ over the set $T_{Att \cup \Delta \cup occ(s)}(Y)$ such that for every $\varphi, \psi \in T_{Att \cup \Delta \cup occ(s)}(Y)$ we define $\varphi \Rightarrow_{N,s} \psi$, if

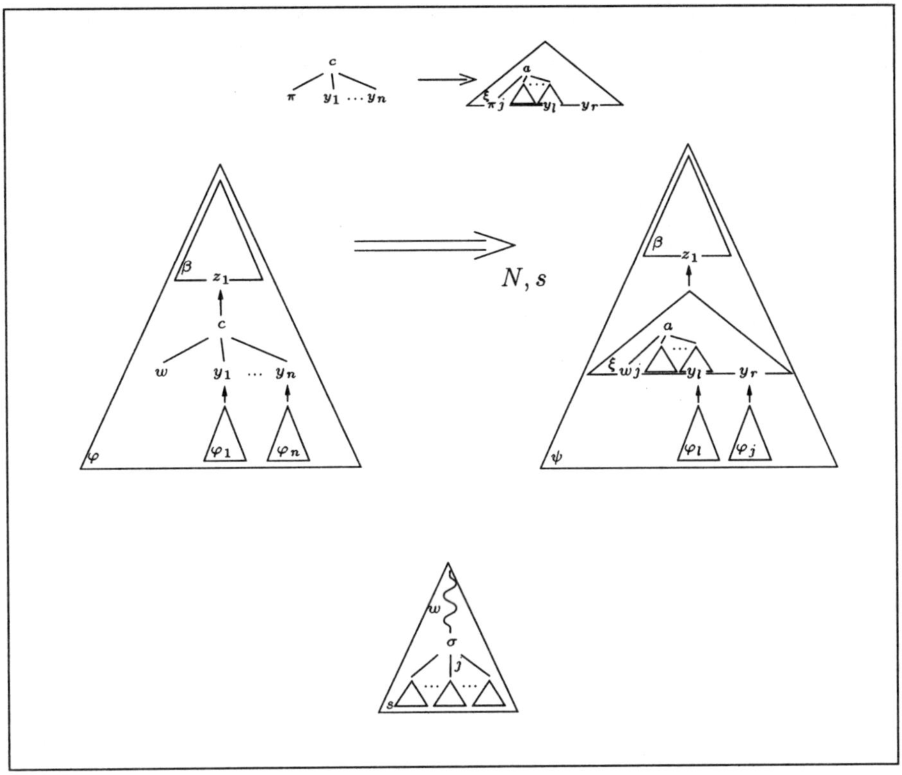

Fig. 7.1. A derivation step induced by N and triggered by a synthesized attribute

- there is a context $\beta \in C_{Att \cup \Delta \cup occ(s) \cup Y, 1}$
- there is an attribute $c \in Att^{(n+1)}$ with $n \geq 0$
- there is a path $w \in occ(s)$
- there are trees $\varphi_1, \ldots, \varphi_n \in T_{Att \cup \Delta \cup occ(s)}(Y)$

such that $\varphi = \beta[c(w, \varphi_1, \ldots, \varphi_n)]$ and one of the following conditions holds.

(1) $c \in Att_{syn}$, $label(s, w) = \sigma$, $\sigma \in \Sigma^{(k)}$ with $k \geq 0$, and
$\psi = \beta[\xi[\pi \leftarrow w][y_1 \leftarrow \varphi_1, \ldots, y_n \leftarrow \varphi_n]]$ where $\xi = rhs(c(\pi), \sigma)$.

(2) $c \in Att_{inh}$, $w = vi$ for some $v \in occ(s)$, $label(s, v) = \sigma$, $\sigma \in \Sigma^{(k)}$ with $k \geq 1$, $1 \leq i \leq k$, and
$\psi = \beta[\xi[\pi \leftarrow v][y_1 \leftarrow \varphi_1, \ldots, y_n \leftarrow \varphi_n]]$, where $\xi = rhs(c(\pi i), \sigma)$.

Example 7.6. Consider the macro attributed tree transducer N_{ssq} of Example 7.3 and the input tree $s = \sigma(\sigma(\alpha, \alpha), \alpha)$. In the following derivation we abbreviate $\Rightarrow_{N_{ssq}, s}$ to $\Rightarrow$.

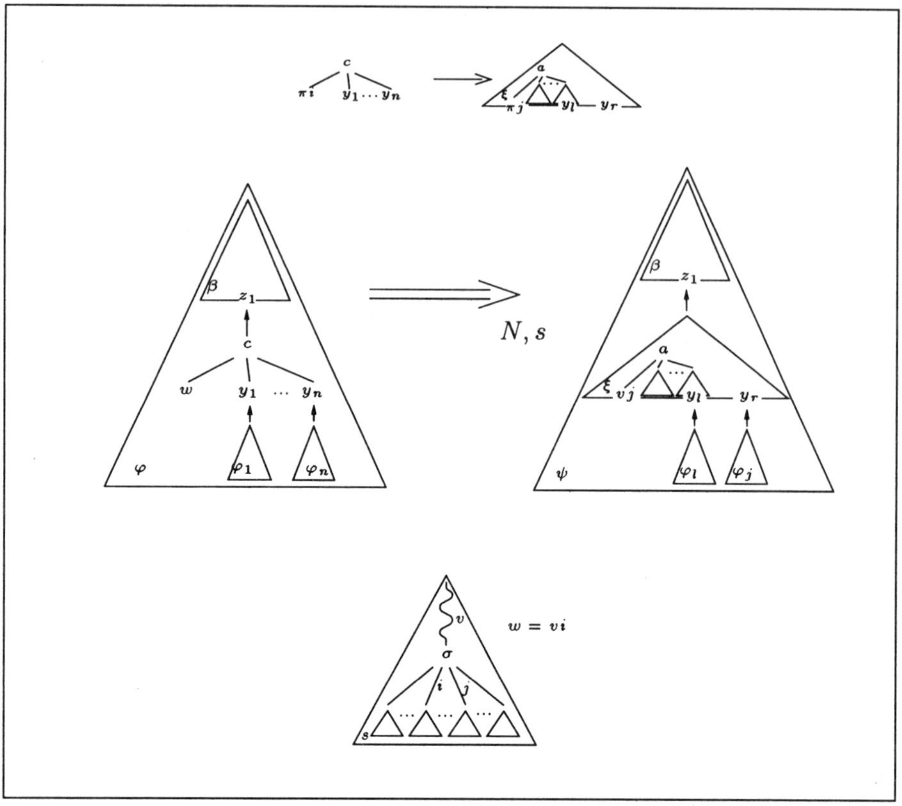

Fig. 7.2. A derivation step induced by N and triggered by an inherited attribute

$$\begin{array}{ll} & sum(\varepsilon, y_1) \\ \Rightarrow & sum(1, sum(2, y_1)) \\ \Rightarrow & sum(11, sum(12, sum(2, y_1))). \end{array}$$

Next we compute $sum(2, y_1)$

$$\begin{array}{ll} & sum(2, y_1) \\ \Rightarrow & squ(2, y_1) \\ \Rightarrow & double(\varepsilon, squ(\varepsilon, y_1)) \end{array}$$

and we call this result t_1. Then we compute $sum(12, t_1)$.

$$\begin{array}{ll} & sum(12, t_1) \\ \Rightarrow & squ(12, t_1) \\ \Rightarrow & double(1, squ(1, t_1)) \\ \Rightarrow & double(1, double(\varepsilon, squ(\varepsilon, t_1))) \\ \Rightarrow & \gamma\gamma double(\varepsilon, double(\varepsilon, squ(\varepsilon, t_1))) \end{array}$$

and we call the result t_2. Finally we compute $sum(11, t_2)$.

$$\begin{array}{ll} & sum(11,t_2) \\ \Rightarrow & squ(11,t_2) \\ \Rightarrow & double(1,squ(1,t_2)) \\ \Rightarrow & double(1,double(\varepsilon,squ(\varepsilon,t_2))) \\ \Rightarrow & \gamma\gamma double(\varepsilon,double(\varepsilon,squ(\varepsilon,t_2))) \end{array}$$

Thus overall N_{ssq} has derived the output tree

$$\gamma\gamma double(\varepsilon,double(\varepsilon,squ(\varepsilon,\gamma\gamma double(\varepsilon,double(\varepsilon,squ(\varepsilon, double(\varepsilon,squ(\varepsilon,y_1)))))))).$$

By means of the environment of N, the inherited attributes at the root of the input tree will be interpreted (in Def. 7.23); roughly speaking, *double* will be replaced by one more γ and *squ* is simply dropped. Then the result of this interpretation is the tree $\gamma^9(\alpha)$ which is the desired result, because

$$\begin{array}{ll} \sum_{w\in occ(s)\wedge label(s,w)\in\Sigma^{(0)}}(length(w))^2 & = \\ \sum_{w\in\{11,12,2\}}(length(w))^2 & = 2^2+2^2+1^2=9. \end{array}$$

□

Definition 7.7. Let s be an input tree of N. The set of *sentential forms of N and s*, denoted by $SF(N,s)$, is the smallest subset SF of $T_{Att\cup\Delta\cup occ(s)}(Y)$ for which the following conditions hold.

(i) For every $c\in Att^{(l+1)}$ with $l\geq 0$, $w\in occ(s)$, and $\varphi_1,\ldots,\varphi_l\in SF$, the term $c(w,\varphi_1,\ldots,\varphi_l)\in SF$.

(ii) For every $\delta\in\Delta^{(l)}$ with $l\geq 0$ and $\varphi_1,\ldots,\varphi_l\in SF$, the term $\delta(\varphi_1,\ldots,\varphi_l)$ is in SF.

(iii) $Y\subseteq SF$. □

In the following, it is also useful to have a notation for the particular set of sentential forms available. These sentential forms occur at the end of derivations and hence, they are called final sentential forms. They consist of output symbols, parameters, and inherited attributes applied to the occurrence ε. In fact later it transpires that it is useful to restrict the set of possible parameters.

Definition 7.8. Let $U\subseteq Y$ be a set of parameters. The *set of final sentential forms of N and U*, denoted by $SF_{fin}(N,U)$, is the smallest subset SF of $T_{Att_{inh}\cup\Delta\cup\{\varepsilon\}}(U)$ such that the following conditions hold.

(i) For every $b\in Att_{inh}^{(l+1)}$ with $l\geq 0$ and $\varphi_1,\ldots,\varphi_l\in SF$, $b(\varepsilon,\varphi_1,\ldots,\varphi_l)\in SF$.

(ii) For every $\delta\in\Delta^{(l)}$ with $l\geq 0$ and $\varphi_1,\ldots,\varphi_l\in SF$, the term $\delta(\varphi_1,\ldots,\varphi_l)$ is in SF.

(iii) $U\subseteq SF$. □

As in the case of the first three formal models of syntax-directed semantics, there is a strong relationship between right-hand sides of rules and sentential forms (see Lemmas 3.7, 4.8, and 5.8).

Lemma 7.9. Let $k, n \geq 0$, $\xi \in RHS(Att_{syn}, Att_{inh}, \Delta, k, n)$, $s \in T_\Sigma$, $w \in occ(s)$ such that, for every $1 \leq i \leq k$, the occurrence wi is in $occ(s)$, and let $\varphi_1 \ldots, \varphi_n \in SF(N, s)$. Then $\xi[\pi \leftarrow w][y_1 \leftarrow \varphi_1, \ldots, y_n \leftarrow \varphi_n] \in SF(N, s)$. □

Here, as for attributed tree transducers, we cannot state a representation of every sentential form as the result of a substitution performed on some right-hand side of a rule. Again this is due to the missing neighborhood relation between the paths which are involved in an arbitrary sentential form; however the relation is necessary for such a representation (see the discussion following Lemma 5.8).

Lemma 7.10. Let s be an input tree of N. Then $SF(N, s)$ is closed under $\Rightarrow_{N,s}$

Proof. The proof of the lemma is by structural induction on sentential forms in $SF(N, s)$. It is very similar to the proof of the corresponding lemma for attributed tree transducers (Lemma 5.9).

□

Clearly, since macro attributed tree transducers are extensions of attributed tree transducers, they share the possibility of cyclic derivations.

Definition 7.11. 1. N is *circular*, if
- there is an $s \in T_\Sigma$,
- there is a sentential form $c(w, t_1, \ldots, t_p) \in SF(N, s)$ with $c \in Att^{(p+1)}$, $p \geq 0$, $w \in occ(s)$, and $t_1, \ldots, t_p \in T_\Delta$,
- there is a context $\beta \in C_{Att \cup \Delta \cup occ(s), 1}$,
- there are $\varphi_1, \ldots, \varphi_p \in SF(N, s)$,

such that $c(w, t_1, \ldots, t_p) \Rightarrow^+_{N,s} \beta[c(w, \varphi_1, \ldots, \varphi_p)]$.

2. N is *noncircular*, if it is not circular. □

Our definition of circularity is pessimistic in the sense that even a macro attributed tree transducer is quoted as circular, if the attribute which is responsible for this property occurs in a parameter position of another attribute and the latter attribute ignores this parameter ("partialness followed by deletion"). We illustrate this phenomenon by an example.

Example 7.12. Let N be a macro attributed tree transducer such that the set R of rules contains the following subsets for the input symbols $\gamma^{(1)}$ and $\alpha^{(0)}$.

- R_γ:
  - $a_0(\pi, y_1) \rightarrow a_0(\pi 1, a(\pi 1))$
  - $a(\pi) \rightarrow \alpha$
  - $b(\pi 1) \rightarrow a(\pi 1)$
- R_α:

- $a_0(\pi, y_1) \to \alpha$
- $a(\pi) \to b(\pi)$

Note that $rhs(a_0(\pi), \gamma)$ does not contain y_1. N is circular, because for the input tree $s = \gamma(\alpha)$, N can derive:

$$a(1) \Rightarrow_{N,s} b(1) \Rightarrow_{N,s} a(1)$$

which is a cycle.

However, for every input tree s, i.e., $s = \gamma^n(\alpha)$ for some $n \geq 0$, there is a terminating derivation:

$$\begin{array}{ll} & a_0(\varepsilon, y_1) \\ \Rightarrow_{N,s} & a_0(1, a(1)) \\ \Rightarrow_{N,s} & a_0(11, a(11)) \\ & \dots \\ \Rightarrow_{N,s} & a_0(1^n, a(1^n)) \\ \Rightarrow_{N,s} & \alpha \end{array}$$

□

This pessimistic point of view is incorporated in the definitions of the dependency graphs in which no distinction is made as to whether an attribute occurs nested in a parameter position or not.

Definition 7.13. Let $\sigma \in \Sigma^{(k)}$ with $k \geq 0$. The *dependency graph of σ for N*, denoted by $D_N(\sigma)$, is the directed graph $(att(\sigma), E(\sigma))$ with $E(\sigma) = \{((c, \pi\eta), (c', \pi\eta')) \in out(\sigma) \times in(\sigma) \mid rhs(c'(\pi\eta'), \sigma)$ contains a subtree of the form $c(\pi\eta, \xi_1, \ldots, \xi_l)\}$ as a set of arcs. □

The dependency graph of an input tree s (Def. 5.13), the composition of a dependency graph with is-graphs of input trees (Def. 5.15), and the is-graph of s for N (Def. 5.15) are defined in the same way as for attributed tree transducers. Moreover, the lemma on which we can base the circularity test for macro attributed tree transducers and the test itself can be taken from the case of attributed tree transducers (Lemma 5.17 and Fig. 5.7, respectively). Hence, we do not repeat these definitions, lemmas, and the algorithm here.

In the rest of this chapter the arbitrary, but fixed, macro attributed tree transducer N is assumed to be noncircular. Then we can prove that, for every input tree s, the derivation relation $\Rightarrow_{N,s}$ is confluent and terminating.

Lemma 7.14. For every input tree s of N, the derivation relation $\Rightarrow_{N,s}$ is locally confluent on $SF(N, s)$.

Proof. The proof is very similar to the proof of the corresponding property of the derivation relation of macro tree transducers (Lemma 4.10) and hence, it is not repeated here. □

As in the case of attributed tree transducers, we define the notions of ω-reducible sentential form and minimal ω-reducible sentential form (Def. 5.21). The following observation about the existence of minimal ω-reducible subtrees corresponds to Observation 5.22.

Observation 7.15. Every ω-reducible sentential form has a minimal ω-reducible subtree. □

Since attributes may have a rank greater than 1, minimal ω-reducible sentential forms of macro attributed tree transducers appear more general than the corresponding forms of attributed tree transducers (Observation 5.23).

Observation 7.16. Every minimal ω-reducible sentential form has the form $c(w, \xi_1, \ldots, \xi_l)$ for some attribute c, occurrence w, and sentential forms $\xi_1, \ldots, \xi_l$; moreover, the $\xi_1, \ldots, \xi_l$ are not ω-reducible.

□

Observation 7.17. If φ is a minimal ω-reducible sentential form, then there is an infinite derivation which starts from φ by application of a rule to the root of φ. □

The proof of the fact that $\Rightarrow_{N,s}$ is terminating proceeds in almost the same way as the proof of the corresponding property of attributed tree transducers (Lemma 5.24). Recall that we are only dealing with noncircular macro attributed tree transducers.

Lemma 7.18. Let s be an input tree of N. Then $\Rightarrow_{N,s}$ is terminating on $SF(N, s)$.

Proof. Again we prove this fact by contradiction. Let N be a noncircular macro attributed tree transducer and assume that there is an $s \in T_\Sigma$ such that $\Rightarrow_{N,s}$ is *not* terminating on $SF(N, s)$. Then there is an ω-reducible sentential form $\varphi \in SF(N, s)$. By Observation 7.15, φ has a minimal ω-reducible subtree, say $\varphi_{0,0}$, and, by Observation 7.16, $\varphi_{0,0}$ has the form $c_0(w_0, \xi_{0,1}, \ldots, \xi_{0,l_0})$. Hence, there is an infinite derivation

$$\varphi_{0,0} \Rightarrow_{N,s} \varphi_{0,1} \Rightarrow_{N,s} \varphi_{0,2} \cdots$$

Clearly, $\varphi_{0,1}$ is also an ω-reducible sentential form and it has a minimal ω-reducible subtree, say $\varphi_{1,0}$. Moreover, there is an infinite derivation

$$\varphi_{1,0} \Rightarrow_{N,s} \varphi_{1,1} \Rightarrow_{N,s} \varphi_{1,2} \cdots$$

Again, $\varphi_{1,1}$ is ω-reducible and it has a minimal ω-reducible subtree $\varphi_{0,2}$. Obviously, this construction of minimal subtrees can be repeated and thereby we obtain two infinite sequences $(\varphi_{i,0})_{i \geq 0}$ and $(\varphi_{i,1})_{i \geq 0}$ such that, for every $i \geq 0$, the following three properties hold:

1. $\varphi_{i,0}$ has the form $c_i(w_i, \xi_{i,1}, \ldots, \xi_{i,l_i})$ for some attribute c_i, some occurrence w_i of s, and some sentential forms $\xi_{i,1}, \ldots, \xi_{i,l_i}$, and $\xi_{i,1}, \ldots, \xi_{i,l_i}$ are not ω-reducible.
2. $\varphi_{i,0} \Rightarrow_{N,s} \varphi_{i,1}$.
3. $\varphi_{i+1,0}$ is a minimal ω-reducible subtree of $\varphi_{i,1}$.

This implies that, for every $i, j \geq 0$ with $j > i$, there is an $(Att \cup \Delta \cup occ(s) \cup Y, 1)$-context $\beta_{i,j}$ such that

$$\varphi_{i,0} \Rightarrow^+_{N,s} \beta_{i,j}[\varphi_{j,0}].$$

Since every $\varphi_{i,0}$ has the form $c_i(w_i, \xi_{i,1}, \ldots, \xi_{i,l_i})$ and since the set $\{c(w) \mid c \in Att, w \in occ(s)\}$ is finite, there must be indices i_0, j_0 with $j_0 > i_0$ such that $c_{i_0} = c_{j_0}$ and $w_{i_0} = w_{j_0}$. Hence, there is a $(Att \cup \Delta \cup occ(s) \cup Y, 1)$-context β such that

$$\begin{aligned} \varphi_{i_0,0} = c_{i_0}(w_{i_0}, \xi_{i_0,1}, \ldots, \xi_{i_0,l_{i_0}}) &\Rightarrow^+_{N,s} \beta[c_{j_0}(w_{j_0}, \xi_{j_0,1}, \ldots, \xi_{j_0,l_{j_0}})] \\ &= \beta[c_{i_0}(w_{i_0}, \xi_{j_0,1}, \ldots, \xi_{j_0,l_{j_0}})]. \end{aligned}$$

Since no rule in this derivation is applied to the trees in the parameter positions of $\varphi_{i_0,0}$ (because $\varphi_{i_0,0}$ is minimal ω-reducible), it follows that for every $t_1, \ldots, t_{l_{i_0}} \in T_\Delta$ there are $\xi_1, \ldots, \xi_{l_{i_0}}$ such that

$$c_{i_0}(w_{i_0}, t_1, \ldots, t_{l_{i_0}}) \Rightarrow^+_{N,s} \beta[c_{i_0}(w_{i_0}, \xi_1, \ldots, \xi_{l_{i_0}})].$$

This is a contradiction to the assumption that N is noncircular. □

Lemma 7.19. For every $s \in T_\Sigma$, the derivation relation $\Rightarrow_{N,s}$ is confluent on $SF(N, s)$.

Proof. This statement follows immediately from the facts that $\Rightarrow_{N,s}$ is locally confluent and terminating (Lemmas 7.14 and 7.18) and from Newman's lemma (Lemma 2.5). □

Lemma 7.20. Let $s \in T_\Sigma$. For every $\varphi \in SF(N, s)$, the normal form $nf(\Rightarrow_{N,s}, \varphi)$ exists.

Proof. The existence of the term $nf(\Rightarrow_{N,s}, \varphi)$ is guaranteed by the fact that $\Rightarrow_{N,s}$ is confluent and terminating on $SF(N, s)$ (Lemmas 7.18 and 7.19) and by Corollary 2.7. □

Corollary 7.21. For every $s \in T_\Sigma$, $c \in Att^{(n+1)}$ with $n \geq 0$, $w \in occ(s)$, and $\varphi_1, \ldots, \varphi_n \in SF(N, s)$,

$$nf(\Rightarrow_{N,s}, c(w, \varphi_1, \ldots, \varphi_n)) = nf(\Rightarrow_{N,s}, c(w, nf(\Rightarrow_{N,s}, \varphi_1), \ldots, nf(\Rightarrow_{N,s}, \varphi_n)))$$

Proof. (See the proof of Lemma 4.14) By Lemma 7.20, $\varphi_i \Rightarrow^*_{N,s} nf(\Rightarrow_{N,s}, \varphi_i)$, for every $1 \leq i \leq n$. Hence, $c(w, \varphi_1, \ldots, \varphi_n) \Rightarrow^*_{N,s} c(w, nf(\Rightarrow_{N,s}, \varphi_1), \ldots, nf(\Rightarrow_{N,s}, \varphi_n))$. Thus our corollary follows from Corollary 2.8. □

The notion of the condensed dependency graph of an input tree s is defined in the same way as for attributed tree transducers (Def. 5.29). Hence, the definition of topological sorting of inside attribute occurrences in Def. 5.30 applies to the situation of a macro attributed tree transducer. This topological sorting is used in the proof that the normal forms of sentential forms are final sentential forms.

Theorem 7.22. For every $\varphi \in SF(N,s)$, the inclusion

$$nf(\Rightarrow_{N,s}, \varphi) \in SF_{fin}(N,Y)$$

holds.

Proof. By simultaneous induction (using Principle 5.28), we can prove that the following predicates K and L hold on T_Σ.

K: For every $a \in Att_{syn}^{(n+1)}$ with $n \geq 0$, and $s \in T_\Sigma$, the inclusion $nf(\Rightarrow_{N,s}, a(\varepsilon, y_1, \ldots, y_n)) \in SF_{fin}(N,Y)$ holds.
L: For every $k \geq 0$ and $\sigma \in \Sigma^{(k)}$, $\xi \in RHS(\sigma)$, and $(s_1, \ldots, s_k) \in (T_\Sigma)^k$, the inclusion $nf(\Rightarrow_{N,\sigma(s_1,\ldots,s_k)}, \xi[\pi \leftarrow \varepsilon]) \in SF_{fin}(N,Y)$ holds.

Since we already have the proof for the corresponding theorem for macro tree transducers (Theorem 4.15) and for attributed tree transducers (Lemma 5.32) in detail, we omit the proof of K and L here. □

Now we can define the tree transformation induced by N.

Definition 7.23. Let $E = (E_1, E_2)$ be the environment of N with $E_1 = (t_1, \ldots, t_r)$.

(a) Let $a \in Att_{syn}^{(n+1)}$ with $n \geq 0$ be a synthesized attribute. The *tree transformation induced by N with a* is the function $\tau_{N,a}^{der} : T_\Sigma \to SF_{fin}(N, Y_n)$ which is defined, for every $s \in T_\Sigma$, by $\tau_{N,a}^{der}(s) = nf(\Rightarrow_{N,s}, a(\varepsilon, y_1, \ldots, y_n))$.
(b) The *tree transformation induced by N* is the function $\tau_N^{der} : T_\Sigma \to T_\Delta$ such that, for every $s \in T_\Sigma$, $\tau_N^{der}(s) = \tilde{E}(\tau_{N,a_0}^{der}(s))$ where $\tilde{E} : SF_{fin}(N, Y_r) \to T_\Delta$ is defined by structural induction as follows.
  (i) For every $b \in Att_{inh}^{(l+1)}$ with $l \geq 0$ and $\varphi_1, \ldots, \varphi_l \in T_{Att_{inh} \cup \Delta}(\{\varepsilon\} \cup Y_r)$,
  $$\tilde{E}(b(\varepsilon, \varphi_1, \ldots, \varphi_l)) = E_2(b)[y_j \leftarrow \tilde{E}(\varphi_j); 1 \leq j \leq l].$$
  (ii) For every $\delta \in \Delta^{(l)}$ with $l \geq 0$ and $\varphi_1, \ldots, \varphi_l \in T_{Att_{inh} \cup \Delta}(\{\varepsilon\} \cup Y_r)$,
  $$\tilde{E}(\delta(\varphi_1, \ldots, \varphi_l)) = \delta(\tilde{E}(\varphi_1), \ldots, \tilde{E}(\varphi_l)).$$

(iii) For every $y_j \in Y_r$, $\tilde{E}(y_j) = t_j$. □

A tree transformation τ is called a *macro attributed tree transformation*, if τ can be induced by some macro attributed tree transducer. The class of macro attributed tree transformations is denoted by MAT.

7.3 Characterization of Macro Attributed Tree Transformations

As for macro attributed tree transformations we provide an inductive characterization. Clearly, this combines somehow the characterizations of Sects. 4.3 and 5.3. We use the principle of definition by simultaneous induction (Principle 5.36) to define sets of functions which characterize the derivation relation of macro attributed tree transducers. In fact, this definition can be seen as an amalgamation of Defs. 4.18 and 5.37. We also refer to the definition of *topsort* in Def. 5.30.

Definition 7.24. Let $<$ be a linear order on Att. We define the set

$$F_{<,N} = \{\tau^{ind}_{<,N,a} : T_\Sigma \to SF_{fin}(N,Y) \mid a \in Att_{syn}\}$$

of functions and the set

$$G_{<,N} = \{\tau^{ind}_{<,N,\sigma,\xi} : (T_\Sigma)^k \to SF_{fin}(N,Y) \,|\, k \geq 0, \sigma \in \Sigma^{(k)}, \xi \in RHS(\sigma)\}$$

of functions by simultaneous induction according to Principle 5.36 as follows.

Application of IB: For every $\sigma \in \Sigma^{(0)}$ and $\xi \in RHS(\sigma)$, we define $\tau^{ind}_{<,N,\sigma,\xi}(())$ by structural induction on ξ.

(i) Since σ has rank 0, no right-hand side of a σ-rule contains a synthesized attribute and hence, case (i) of Def. 7.1 cannot occur.

(ii) If $\xi = b(\pi, \xi_1, \ldots, \xi_l)$ for some $b \in Att^{(l+1)}_{inh}$, $l \geq 0$, and $\xi_1, \ldots, \xi_l \in RHS(Att_{syn}, Att_{inh}, \Delta, 0, n)$, then we define

$$\tau^{ind}_{<,N,\sigma,\xi}(()) = b(\varepsilon, \tau^{ind}_{<,N,\sigma,\xi_1}(()), \ldots, \tau^{ind}_{<,N,\sigma,\xi_l}(())).$$

(iii) If $\xi = \delta(\xi_1, \ldots, \xi_l)$ for some $\delta \in \Delta^{(l)}$ with $l \geq 0$ and $\xi_1, \ldots, \xi_l \in RHS(Att_{syn}, Att_{inh}, \Delta, 0, n)$, then we define

$$\tau_{<,N,\sigma,\xi}(()) = \delta(\tau^{ind}_{<,N,\sigma,\xi_1}(()), \ldots, \tau^{ind}_{<,N,\sigma,\xi_l}(())).$$

(iv) If $\xi = y_j$, then $\tau^{ind}_{<,N,\sigma,\xi}(()) = y_j$.

Application of IS1: For every $a \in Att_{syn}^{(n+1)}$ with $n \geq 0$ and $s = \sigma(s_1, \ldots, s_k) \in T_\Sigma$, we define

$$\tau^{ind}_{<,N,a}(s) = \tau^{ind}_{<,N,\sigma,rhs(a(\pi),\sigma)}(\omega),$$

where $\omega = (s_1, \ldots, s_k)$.

Application of IS2: Let $k \geq 1$, $\sigma \in \Sigma^{(k)}$, and $s_1, \ldots, s_k \in T_\Sigma$. Let $w = topsort(N, <, s)$ with $s = \sigma(s_1, \ldots, s_k)$. By (finite) mathematical induction on ν with $1 \leq \nu \leq length(w)$, we define $\tau^{ind}_{<,N,\sigma,rhs(w(\nu),\sigma)}(\omega)$, where $\omega = (s_1, \ldots, s_k)$.

Induction base $\nu = 1$: The value $\tau^{ind}_{<,N,\sigma,rhs(w(1),\sigma)}(\omega)$ is defined by structural induction on $rhs(w(1), \sigma)$. Let SUB represent $sub(rhs(w(1), \sigma))$.

(i) Let $d(\pi i, \xi_1, \ldots, \xi_l) \in SUB$ for some $d \in Att_{syn}^{(l+1)}$ with $l \geq 0$ and $1 \leq i \leq k$, and $\xi_1, \ldots, \xi_l \in SUB$. Since $\nu = 1$, the value $\tau^{ind}_{<,N,d}(s_i) \in T_\Delta(Y_l)$, in particular; this value does not contain inherited attributes. Thus we can define

$$\tau^{ind}_{<,N,\sigma,d(\pi i,\xi_1,\ldots,\xi_l)}(\omega) = \tau^{ind}_{<,N,d}(s_i)[y_j \leftarrow \tau^{ind}_{<,N,\sigma,\xi_j}(\omega); 1 \leq j \leq l]$$

(ii) Let $b(\pi, \xi_1, \ldots, \xi_l) \in SUB$ for some $b \in Att_{inh}^{(l+1)}$ with $l \geq 0$ and $1 \leq i \leq k$, and $\xi_1, \ldots, \xi_l \in SUB$. Then define

$$\tau^{ind}_{<,N,\sigma,b(\pi,\xi_1,\ldots,\xi_l)}(\omega) = b(\omega, \tau^{ind}_{<,N,\sigma,\xi_1}(\omega), \ldots, \tau^{ind}_{<,N,\sigma,\xi_l}(\omega)).$$

(iii) Let $\delta(\xi_1, \ldots, \xi_l) \in SUB$ for some $\delta \in \Delta^{(l)}$, $l \geq 0$, $\xi_1, \ldots, \xi_l \in SUB$. Then define

$$\tau^{ind}_{<,N,\sigma,\delta(\xi_1,\ldots,\xi_l)}(\omega) = \delta(\tau^{ind}_{<,N,\sigma,\xi_1}(\omega), \ldots, \tau^{ind}_{<,N,\sigma,\xi_l}(\omega)).$$

(iv) Let $y_j \in SUB$. Then define $\tau^{ind}_{<,N,\sigma,y_j}(\omega) = y_j$.

Induction step $\nu \to \nu + 1$: Assume that for every κ with $1 \leq \kappa \leq \nu$, the value $\tau^{ind}_{<,N,\sigma,rhs(w(\kappa),\sigma)}(\omega)$ is already defined. Then we define the value $\tau^{ind}_{<,N,\sigma,rhs(w(\nu+1),\sigma)}(\omega)$ by structural induction on $rhs(w(\nu+1), \sigma)$. Let SUB abbreviate the set $sub(rhs(w(\nu + 1), \sigma))$.

(i) Let $d(\pi i, \xi_1, \ldots, \xi_l) \in SUB$ for some $d \in Att_{syn}^{(l)}$ with $l \geq 0$ and $1 \leq i \leq k$, and $\xi_1, \ldots, \xi_l \in SUB$. Then define

$$\tau^{ind}_{<,N,\sigma,d(\pi i,\xi_1,\ldots,\xi_l)}(\omega) = \phi_{\omega,i,\tilde{\xi}}(\tau^{ind}_{<,N,d}(s_i))$$

where $\tilde{\xi}$ represents $(\xi_1, \ldots, \xi_l)$ and $\phi_{\omega,i,\tilde{\xi}}$ is defined by structural induction on trees in $SF_{fin}(N, Y_l)$ as follows (where $SF_{fin}(N, Y_l)$ is represented by S):

(a) Let $b(\varepsilon, \varphi_1, \ldots, \varphi_l) \in S$ for some $b \in Att_{inh}^{(l)}$ with $l \geq 0$ and $1 \leq i \leq k$, and $\varphi_1, \ldots, \varphi_l \in S$. Then define
$\phi_{\omega,i,\tilde{\xi}}(b(\varepsilon, \varphi_1, \ldots, \varphi_l)) =$
$\tau^{ind}_{<,N,\sigma,rhs(b(\pi i),\sigma)}(\omega)[y_j \leftarrow \phi_{\omega,i,\tilde{\xi}}(\varphi_j); 1 \leq j \leq l]$.

(b) Let $\delta(\varphi_1, \ldots, \varphi_l) \in S$ for some $\delta \in \Delta^{(l)}$, $l \geq 0$, $\varphi_1, \ldots, \varphi_l \in S$. Then define

$$\phi_{\omega,i,\tilde{\xi}}(\delta(\varphi_1, \ldots, \varphi_l)) = \delta(\phi_{\omega,i,\tilde{\xi}}(\varphi_1), \ldots, \phi_{\omega,i,\tilde{\xi}}(\varphi_l)).$$

(c) Let $y_j \in S$. Then define

$$\phi_{\omega,i,\tilde{\xi}}(y_j) = \tau^{ind}_{<,N,\sigma,\xi_j}(\omega).$$

(ii),(iii),(iv) These cases are the same as the corresponding cases of the induction base. □

As in Lemma 5.38, the order $<$ does not matter here.

Lemma 7.25. Let $<_1$ and $<_2$ be two linear orders on Att. Then

- for every $a \in Att_{syn}$, $\tau^{ind}_{<_1,N,a} = \tau^{ind}_{<_2,N,a}$ and
- for every $k \geq 0$, $\sigma \in \Sigma^{(k)}$, and $\xi \in RHS(\sigma)$, $\tau^{ind}_{<_1,N,\sigma,\xi} = \tau^{ind}_{<_2,N,\sigma,\xi}$. □

Thus, in the following, we will omit the linear order on the set of attributes from the notation of the τ^{ind}-functions.

Again we show the correctness and completeness of the τ^{ind}-function with respect to the derivation relation. For this purpose we define, for every $\sigma \in \Sigma^{(k)}$ with $k \geq 0$ and $\xi \in RHS(\sigma)$, the function $\tau^{der}_{N,\sigma,\xi} : (T_\Sigma)^k \to SF_{fin}(N, Y)$ such that $\tau^{der}_{N,\sigma,\xi}((s_1, \ldots, s_k)) = nf(\Rightarrow_{N,\sigma(s_1,\ldots,s_k)}, \xi[\pi \leftarrow \varepsilon])$.

Theorem 7.26. 1. For every $s \in T_\Sigma$ and $a \in Att_{syn}^{(n+1)}$ with $n \geq 0$, $\tau^{der}_{N,a}(s) = \tau^{ind}_{N,a}(s)$.
2. For every $k \geq 0$, $\sigma \in \Sigma^{(k)}$, $\xi \in RHS(\sigma)$, and $\omega = (s_1, \ldots, s_k) \in (T_\Sigma)^k$, $\tau^{der}_{N,\sigma,\xi}(\omega) = \tau^{ind}_{N,\sigma,\xi}(\omega)$.

Proof. The two statements are proved by the established technique of simultaneous induction using Principle 5.28. An example of a proof according to this principle is shown in Lemma 5.32. Since no unexpected conditions arise in the proof of the two statements and the proof technique is practically a combination of the proofs for macro tree transducers and attributed tree transducers, we do not give the proof here. □

7.4 Composition and Decomposition Results

First we will prove a decomposition result for macro attributed tree transducers which generalizes the decomposition of MAC into TOP and $YIELD$ (Theorem 4.37). In fact, the two underlying lemmas are straightforward generalizations of the corresponding lemmas for the decomposition of macro tree transducers (Lemmas 4.34 and 4.36).

Lemma 7.27. $MAT \subseteq ATT \circ YIELD$.

Proof. (See Lemma 4.34) Let $N = (Att, \Sigma, \Delta, a_0, R, E)$ be a noncircular macro attributed tree transducer with $E = (E_1, E_2)$ and $E_1 = (t_1, \ldots, t_r)$ and $r + 1 = rank(q_0)$. Denote the number $max\{n \mid Att^{(n+1)} \cup \Delta^{(n)} \neq \emptyset\}$ by mx.

We construct an attributed tree transducer $A = (Att', \Sigma, \Gamma, a_0', R', E')$, a function $g : \Gamma^{(0)} \to T_\Delta(Z_{mx})$, and a tuple $\tilde{t} \in (T_\Delta)^{mx}$ such that $\tau_N = \tau_A \circ yield_{g,\tilde{t}}$.

- Define $Att' = (Att')_{syn} \cup (Att')_{inh}$ where $(Att')_{syn} = \{a' \mid a \in Att_{syn}\}$ and $(Att')_{inh} = \{b' \mid b \in Att_{inh}\}$. Every attribute in Att' has rank 1.
- Define Γ to be the ranked alphabet such that $\Gamma^{(0)} = \{\delta' \mid \delta \in \Delta\} \cup \{\alpha_1, \ldots, \alpha_{mx}\}$ where the α's are new symbols, and, for every k such that $Att^{(k)} \neq \emptyset$ or $\Delta^{(k-1)} \neq \emptyset$, define $\Gamma^{(k)} = \{c_k\}$ and $\Gamma^{(k)} = \emptyset$ otherwise.
- Define the function $g : \Gamma^{(0)} \to T_\Delta(Z_{mx})$ by $g(\delta') = \delta(z_1, \ldots, z_k)$ if $\delta \in \Delta^{(k)}$, and $g(\alpha_i) = z_i$ for every $i \in \{1, \ldots, mx\}$.
- Define $\tilde{t} = E$ if $mx = r$, and $\tilde{t} = (t_1, \ldots, t_r, z_{r+1}, \ldots, z_{mx})$ if $mx > r$.
- For the construction of the rules of A we need the set $COMB = \{COMB_{k,n} \mid k, n \geq 0\}$ of functions $COMB_{k,n} : RHS(Att_{syn}, Att_{inh}, \Delta, k, n) \to RHS((Att')_{syn}, (Att')_{inh}, \Gamma, k)$ that are defined inductively on the structure of their arguments.
  (i) For every $a \in Att_{syn}^{(l+1)}$ with $l \geq 0$, $1 \leq i \leq k$ and $\xi_1, \ldots, \xi_l \in RHS(Att_{syn}, Att_{inh}, \Delta, k, n)$, we define $COMB_{k,n}(a(\pi i, \xi_1, \ldots, \xi_l)) = c_{l+1}(a'(\pi i), COMB_{k,n}(\xi_1), \ldots, COMB_{k,n}(\xi_l))$.
  (ii) For every $b \in Att_{inh}^{(l+1)}$ with $l \geq 0$ and $\xi_1, \ldots, \xi_l \in RHS(Att_{syn}, Att_{inh}, \Delta, k, n)$, we define $COMB_{k,n}(b(\pi, \xi_1, \ldots, \xi_l)) = c_{l+1}(b'(\pi), COMB_{k,n}(\xi_1), \ldots, COMB_{k,n}(\xi_l))$.
  (iii) For every $\delta \in \Delta^{(l)}$ with $l \geq 0$ and $\xi_1, \ldots, \xi_l \in RHS(Att_{syn}, Att_{inh}, \Delta, k, n)$, we define $COMB_{k,n}(\delta(\xi_1, \ldots, \xi_l)) = c_{l+1}(\delta', COMB_{k,n}(\xi_1), \ldots, COMB_{k,n}(\xi_l))$.
  (iv) For every $1 \leq j \leq n$, we define $COMB_{k,n}(y_j) = \alpha_j$.

  Now we define R' to be the set $\bigcup_{\sigma \in \Sigma} R'_\sigma$; for every $\sigma \in \Sigma^{(k)}$ with $k \geq 0$ we define:

$$R'_\sigma = \{a'(\pi) \to COMB_{k,n}(\xi) \mid (a(\pi, y_1, \ldots, y_n) \to \xi) \in R_\sigma\}$$
$$\cup \{b'(\pi i) \to COMB_{k,n}(\xi) \mid (b(\pi i, y_1, \ldots, y_n) \to \xi) \in R_\sigma\}.$$

- Define $E' : (Att')_{inh} \to T_\Gamma$ such that, for every $b \in Att_{inh}^{(n+1)}$ and $n \geq 0$, we let $E'(b') = COMB_{0,n}(E_2(b))$.

Due to the definition of the dependency graphs for macro attributed tree transducers (Def. 7.13) and for attributed tree transducers (Def. 5.12), it is clear that A is also noncircular.

We extend the function g to

$$g_\varepsilon : \Gamma^{(0)} \cup \{b'(\varepsilon) \mid b \in Att_{inh}\} \to T_{Att_{inh} \cup \Delta}(\{\varepsilon\} \cup Z)$$

such that for every $b \in Att_{inh}^{(n+1)}$ with $n \geq 0$, $g_\varepsilon(b'(\varepsilon)) = b(\varepsilon, z_1, \ldots, z_n)$.

The correctness of the construction can be shown by proving the following statements by simultaneous induction where we represent the substitution $[z_1 \leftarrow y_1, \ldots, z_n \leftarrow y_n]$ by $[z \leftarrow y]_n$:

K: For every $n \geq 0$, $a \in Att_{syn}^{(n+1)}$, and $s \in T_\Sigma$, the relation $yield_{g_\varepsilon}(\tau_{A,a'}(s)) \in SF(N, Z_n)$ holds and the equality $\tau_{N,a}(s) = yield_{g_\varepsilon}(\tau_{A,a'}(s))[z \leftarrow y]_n$ is true.

L: For every $k \geq 0$, $\sigma \in \Sigma^{(k)}$, $\xi \in RHS(\sigma)$, and $\omega \in (T_\Sigma)^k$, the relation $yield_{g_\varepsilon}(\tau_{A,\sigma,COMB_{k,n}(\xi)}(\omega)) \in SF(N, Z_n)$ holds and the equality $\tau_{N,\sigma,\xi}(\omega) = yield_{g_\varepsilon}(\tau_{A,\sigma,COMB_{k,n}(\xi)}(\omega))[z \leftarrow y]_n$ is true.

These statements can be proved by straightforward simultaneous induction and hence, it is left to the reader. The correctness of the construction conforms with that in the proof of Lemma 4.34. □

Lemma 7.28. $ATT \circ YIELD \subseteq MAT$

Proof. Let $A = (Att, \Sigma, \Gamma, a_0, R, E)$ be an attributed tree transducer, $g : \Gamma^{(0)} \to T_\Delta(Z_r)$ such that for every $\alpha \in \Gamma$, $g(\alpha)$ is linear in Z, and $\tilde{t} = (t_1, \ldots, t_r)$ for some $r \geq 0$. We construct a macro attributed tree transducer $N = (Att', \Sigma, \Delta, a_0', R', E')$ such that $\tau_A \circ yield_{g,\tilde{t}} = \tau_N$.

- Define $Att' = (Att')_{syn} \cup (Att')_{inh}$ and $(Att')_{syn} = \{a' \mid a \in Att_{syn}\}$ and $(Att')_{inh} = \{b' \mid b \in Att_{inh}\}$; moreover, every attribute has rank $r+1$, i.e., $Att' = (Att')^{(r+1)}$.
- Extend g to $g_\pi : \Gamma^{(0)} \cup Att(\{\pi, \pi 1, \ldots, \pi\kappa\}) \to T_{\Delta \cup Att'}(Z_r \cup \{\pi, \pi 1, \ldots, \pi\kappa\})$ where $\kappa = max\{k \mid \Sigma^{(k)} \neq \emptyset\}$ by defining, for every $c \in Att$ and $\eta \in \{\varepsilon\} \cup \{1, \ldots, \kappa\}$:

$$g_\pi(c(\pi\eta)) = c'(\pi\eta, z_1, \ldots, z_r).$$

(Clearly, for every $\alpha \in \Gamma^{(0)}$ we have $g'(\alpha) = g(\alpha)$.)
- R' is the set $\bigcup_{\sigma \in \Sigma} (R')_\sigma$ where $(R')_\sigma$ is the set of rules such that, if, for some $c \in Att$ and $\eta \in \{\varepsilon\} \cup \{1, \ldots, \kappa\}$, the rule $c(\pi\eta) \to \xi$ is in R_σ, then the rule

$$c'(\pi\eta, y_1, \ldots, y_r) \to yield_{g_\pi}(\xi)[z_1 \leftarrow y_1, \ldots, z_r \leftarrow y_r]$$

is in $(R')_\sigma$.

- $E' = (E_1, E_2)$ where $E_1 = (t_1, \ldots, t_r)$ and $E_2 : (Att')_{inh} \to T_\Delta(Y)$ is the function such that, for every $b' \in Att_{inh}^{(r+1)}$, we define $E_2(b') = yield_g(E(b))[z_1 \leftarrow y_1, \ldots, z_r \leftarrow y_r]$.

The correctness of the construction is proved by showing the following two statements by simultaneous induction in which we represent the substitution $[z_1 \leftarrow y_1, \ldots, z_r \leftarrow y_r]$ by $[z \leftarrow y]_r$. As in the proof of the decomposition result $MAT \subseteq ATT \circ YIELD$ (Lemma 7.27), we need an extension of g which can handle objects of the form $b(\varepsilon)$. Thus, we extend g to $g_\varepsilon : \Gamma^{(0)} \cup \{b(\varepsilon) \mid b \in Att_{inh}\} \to T_{(Att')_{inh} \cup \Delta}(\{\varepsilon\} \cup Z)$ such that for every $b \in Att_{inh}$:

$$g_\varepsilon(b(\varepsilon)) = b'(\varepsilon, z_1, \ldots, z_r).$$

K: For every $a \in Att_{syn}$ and $s \in T_\Sigma$, the equality $yield_{g_\varepsilon}(\tau_{A,a}(s))[z \leftarrow y]_r = \tau_{N,a'}(s)$ holds.

L: For every $k \geq 0$, $\sigma \in \Sigma^{(k)}$, $\xi \in RHS(\sigma)$, and $\omega \in (T_\Sigma)^k$, the equality $yield_{g_\varepsilon}(\tau_{A,\sigma,\xi}(\omega))[z \leftarrow y]_r = \tau_{N,\sigma,\xi'}(\omega)$ holds where $\xi' = yield_{g_\pi}(\xi)[z \leftarrow y]_r$.

□

From Lemmas 7.27 and 7.28, we can derive the following characterization of MAT.

Theorem 7.29. $MAT = ATT \circ YIELD$.

This characterization generalizes the characterization of MAC in terms of the composition of TOP and $YIELD$ (see Theorem 4.37). Intuitively, it is clear that $MAT = ATT \circ YIELD$ must be true, because on both sides of the equation the model which handles the customary replacement (i.e., macro tree transducers or top-down tree transducers) is generalized from a recursive descent device to the corresponding tree-walking device (i.e., attributed tree transducers and macro attributed tree transducers, respectively).

From the characterization of MAT we can draw some conclusions.

Corollary 7.30. $MAT = ATT \circ ATT$.

Proof.

	MAT	
$=$	$ATT \circ YIELD$	(Theorem 7.29)
$\subseteq$	$ATT \circ ATT$	(Corollary 6.24)
$\subseteq$	$ATT \circ MAC$	(Lemma 6.1)
$=$	$ATT \circ TOP \circ YIELD$	(Theorem 4.37)
$=$	$ATT \circ YIELD$	(Theorem 5.47)

□

Corollary 7.31. $MAT \circ TOP = MAT$.

Proof.

$$\begin{array}{lll} & MAT \circ TOP & \\ = & ATT \circ ATT \circ TOP & \text{(Corollary 7.30)} \\ = & ATT \circ ATT & \text{(Theorem 5.47)} \end{array}$$

□

Corollary 7.32. $HOM \circ MAT = MAC \circ MAC$

Proof.

$$\begin{array}{lll} & HOM \circ MAT & \\ = & HOM \circ ATT \circ YIELD & \text{(Theorem 7.29)} \\ = & MAC \circ YIELD & \text{(Theorem 6.25)} \\ \subseteq & MAC \circ MAC & \text{(Lemma 4.32)} \\ = & HOM \circ sl\text{-}MAC \circ MAC & \text{(Corollary 4.39)} \\ \subseteq & HOM \circ 1v\text{-}ATT \circ MAC & \text{(Note 6.9, Theorem 6.23)} \\ = & HOM \circ 1v\text{-}ATT \circ TOP \circ YIELD & \text{(Theorem 4.37)} \\ \subseteq & HOM \circ ATT \circ TOP \circ YIELD & (1v\text{-}ATT \subseteq ATT) \\ = & HOM \circ ATT \circ YIELD & \text{(Theorem 5.47)} \\ \subseteq & HOM \circ ATT \circ ATT & \text{(Corollary 6.24)} \\ = & HOM \circ MAT & \text{(Corollary 7.30)} \end{array}$$

□

7.5 Height Property

In the chapters concerning top-down tree transducers, macro tree transducers, and attributed tree transducers we have already shown an approximation of the height of an output tree t with respect to the height or size of the corresponding input tree s:

- for top-down tree transducers, there is a constant c such that $height(t) \leq c \cdot height(s)$ (Lemma 3.27),
- for macro tree transducers, there is a constant c such that $height(t) \leq c^{height(s)}$ (Lemma 4.22), and
- for attributed tree transducers, there is a constant c such that $height(t) \leq c \cdot size(s)$ (Lemma 5.40).

In Corollary 7.30 it was shown that the class MAT is characterized by the two-fold composition of the class ATT, i.e., $MAT = ATT \circ ATT$. We can use this fact in order to derive an approximation of the height of the output trees for macro attributed tree transducers.

Lemma 7.33. Let N be a macro attributed tree transducer. There is a constant $c > 0$ such that, for every $s \in T_\Sigma$, $height(\tau_N(s)) \leq 2^{c \cdot size(s)}$.

Proof. By Corollary 7.30, there are two attributed tree transducers A_1 and A_2 such that $\tau_M = \tau_{A_1} \circ \tau_{A_2}$. Let $s \in T_\Sigma$ and $t \in T_\Delta$ such that $\tau_N(s) = t$. Then there is a t' such that $\tau_{A_1}(s) = t'$ and $\tau_{A_2}(t') = t$. By Lemma 5.40, there are constants c and d such that $height(t') \leq c \cdot size(s)$ and $height(t) \leq d \cdot size(t')$. Thus, $height(t) \leq d \cdot size(t') \leq d \cdot 2^{b \cdot height(t')}$ for some constant $b > 0$, according to Lemma 2.13. Then, $d \cdot 2^{b \cdot height(t')} \leq d \cdot 2^{b \cdot c \cdot size(s)} \leq 2^{e \cdot size(s)}$ for some constant e. □

7.6 Bibliographic Notes

Most of the material presented in this chapter has been taken from [KV94b]. Example 7.3 is due to [Hou94].

In [Küh97] the syntactic single use restriction of [GG84, Gie88] is considered for macro attributed tree transducers. It transpires that the resulting class of tree transformations is a strict subclass of ATT.

8. Two Examples

In this chapter we will illustrate the usefulness of the formal models presented by specifying two non-trivial functions in a syntax-directed way via tree transducers. Clearly, our formal models are rather archaic, e.g., they abstract from many-sortedness and from concrete computations in particular semantic domains. Thus, in order to provide a more realistic specification language, we will enrich macro attributed tree transducers by these two features. This yields the formal model of so called many-sorted macro attributed tree transducer, and the two functions will be formalized by means of such transducers. The first function is the contextual analysis of a small, block-structured programming language. The second function is the insertion function in 2-3 trees.

8.1 The Specification Language

Here we superpose the many-sortedness and the possibility of concrete calculations in semantic domains on to the concept of macro attributed tree transducers; this leads to the concept of many-sorted macro attributed tree transducers. The formalization is straightforward, although slightly technical. However, for the sake of completeness, we have decided to include this formalization in the book. First we need some notions concerning sorted sets and many-sorted algebras.

Let S be a set of *sorts* (or *types*); as examples of sorts in this context we mention *integer, boolean, real, list of ..., environment.* An *S-sorted set* is a tuple $(A, sort)$ where A is a set and $sort : A \to S$ is a function which associates a sort with every element in A. For every $\kappa \in S$, we represent by A^κ the set of elements of A which have sort κ. An *S-ranked alphabet* is a finite $S^* \times S$-sorted set.

Let $(\Sigma, sort)$ be an S-ranked alphabet and A be an S-sorted set. The set of *S-sorted trees over Σ indexed by A*, denoted by $T_\Sigma(A)$, is the smallest S-sorted set T such that the following conditions hold.

(i) For every $\kappa \in S$, the set $\Sigma^{(\varepsilon,\kappa)} \cup A^\kappa \subseteq T^\kappa$.
(ii) For every $\sigma \in \Sigma^{(\kappa_1\ldots\kappa_k,\kappa)}$ with $\kappa, \kappa_1, \ldots, \kappa_k \in S$ and trees $t_1 \in T^{\kappa_1}, \ldots, t_k \in T^{\kappa_k}$, the tree $\sigma(t_1, \ldots, t_k) \in T^\kappa$.

We note that, if S is a singleton, then an S-ranked alphabet $(\Sigma, sort)$ is a usual ranked alphabet, and the set of S-sorted trees over Σ is the set of usual trees over Σ.

Let $(\Sigma, sort)$ be an S-ranked alphabet. A *many-sorted Σ-algebra $\mathcal{A}$* is a pair (A, α) where A is an S-sorted set such that, for every $\kappa \in S$, the set $A^\kappa \neq \emptyset$, and α, called the *interpretation function of $\mathcal{A}$*, is a function such that

(i) for every $\sigma \in \Sigma^{(\varepsilon,\kappa)}$, $\alpha(\sigma) \in A^\kappa$ and
(ii) for every $\sigma \in \Sigma^{(\kappa_1 \ldots \kappa_k, \kappa)}$, $\alpha(\sigma)$ is a function of type $A^{\theta_1} \times \ldots \times A^{\theta_k} \to A^\theta$.

We also have to associate sorts with the variables which occur in a many-sorted macro attributed tree transducer. For this purpose, we fix the sets IS and OS of *input sorts* and *output sorts*, respectively. Then, let $Y = \{y_{i,\kappa} \mid i \geq 0, \kappa \in OS\}$ be a sorted set of variables such that $sort(y_{i,\kappa}) = \kappa$. Moreover, let $\Pi = \{\pi_\iota \mid \iota \in IS\}$ be the IS-sorted set of path variables such that $sort(\pi_\iota) = \iota$.

Definition 8.1. A *many-sorted macro attributed tree transducer* is a tuple $N = (Att, \Sigma, \Delta, \mathcal{A}, a_0, R, E)$ where

- Att is an $(IS \cdot OS^*) \times OS$-sorted set, called the set of *attributes*; thus, every element $c \in Att$ has a sort of the form $(\iota\kappa_1 \ldots \kappa_r, \kappa_0)$ for some $\iota \in IS$ and $\kappa_0, \kappa_1, \ldots, \kappa_r \in OS$; Att is partitioned into the disjoint sets Att_{syn} and Att_{inh}, of which the elements are called *synthesized attributes* and *inherited attributes*, respectively, such that $Att \cap (\Sigma \cup \Delta) = \emptyset$.
- Σ is an IS-ranked alphabet, called the *input alphabet*, and Δ is an OS-ranked alphabet, called the *output alphabet*.
- $\mathcal{A} = (A, \alpha)$ is a many-sorted Δ-algebra; thus. in particular, A is an OS-sorted set.
- $a_0 \in Att_{syn}^{(\iota_0\kappa_{0,1} \ldots \kappa_{0,r}, \kappa_0)}$ for some $\iota_0 \in IS$ and $\kappa_{0,1}, \ldots, \kappa_{0,r}, \kappa_0 \in OS$ is a designated attribute, called the *initial attribute*.
- R is a set $\bigcup_{\sigma \in \Sigma} R_\sigma$ such that for every $\sigma \in \Sigma$, the set R_σ of rules satisfies the following conditions.

(1) For every $a \in Att_{syn}^{(\iota\kappa_1 \ldots \kappa_n, \kappa)}$ and $\sigma \in \Sigma^{\tilde{\iota}}$ with $\tilde{\iota} = (\iota_1 \ldots \iota_k, \iota)$, the set R_σ contains exactly one rule of the form

$$a(\pi_\iota, y_{1,\kappa_1}, \ldots, y_{n,\kappa_n}) \to \xi$$

where $\xi \in RHS(Att_{syn}, Att_{inh}, \Delta, \tilde{\iota}, \tilde{\kappa})^\kappa$ and $\tilde{\kappa} = (\kappa_1 \ldots \kappa_n, \kappa)$.

(2) For every $b \in Att_{inh}^{(\iota_i\kappa_1 \ldots \kappa_n, \kappa)}$, $\sigma \in \Sigma^{\tilde{\iota}}$ with $\tilde{\iota} = (\iota_1 \ldots \iota_k, \iota)$, and $1 \leq i \leq k$, the set R_σ contains exactly one rule of the form

$$b(\pi_\iota i, y_{1,\kappa_1}, \ldots, y_{n,\kappa_n}) \to \xi$$

where $\xi \in RHS(Att_{syn}, Att_{inh}, \Delta, \tilde{\iota}, \tilde{\kappa})^\kappa$ and $\tilde{\kappa} = (\kappa_1 \ldots \kappa_n, \kappa)$.

Let $\tilde{\iota} = (\iota_1 \ldots \iota_k, \iota) \in IS^* \times IS$ and $\tilde{\kappa} = (\kappa_1 \ldots \kappa_n, \kappa) \in OS^* \times OS$. The set $RHS(Att_{syn}, Att_{inh}, \Delta, \tilde{\iota}, \tilde{\kappa})$ of *right-hand sides over Att_{syn}, Att_{inh}, Δ, $\tilde{\iota}$, and $\tilde{\kappa}$* is the smallest OS-sorted set RHS satisfying the following conditions.

(i) For every $a \in Att_{syn}^{(\iota_i\theta_1\ldots\theta_l,\theta)}$ with $1 \leq i \leq k$, $l \geq 0$, and $\xi_1 \in RHS^{\theta_1}$, ..., $\xi_l \in RHS^{\theta_l}$, the term $a(\pi_\iota i, \xi_1, \ldots, \xi_l)$ is in RHS^θ.

(ii) For every $b \in Att_{inh}^{(\iota\theta_1\ldots\theta_l,\theta)}$ and $\xi_1 \in RHS^{\theta_1}$, ..., $\xi_l \in RHS^{\theta_l}$, the term $b(\pi_\iota, \xi_1, \ldots, \xi_l)$ is in RHS^θ.

(iii) For every $\delta \in \Delta^{(\theta_1\ldots\theta_l,\theta)}$ and $\xi_1 \in RHS^{\theta_1}$, ..., $\xi_l \in RHS^{\theta_l}$, the term $\delta(\xi_1, \ldots, \xi_l)$ is in RHS^θ.

(iv) For every $\theta \in OS$, $Y^\theta \subseteq RHS^\theta$.

- $E = (E_1, E_2)$ is the *environment* where $E_1 = (t_1, \ldots, t_r)$ and $t_i \in T_\Delta^{\kappa_{0,i}}$ (recall that the $\kappa_{0,i}$'s are the argument sorts of the initial attribute) and $E_2 : Att_{inh} \to T_\Delta(Y)$ is a function such that for every $b \in Att_{inh}^{(\iota\kappa_1\ldots\kappa_k,\kappa)}$, $E_2(b) \in T_\Delta(\{y_{1,\kappa_1}, \ldots, y_{k,\kappa_k}\})^\kappa$. □

The notions and notations which were fixed for macro attributed tree transducers after Note 7.4, translate to the concept of many-sorted macro attributed tree transducers. In the following, let $N = (Att, \Sigma, \Delta, \mathcal{A}, a_0, R, E)$ be an arbitrary, but fixed, many-sorted macro attributed tree transducer.

To define the derivation relation consistently, we also have to associate a sort with every occurrence $w \in occ(s)$ of a tree s. Intuitively, the sort of w is the result sort of the label of w. Formally, if s is a tree over Σ, $w \in occ(s)$, and $label(s, w) = \sigma \in \Sigma^{(\iota_1\ldots\iota_k,\iota)}$, then $sort(w) = \iota$.

The derivation relation of a many-sorted macro attributed tree transducer is defined in the same way as for a macro attributed tree transducer except that an additional derivation step may consist of a calculation in the algebra $\mathcal{A}$.

Definition 8.2. Let $s \in T_\Sigma^\iota$ for some $\iota \in IS$. The *derivation relation induced by N on s* is the binary relation $\Rightarrow_{N,s}$ over the set T_Ψ where $\Psi = Att \cup \Delta \cup A \cup occ(s) \cup Y$, such that for every $\varphi, \psi \in T_\Psi$, we define $\varphi \Rightarrow_{N,s} \psi$ if there is a context $\beta \in C_{\Psi,1}$ and one of the following three conditions holds.

(1)
- There is a synthesized attribute $a \in Att_{syn}^{(\iota\kappa_1\ldots\kappa_n,\kappa)}$,
- there is an occurrence $w \in occ(s)^\iota$ with $label(s, w) = \sigma \in \Sigma^{(\iota_1\ldots\iota_k,\iota)}$, and
- there are trees $\varphi_1 \in T_\Psi^{\kappa_1}$, ..., $\varphi_n \in T_\Psi^{\kappa_n}$

such that $\varphi = \beta[a(w, \varphi_1, \ldots, \varphi_n)]$ and $\psi = \beta[rhs(a(\pi_\iota), \sigma)[\pi_\iota \leftarrow w][y_{1,\kappa_1} \leftarrow \varphi_1, \ldots, y_{n,\kappa_n} \leftarrow \varphi_n]]$.

(2)
- There is an inherited attribute $b \in Att_{inh}^{(\iota_i\kappa_1\ldots\kappa_n,\kappa)}$,
- there is an occurrence $w \in occ(s)^{\iota_i}$ with $w = vi$ for some $v \in occ(s)^\iota$, $label(s, v) = \sigma \in \Sigma^{(\iota_1\ldots\iota_k,\iota)}$, $1 \leq i \leq k$, and
- there are trees $\varphi_1 \in T_\Psi^{\kappa_1}$, ..., $\varphi_n \in T_\Psi^{\kappa_n}$

such that $\varphi = \beta[b(w, \varphi_1, \ldots, \varphi_n)]$
and $\psi = \beta[rhs(b(\pi_\iota i), \sigma)[\pi_\iota \leftarrow v][y_{1,\kappa_1} \leftarrow \varphi_1, \ldots, y_{n,\kappa_n} \leftarrow \varphi_n]]$.

(3) • There is a $\delta \in \Delta^{(\kappa_1 \ldots \kappa_n, \kappa)}$, and
• there are elements $v_1 \in A^{\kappa_1}$, ..., $v_n \in A^{\kappa_n}$
such that $\varphi = \beta[\delta(v_1, \ldots, v_n)]$
and $\psi = \beta[\alpha(\delta)(v_1, \ldots, v_n)]$. □

It is clear from Condition (3) of Def. 8.2 that the interpretation of an element δ is strict in the sense that all its arguments have to be evaluated first.

Since we will not be proving theoretical results about many-sorted macro attributed tree transducers here, but rather using them to show their practical relevance, we will also not be reconsidering the topics about circularity, confluency, termination, and inductive characterization. These notions and the corresponding results can be generalized in a straightforward manner from Sects. 7.2–7.4.

8.2 Contextual Analysis

It is well known that contextual constraints of an imperative, block-structured programming language, like e.g. the scope rule and the type rule, cannot be expressed by a context-free grammar. Such constraints are said to be part of the *static semantics* of a language. Recall that the scope rule indicates the scope of a declared variable, and the type rule defines the way in which typed expressions may be combined. The *contextual analysis* checks whether the contextual constraints are obeyed by a given program.

In this section we will formalize the contextual analysis by means of a many-sorted macro attributed tree transducer. The programming language concerned is a very simple imperative, block-structured programming language in the spirit of ALGOL 60; the language is called CHECK. It only allows the declaration of variables of type either boolean or integer, and the assignment of variables (see Fig. 8.1 for a CHECK-program). For the sake of simplicity, we will restrict CHECK-programs to have at most two variables a and b.

The scope and type rules of CHECK are very easy: the scope of a variable is the entire block in which it is declared, and the types of the variables involved in an assignment must be equal. Thus, in the program in Fig. 8.1, the scope of the integer variable b is statement $stat_2$, and the scope of the boolean variable b is $stat_1$, i.e., the **begin-end** block, and $stat_3$ (in fact, only $stat_3$, because b is redeclared inside $stat_1$). The statement $stat_2$ does not yield a type conflict, i.e, it matches with the type rule, whereas $stat_3$ yields a type conflict, because b has type boolean and a has type integer. Let us now define the context-free part of CHECK. We have indexed the abbreviations of productions by the corresponding types.

```
program
    var b: bool; var a: int;          {decl_1}
    begin                             {stat_1}
        var b: int;                   {decl_2}
        b := a                        {stat_2}
    end;
    b := a                            {stat_3}
end
```

Fig. 8.1. A CHECK-program P

Definition 8.3. The *context-free part of CHECK-programs* is generated by a context-free grammar which is specified by the following set Σ_{CHECK} of productions:

$$
\begin{array}{ll}
r_1^{(DS,P)}: & P \to \textbf{program } D; S \textbf{ end.} \\
r_2^{(IT,D)}: & D \to \textbf{var } I : T \\
r_3^{(DD,D)}: & D \to D; D \\
r_4^{(II,S)}: & S \to I := I \\
r_5^{(DS,S)}: & S \to \textbf{begin } D; S \textbf{ end} \\
r_6^{(SS,S)}: & S \to S; S \\
r_7^{(\varepsilon,I)}: & I \to a \\
r_8^{(\varepsilon,I)}: & I \to b \\
r_9^{(\varepsilon,T)}: & T \to int \\
r_{10}^{(\varepsilon,T)}: & T \to bool
\end{array}
$$

We view the set Σ_{CHECK} of productions as an IS_{CHECK}-ranked alphabet where $IS_{CHECK} = \{P, D, S, I, T\}$. Intuitively, P, D, S, I, and T stand for program, declaration, statement, identifier, and type. □

In order to describe environments, i.e., the bindings of variables and types, and their maintenance formally, we construct a many-sorted algebra which determines several appropriate semantic domains.

Definition 8.4. The *semantic domain for the contextual analysis of CHECK-programs* is the many-sorted Δ_{CHECK}-algebra $\mathcal{A}_{CHECK} = (A, \alpha)$ defined as follows.

- OS_{CHECK} is the set $\{i, t, e, b\}$ of sorts.
- A is the OS_{CHECK}-sorted set $A^i \cup A^t \cup A^e \cup A^b$, where

- $A^i = \{\widehat{a}, \widehat{b}\}$ is the semantic domain of *identifiers*,
- $A^t = \{\widehat{int}, \widehat{bool}, \widehat{undef}\}$ is the semantic domain of *types*,
- $A^e = \{\widehat{\rho} \mid \widehat{\rho} : A^i \to A^t\}$ is the semantic domain of *environments*, and
- $A^b = \{\widehat{true}, \widehat{false}\}$ is the semantic domain of *boolean values*.

- Δ_{CHECK} is the OS-ranked alphabet $\{look\text{-}up^{(ei,t)}, update^{(eit,e)}, \rho_0^{(\varepsilon,e)}, and^{(bb,b)}, ty\text{-}eq^{(tt,b)}\} \cup \Delta_{CHECK,0}$ with $\Delta_{CHECK,0} = \{a^{(\varepsilon,i)}, b^{(\varepsilon,i)}, int^{(\varepsilon,t)}, bool^{(\varepsilon,t)}, undef^{(\varepsilon,t)}\}$.
- For every $\widehat{i}, \widehat{i_1} \in A^i$, $\widehat{t_1}, \widehat{t_2} \in A^t$, $\widehat{\rho} \in A^e$, and $\widehat{b_1}, \widehat{b_2} \in A^b$, define α as follows:

$$\begin{aligned}
\alpha(look\text{-}up)(\widehat{\rho}, \widehat{i}) &= \widehat{\rho}(\widehat{i}) \\
\alpha(update)(\widehat{\rho}, \widehat{i_1}, \widehat{t}) &= \lambda\widehat{i}.\ \text{if } \widehat{i_1} = \widehat{i} \text{ then } \widehat{t} \text{ else } \widehat{\rho}(\widehat{i}) \\
\alpha(\rho_0)() &= \lambda\widehat{i}.\widehat{undef} \\
\alpha(and)(\widehat{b_1}, \widehat{b_2}) &= \widehat{true} \text{ iff } (\widehat{b_1} \text{ is } \widehat{true} \text{ and } \widehat{b_2} \text{ is } \widehat{true}) \\
\alpha(ty\text{-}eq)(\widehat{t_1}, \widehat{t_2}) &= \widehat{true} \text{ iff } (\widehat{t_1} = \widehat{t_2}) \text{ and } \widehat{t_1}, \widehat{t_2} \in \{\widehat{int}, \widehat{bool}\}
\end{aligned}$$

and, for every symbol $\delta \in \Delta_{CHECK,0}$, define $\alpha(\delta)() = \widehat{\delta}$. □

Now we construct a many-sorted macro attributed tree transducer N which performs the contextual analysis of CHECK-programs. The transducer N uses the synthesized attribute *Envir-syn* and the inherited attribute *Envir-inh* to traverse depth-first left-to-right through a declaration and to accumulate the bindings between identifiers and types in a parameter position. After having accumulated the current environment, N distributes this environment to every statement of the program; the distribution is done in a strict recursive descent, i.e., there is no tree-walking involved in the statement part of the program.

Definition 8.5. Define the many-sorted macro attributed tree transducer $N = (Att, \Sigma_{CHECK}, \Delta_{CHECK}, \mathcal{A}_{CHECK}, Check\text{-}prog, R, E)$ as follows:

- $Att = Att_{syn} \cup Att_{inh}$ where $Att_{syn} = \{Check\text{-}prog^{(P,b)}, Check^{(Se,b)}, Envir\text{-}syn^{(De,e)}, Id^{(I,i)}, Ty^{(T,t)}\}$ and $Att_{inh} = \{Envir\text{-}inh^{(De,e)}\}$
- Σ_{CHECK} contains the following elements (as in Def. 8.3) which are grouped according to their result sort:

$r_1^{(DS,P)}$,
$r_2^{(IT,D)}$, $r_3^{(DD,D)}$,
$r_4^{(II,S)}$, $r_5^{(DS,S)}$, $r_6^{(SS,S)}$,
$r_7^{(\varepsilon,I)}$, $r_8^{(\varepsilon,I)}$,
$r_9^{(\varepsilon,T)}$, $r_{10}^{(\varepsilon,T)}$

- Δ_{CHECK} contains the following elements (as in Def. 8.4):

$update^{(eit,e)}$, $\rho_0^{(\varepsilon,e)}$,
$look\text{-}up^{(ei,t)}$,

$and^{(bb,b)}$, $ty\text{-}eq^{(tt,b)}$,
$a^{(\varepsilon,i)}$, $b^{(\varepsilon,i)}$,
$int^{(\varepsilon,t)}$, $bool^{(\varepsilon,t)}$

- $\mathcal{A}_{CHECK}$ is the semantic domain for the contextual analysis of *CHECK*-programs (Def. 8.4).
- R contains the sets R_{r_1}, ..., $R_{r_{10}}$:

$$
\begin{array}{llcl}
R_{r_1}: & Check\text{-}prog(\pi_P) & \rightarrow & Check(\pi_P 2, Envir\text{-}syn(\pi_P 1, \rho_0()))\\
 & Envir\text{-}inh(\pi_P 1, y_{1,e}) & \rightarrow & y_{1,e}\\
R_{r_2}: & Envir\text{-}syn(\pi_D, y_{1,e}) & \rightarrow & Envir\text{-}inh(\pi_D, update(y_{1,e}, Id(\pi_D 1),\\
 & & & Ty(\pi_D 2)))\\
R_{r_3}: & Envir\text{-}syn(\pi_D, y_{1,e}) & \rightarrow & Envir\text{-}syn(\pi_D 2, y_{1,e})\\
 & Envir\text{-}inh(\pi_D 2, y_{1,e}) & \rightarrow & Envir\text{-}syn(\pi_D 1, y_{1,e})\\
 & Envir\text{-}inh(\pi_D 1, y_{1,e}) & \rightarrow & Envir\text{-}inh(\pi_D, y_{1,e})\\
R_{r_4}: & Check(\pi_S, y_{1,e}) & \rightarrow & ty\text{-}eq(look\text{-}up(y_{1,e}, Id(\pi_S 1)),\\
 & & & look\text{-}up(y_{1,e}, Id(\pi_S 2)))\\
R_{r_5}: & Check(\pi_S, y_{1,e}) & \rightarrow & Check(\pi_S 2, Envir\text{-}syn(\pi_S 1, y_{1,e}))\\
 & Envir\text{-}inh(\pi_S 1, y_{1,e}) & \rightarrow & y_{1,e}\\
R_{r_6}: & Check(\pi_S, y_{1,e}) & \rightarrow & and(Check(\pi_S 1, y_{1,e}),\\
 & & & Check(\pi_S 2, y_{1,e}))\\
R_{r_7}: & Id(\pi_I) & \rightarrow & a()\\
R_{r_8}: & Id(\pi_I) & \rightarrow & b()\\
R_{r_9}: & Ty(\pi_T) & \rightarrow & int()\\
R_{r_{10}}: & Ty(\pi_T) & \rightarrow & bool()
\end{array}
$$

- $E = (E_1, E_2)$ with $E_1 = ()$, because the initial attribute *Check-prog* does not have context parameters, and E_2 is an arbitrary function of appropriate type; in fact, it does not matter how E_2 is defined, because *Check-Prog* does not depend on the values of the inherited attributes at the root of the input tree.

Now let us consider again the CHECK-program P of Fig. 8.1. The abstract syntax tree t_P of P (Fig. 8.2) is the tree:

$$t_P = r_1^{(DS,P)}(decl_1, r_6^{(SS,S)}(stat_1, stat_2))$$

where

$$
\begin{aligned}
decl_1 &= r_3^{(DD,D)}(r_2^{(IT,D)}(r_8^{(\varepsilon,I)}, r_{10}^{(\varepsilon,T)}), r_2^{(IT,D)}(r_7^{(\varepsilon,I)}, r_9^{(\varepsilon,T)}))\\
stat_1 &= r_5^{(DS,S)}(decl_2, stat_2)
\end{aligned}
$$

$$decl_2 = r_2^{(IT,D)}(r_8^{(\varepsilon,I)}, r_9^{(\varepsilon,T)})$$
$$stat_2 = r_4^{(II,S)}(r_8^{(\varepsilon,I)}, r_7^{(\varepsilon,I)})$$
$$stat_3 = stat_2.$$

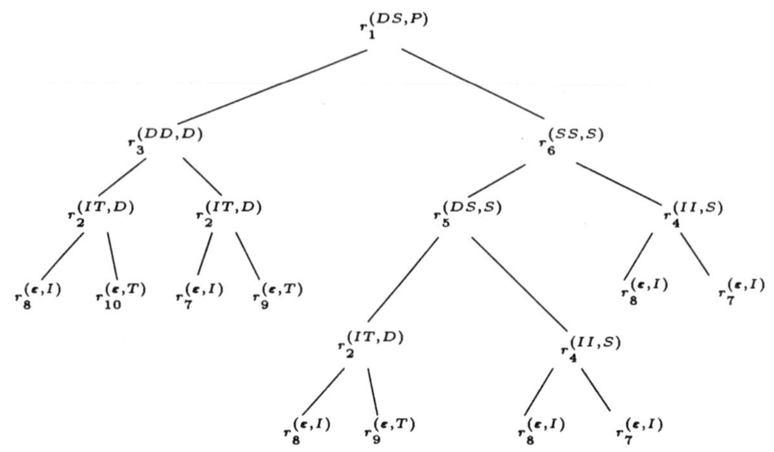

Fig. 8.2. The abstract syntax tree t_P of P

Finally, we show the derivation of N with the input tree t_P, where we represent $\Rightarrow_{N,t_P}$ by $\Rightarrow$.

$$\begin{array}{ll}
 & Check\text{-}prog(\varepsilon) \\
\Rightarrow & Check(2, Envir\text{-}syn(1, \rho_0())) \\
\Rightarrow & Check(2, Envir\text{-}syn(12, \rho_0())) \\
\Rightarrow & Check(2, Envir\text{-}inh(12, update(\rho_0(), Id(121), Ty(122)))) \\
\Rightarrow^2 & Check(2, Envir\text{-}inh(12, update(\rho_0(), a(), int()))) \\
\Rightarrow^3 & Check(2, Envir\text{-}inh(12, update(\lambda\widehat{i}.\widehat{undef}, \widehat{a}, \widehat{int}))) \\
\Rightarrow & Check(2, Envir\text{-}inh(12, \widehat{\rho_1})) \\
 & \text{where } \widehat{\rho_1} = \lambda\widehat{i}.\ \text{if } \widehat{i} = \widehat{a} \text{ then } \widehat{int} \text{ else } \widehat{undef} \\
\Rightarrow & Check(2, Envir\text{-}syn(11, \widehat{\rho_1})) \\
\Rightarrow & Check(2, Envir\text{-}inh(11, update(\widehat{\rho_1}, Id(111), Ty(112))))
\end{array}$$

$$\begin{aligned}
&\Rightarrow^5 && Check(2, Envir\text{-}inh(11, \widehat{\rho_2}))\\
& && \text{where } \widehat{\rho_2} = \lambda\widehat{i}.\ \text{case } \widehat{i} \text{ of } \widehat{a} : \widehat{int}; \widehat{b} : \widehat{bool}; otherwise : \widehat{undef}\\
&\Rightarrow && Check(2, Envir\text{-}inh(1, \widehat{\rho_2}))\\
&\Rightarrow && Check(2, \widehat{\rho_2})\\
&\Rightarrow && and(Check(21, \widehat{\rho_2}), Check(22, \widehat{\rho_2}))\\
&\Rightarrow^2 && and(Check(212, Envir\text{-}syn(211, \widehat{\rho_2})),\\
& && \qquad ty\text{-}eq(look\text{-}up(\widehat{\rho_2}, Id(221)), look\text{-}up(\widehat{\rho_2}, Id(222))))\\
&\Rightarrow^{7+4} && and(Check(212, \widehat{\rho_3}), ty\text{-}eq(look\text{-}up(\widehat{\rho_2}, \widehat{b}), look\text{-}up(\widehat{\rho_2}, \widehat{a})))\\
& && \text{where } \widehat{\rho_3} = \lambda\widehat{i}.\ \text{case } \widehat{i} \text{ of } \widehat{a} : \widehat{int}; \widehat{b} : \widehat{int}; otherwise : \widehat{undef}\\
&\Rightarrow^{5+2} && and(ty\text{-}eq(look\text{-}up(\widehat{\rho_3}, \widehat{b}), look\text{-}up(\widehat{\rho_3}, \widehat{a})), ty\text{-}eq(\widehat{bool}, \widehat{int}))\\
&\Rightarrow^{2+1} && and(ty\text{-}eq(\widehat{int}, \widehat{int}), \widehat{false})\\
&\Rightarrow^2 && \widehat{false}.
\end{aligned}$$

Hence, the many-sorted macro attributed tree transducer N computes the answer $\widehat{false}$ for the contextual analysis of the CHECK-program P.

8.3 Insertion into 2-3 Trees

This second example concerns the insertion of information into 2-3 trees. Such data structures are well suited for storing and retrieving information in a very efficient way. For our consideration, it is convenient and quite sufficient to identify a piece of information by its key; a key is assumed to be a natural number.

A 2-3 tree is a balanced search tree in which the nodes may have rank two, three, or zero. A node with rank two carries one key, a node with rank three carries two keys, and a leaf can carry either one or two keys. We assume that every key occurs at no more than one node (for an example see Fig. 8.3(a)).

A search tree implies the following. If n is a node of rank three and n carries the keys z_1 and z_2 with $z_1 < z_2$, then

- every key which occurs in the first subtree of n is less than z_1,
- every key which occurs in the second subtree of n is greater than z_1 and less than z_2, and
- every key which occurs in the third subtree of n is greater than z_2,

and if n is a node of rank two and n carries the key z, then

- every key which occurs in the first subtree of n is less than z, and
- every key which occurs in the second subtree of n is greater than z.

Moreover, in a 2-3 tree all maximal paths have the same length; in other words, the leaves occur at the same depth.

Here we represent 2-3 trees as trees over an appropriate ranked alphabet (see Fig. 8.3 (b), where the dotted circles indicate the corresponding nodes in (a)). Since the keys are natural numbers and since we do not want to restrict the set of possible keys, we will construct a sorted ranked alphabet which is infinite. Clearly, this extends the concept of many-sorted macro attributed tree transducers again. However, this extension is justified, because it is not exploited to code extra transformational power into the concept.

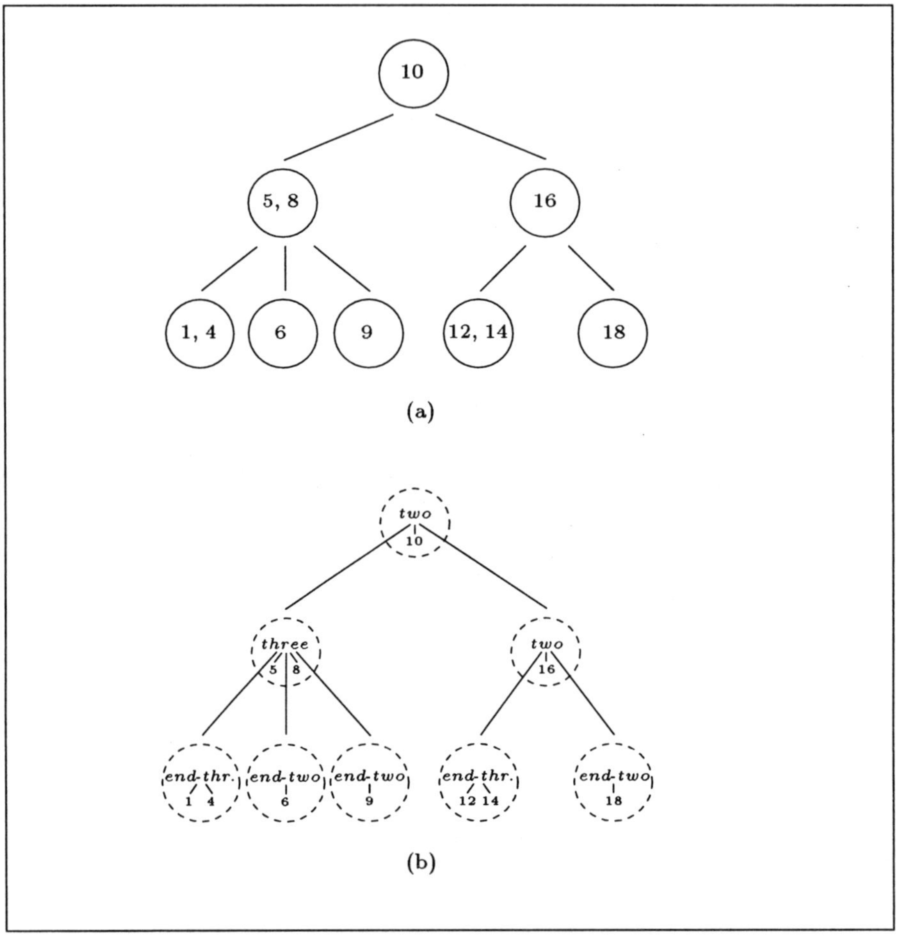

Fig. 8.3. (a) A 2-3-tree and (b) its representation

Definition 8.6. 1. The set of *input sorts for 2-3 trees* is the set $IS_{2,3} = \{t, n\}$; the input sorts t and n stand for *tree* and *natural number*, respectively.

2. The *ranked alphabet for 2-3 trees* is the $IS_{2,3}$-ranked alphabet $\Sigma_{2,3} = \{two^{(tnt,t)}, three^{(tntnt,t)}, end\text{-}two^{(n,t)}, end\text{-}three^{(nn,t)}\} \cup \{z^{(\varepsilon,n)} \mid z \text{ is a natural number}\}$.
3. The *set of 2-3 trees*, denoted by $T_{2,3}$, is the set $\{t \in T_{\Sigma_{2,3}} \mid$ there is a k such that every path w of t with $label(t,w) \in \{end\text{-}two, end\text{-}three\}$ has length $k\}$. □

Definition 8.7. The *semantic domain for 2-3 trees* is the many-sorted $\Delta_{2,3}$-algebra $\mathcal{A}_{2,3} = (A, \alpha)$ defined as follows:

- The set $OS_{2,3}$ of output sorts is the set $\{t, n, b\}$; the sorts t and n have the same meaning as in the set $IS_{2,3}$ of input sorts and b stands for *boolean*.
- The output alphabet $\Delta_{2,3}$ is the $OS_{2,3}$-ranked alphabet $\Sigma_{2,3} \cup \{$**if-then-else**$^{(btt,t)}, <^{(nn,b)}\}$. We will write expressions, which include **if-then-else** or $<$, in the usual infix-notation.
- Let $A = A^t \cup A^n \cup A^b$ where
  - $A^t = T_{2,3}$, i.e., the set of 2-3 trees
  - $A^n = \mathbb{N}$, i.e., the set of natural numbers
  - $A^b = \{true, false\}$, i.e., the set of boolean values.
- For every $t_1, t_2, t_3 \in A^t$, $z_1, z_2 \in A^n$, define α as follows:

$$\begin{aligned}
\alpha(two)(t_1, z_1, t_2) &= two(t_1, z_1, t_2)\\
\alpha(three)(t_1, z_1, t_2, z_2, t_3) &= three(t_1, z_1, t_2, z_2, t_3)\\
\alpha(end\text{-}two)(z_1) &= end\text{-}two(z_1)\\
\alpha(end\text{-}three)(z_1, z_2) &= end\text{-}three(z_1, z_2)\\
\alpha(\textbf{if-then-else})(true, t_1, t_2) &= t_1\\
\alpha(\textbf{if-then-else})(false, t_1, t_2) &= t_2\\
\alpha(<)(z_1, z_2) &= true \text{ iff } z_1 \text{ is less than } z_2.
\end{aligned}$$

□

Condition (3) of Def. 8.2 stated that every symbol $\delta \in \Delta$ is interpreted in a strict way. In particular, for **if-then-else** this is not desirable because the values of both cases (**then** and **else**) have to be evaluated before a decision in favor of one of them is performed. Hence, we add a fourth condition to the definition of the rewrite relation of many-sorted macro attributed tree transducers, which handles the non-strict interpretation of **if-then-else**.

Definition 8.8. (Modification of Def. 8.2) Let $s \in T_\Sigma^\iota$. The *derivation relation induced by N on s* is the binary relation $\Rightarrow_{N,s}$ over the set T_Ψ where $\Psi = Att \cup \Delta \cup A \cup occ(s) \cup Y$, such that for every $\varphi, \psi \in T_\Psi$ we define $\varphi \Rightarrow_{N,s} \psi$ if there is a context $\beta \in C_{\Psi,1}$ and one of the following four conditions holds.

(1) As in Def. 8.2
(2) As in Def. 8.2
(3) As in Def. 8.2 except that $\delta \neq$ **if-then-else**
(4) There is an **if-then-else** $\in \Delta^{(b\kappa\kappa,\kappa)}$, there are $v \in A^b$ and $\varphi_{true} \in T_\Psi^\kappa$, $\varphi_{false} \in T_\Psi^\kappa$ where $\Psi = Att \cup \Delta \cup A \cup occ(s) \cup Y$ such that $\varphi = \beta[\textbf{if } v \textbf{ then } \varphi_{true} \textbf{ else } \varphi_{false}]$, and $\psi = \beta[\varphi_v]$. □

Now let us explain briefly the way in which a new natural number z is inserted into a 2-3 tree s. We assume that z does not occur in s. This insertion proceeds in two phases.

First, the attribute *insert* is called at the root of s with the parameter z, i.e., the rewriting starts with the sentential form $insert(\varepsilon, z)$. By moving from the root of s towards a node which is labeled either by *end-two* or *end-three*, the attribute *insert* searches for the so-called *insertion point of z in s*. Informally, the insertion point of z in s is the uniquely determined node n of s such that every key which occurs left of n is less than z, and every key which occurs right of n is greater than z. We formulate this search procedure in the form of sets $R_{two}^{insertion}$, $R_{three}^{insertion}$, $R_{end-two}^{insertion}$, and $R_{end-three}^{insertion}$ of rules of a many-sorted macro attributed tree transducer where the auxiliary synthesized attribute *id* computes the identity of its argument (for the sake of brevity we have omitted the sorts). In Fig. 8.4 the result of this search process is illustrated.

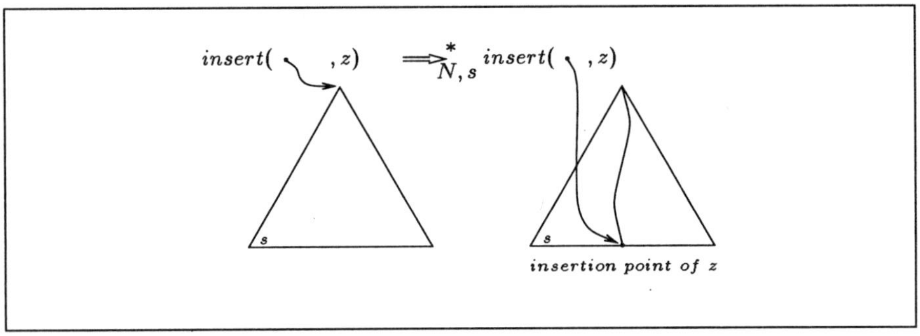

Fig. 8.4. Result of the search process

$R_{two}^{insertion}$:

$insert(\pi, y_1) \quad \rightarrow \quad$ **if** $y_1 < id(\pi 2)$
then $insert(\pi 1, y_1)$
else $insert(\pi 3, y_1)$

$R^{insertion}_{three}$:

$$
\begin{array}{lll}
insert(\pi, y_1) & \rightarrow & \textbf{if } y_1 < id(\pi 2) \\
& & \textbf{then } insert(\pi 1, y_1) \\
& & \textbf{else if } y_1 < id(\pi 4) \\
& & \quad \textbf{then } insert(\pi 3, y_1) \\
& & \quad \textbf{else } insert(\pi 5, y_1)
\end{array}
$$

$R^{insertion}_{end\text{-}two}$:

$$
\begin{array}{lll}
insert(\pi, y_1) & \rightarrow & \textbf{if } y_1 < id(\pi 1) \\
& & \textbf{then } subst(\pi, end\text{-}three(y_1, id(\pi 1))) \\
& & \textbf{else } subst(\pi, end\text{-}three(id(\pi 1), y_1))
\end{array}
$$

$R^{insertion}_{end\text{-}three}$:

$$
\begin{array}{lll}
insert(\pi, y_1) & \rightarrow & \textbf{if } y_1 < id(\pi 1) \\
& & \textbf{then } re(\pi, end\text{-}two(y_1), id(\pi 1), end\text{-}two(id(\pi 2))) \\
& & \textbf{else} \\
& & \quad \textbf{if } y_1 < id(\pi 2) \\
& & \quad \textbf{then } re(\pi, end\text{-}two(id(\pi 1)), y_1, end\text{-}two(id(\pi 2))) \\
& & \quad \textbf{else } re(\pi, end\text{-}two(id(\pi 1)), id(\pi 2), end\text{-}two(y_1))
\end{array}
$$

As the rules in $R^{insertion}_{two}$ and $R^{insertion}_{three}$ show, the attribute *insert* moves down from the root by selecting an appropriate subtree each time it visits a node, until it reaches a node labeled either by *end-two* or *end-three*. Now the second phase starts. If *insert* has reached a node which is labeled by *end-two*, then, intuitively, the number z can be added to this node by transforming the *end-two*-node into an *end-three*-node and filling in z at the appropriate free place. Then, the new 2-3 tree has to be generated; this is done by the inherited attribute *subst* which stands for substitution. In Fig. 8.5 the computation of *subst* is illustrated where $s[w \leftarrow x]$ denotes the tree which is obtained from s by replacing the subtree at occurrence w by some variable x. Now we show the rules for *subst*.

R^{subst}_{two}:

$$
\begin{array}{lll}
subst(\pi 1, y_1) & \rightarrow & subst(\pi, two(y_1, id(\pi 2), id(\pi 3))) \\
subst(\pi 3, y_1) & \rightarrow & subst(\pi, two(id(\pi 1), id(\pi 2), y_1))
\end{array}
$$

R^{subst}_{three}:

$$
\begin{array}{lll}
subst(\pi 1, y_1) & \rightarrow & subst(\pi, three(y_1, id(\pi 2), id(\pi 3), id(\pi 4), id(\pi 5))) \\
subst(\pi 3, y_1) & \rightarrow & subst(\pi, three(id(\pi 1), id(\pi 2), y_1, id(\pi 4), id(\pi 5))) \\
subst(\pi 5, y_1) & \rightarrow & subst(\pi, three(id(\pi 1), id(\pi 2), id(\pi 3), id(\pi 4), y_1))
\end{array}
$$

If *insert* has reached a node which is labeled by *end-three*, then there is no place in which to insert z at this node. Instead, the 2-3 tree above the

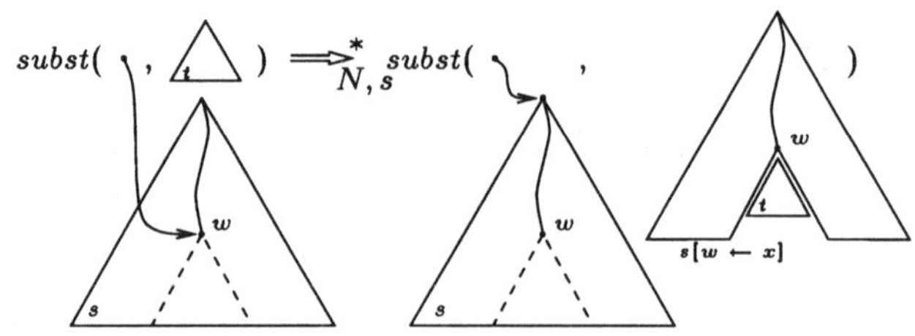

Fig. 8.5. Result of the substitution process

insertion point for z has to be rearranged. This rearrangement is performed by the inherited attribute re. The rules for re are:

R^{re}_{two}:

$$\begin{array}{lcl} re(\pi 1, y_1, y_2, y_3) & \rightarrow & subst(\pi, three(y_1, y_2, y_3, id(\pi 2), id(\pi 3))) \\ re(\pi 3, y_1, y_2, y_3) & \rightarrow & subst(\pi, three(id(\pi 1), id(\pi 2), y_1, y_2, y_3)) \end{array}$$

R^{re}_{three}:

$$\begin{array}{lcl} re(\pi 1, y_1, y_2, y_3) & \rightarrow & re(\pi, two(y_1, y_2, y_3), id(\pi 2), \\ & & two(id(\pi 3), id(\pi 4), id(\pi 5))) \\ re(\pi 3, y_1, y_2, y_3) & \rightarrow & re(\pi, two(id(\pi 1), id(\pi 2), y_1), y_2, \\ & & two(y_3, id(\pi 4), id(\pi 5))) \\ re(\pi 5, y_1, y_2, y_3) & \rightarrow & re(\pi, two(id(\pi 1), id(\pi 2), id(\pi 3)), id(\pi 4), \\ & & two(y_1, y_2, y_3)) \end{array}$$

Finally, we show the rules for the synthesized attribute id which computes the identity of its argument.

R^{id}_{two}:

$$id(\pi) \rightarrow two(id(\pi 1), id(\pi 2), id(\pi 3))$$

R^{id}_{three}:

$$id(\pi) \rightarrow three(id(\pi 1), id(\pi 2), id(\pi 3), id(\pi 4), id(\pi 5))$$

$R^{id}_{end-two}$:

$$id(\pi) \rightarrow end\text{-}two(id(\pi 1))$$

$R^{id}_{end-three}$:

$$id(\pi) \rightarrow end\text{-}three(id(\pi 1), id(\pi 2))$$

R^{id}_{z}:

$$id(\pi) \rightarrow z$$

Note that there is an infinite number of rules of the form $id(\pi) \rightarrow z$. Such rules are used to transform the syntactic object z into a semantic object $z \in \mathbb{N}$.

Now the following question arises: how does the number z which has to be inserted come into the macro attributed tree transducer? For the sake of convenience, we solve this problem by adding an extra root-marker $root^{(t,t)}$ on top of the original input tree and, for every number z, we specify a particular macro attributed tree transducer. For two different numbers z_1 and z_2 the two corresponding transducers only differ in one rule in R_{root}. Here again we should point out to the enquiring reader that we wanted only to show the usefulness of macro attributed tree transducers as a *computation model*; we did not claim that this is a perfect specification language with corresponding input/output features.

Definition 8.9. Let z be a natural number. Construct the many-sorted macro attributed tree transducer $N = (Att, \Sigma, \Delta, \mathcal{A}_{2,3}, in, R, -)$ for the insertion of z as follows.

- $Att = Att_{syn} \cup Att_{inh}$ where
  $Att_{syn} = \{in^{(t,t)}, insert^{(tn,t)}, id^{(t,t)}\}$ and $Att_{inh} = \{subst^{(tt,t)}, re^{(ttnt,t)}\}$
- $\Sigma = \{root^{(t,t)}\} \cup \Sigma_{2,3}$ and $\Delta = \Delta_{2,3}$ where the sorted ranked alphabets $\Sigma_{2,3}$ and $\Delta_{2,3}$ are defined as in Defs. 8.6 and 8.7, respectively.
- $\mathcal{A}_{2,3}$ is the semantic domain as defined in Definition 8.7.
- $R = R_{root} \cup R_{two} \cup R_{three} \cup R_{end-two} \cup R_{end-three} \cup R_z$
  and the subsets are defined by

R_{root}:

$$\begin{array}{lll} in(\pi) & \rightarrow & insert(\pi, z) \\ re(\pi 1, y_1, y_2, y_3) & \rightarrow & two(y_1, y_2, y_3) \end{array}$$

$R_{two} = R_{two}^{insertion} \cup R_{two}^{subst} \cup R_{two}^{re} \cup R_{two}^{id}$
$R_{three} = R_{three}^{insertion} \cup R_{three}^{subst} \cup R_{three}^{re} \cup R_{three}^{id}$
$R_{end-two} = R_{end-two}^{insertion} \cup R_{end-two}^{id}$
$R_{end-three} = R_{end-three}^{insertion} \cup R_{end-three}^{id}$
$R_z = R_z^{id}$ □

Finally, we show a computation induced by N. Consider the tree $t = root(three(end\text{-}two(1), 2, end\text{-}two(3), 4, end\text{-}three(5, 6)))$ and the number $z = 7$ to be inserted. Then N computes as follows where we abbreviate $\Rightarrow_{N,t}$ to $\Rightarrow$:

$$\begin{aligned}
& \mathit{in}(\varepsilon) \\
\Rightarrow\ & \mathit{insert}(1,7) \\
\Rightarrow^*\ & \mathit{insert}(15,7) \\
\Rightarrow^*\ & \mathit{re}(15, \mathit{end\text{-}two}(\mathit{id}(151)), \mathit{id}(152), \mathit{end\text{-}two}(7)) \\
\Rightarrow^2\ & \mathit{re}(15, \mathit{end\text{-}two}(5), 6, \mathit{end\text{-}two}(7)) \\
\Rightarrow\ & \mathit{re}(1, \mathit{two}(\mathit{id}(11), \mathit{id}(12), \mathit{id}(13)), \mathit{id}(14), \mathit{two}(\mathit{end\text{-}two}(5), 6, \\
& \mathit{end\text{-}two}(7))) \\
\Rightarrow^2\ & \mathit{re}(1, \mathit{two}(\mathit{end\text{-}two}(1), 2, \mathit{end\text{-}two}(3)), 4, \mathit{two}(\mathit{end\text{-}two}(5), 6, \\
& \mathit{end\text{-}two}(7))) \\
\Rightarrow\ & \mathit{two}(\mathit{two}(\mathit{end\text{-}two}(1), 2, \mathit{end\text{-}two}(3)), 4, \mathit{two}(\mathit{end\text{-}two}(5), 6, \\
& \mathit{end\text{-}two}(7)))
\end{aligned}$$

8.4 Bibliographic Notes

The approach of many-sorted tree transducers has been formalized in [Vog91] and the embedding of computations in particular domains has been studied in [FHVV93]. This enrichment leads to the concept of the *many-sorted macro attributed tree transducer*.

The example of contextual analysis is taken from [Vog91] where this analysis has been described in terms of top-down tree transducers, macro tree transducers, high-level tree transducers, and modular tree transducers. The example of insertion into 2-3 trees is due to [Küh95a].

In [Ind85, Ind88] the code generation for an easy programming language has been expressed as a system of recursively defined functions. In fact, this system can be viewed as a many-sorted macro tree transducer.

In [BEMV97] a language-based environment is described. The language is a functional programming language called *NoName* that supports the tree transducers investigated in this book, in particular, the many-sorted macro attributed tree transducers. The software environment contains a syntax-directed editor for *NoName*-programs [Ern95], a link to an Ingres database system which contains standard tree transformations [Mül95], and a transformation system which transforms the primitive recursion on trees into iteration [MV95a, MV95b].

Bibliography

[ABB97] J. Autebert, J. Berstel, and L. Boasson. Context-free languages and push-down automata. In G. Rozenberg and A. Salomaa (eds.), *Handbook of Formal Languages, Vol. 1: Word, Language, Grammar*, volume 1, pages 111–174. Springer-Verlag, Berlin, 1997.

[AM91] H. Alblas and B. Melichar (eds.). *Attribute grammars, applications and systems.* Lecture Notes in Computer Science, 545. Springer-Verlag, Berlin, 1991.

[ASU86] A.V. Aho, R. Sethi, and J.D. Ullman. *Compilers – principles, techniques, and tools.* Addison-Wesley, Reading, MA, 1986.

[AU69a] A. V. Aho and J. D. Ullman. Properties of syntax-directed translations. *J. Comput. Syst. Sci.*, 3:319–334, 1969.

[AU69b] A. V. Aho and J. D. Ullman. Syntax-directed translations and the push-down assembler. *J. Comput. System Sci.*, 3:37–56, 1969.

[AU71] A.V. Aho and J.D. Ullman. Translations on a context-free grammar. *Inform. and Control*, 19:439–475, 1971.

[AU73] A.V. Aho and J.D. Ullman. *The theory of parsing, translation and compiling, Vols. I and II.* Prentice Hall, Englewood Cliffs, NJ, 1973.

[Bak78] B. S. Baker. Tree transducers and tree languages. *Inform. and Control*, 37:241–266, 1978.

[Bak79] B. S. Baker. Composition of top-down and bottom-up tree transductions. *Inform. and Control*, 41:186–213, 1979.

[Bar81] M. Bartha. An algebraic definition of attributed transformations. In F. Gécseg (ed.), *Fundamentals of Computation Theory*, Lecture Notes in Computer Science, 117, pages 51–60. Springer Publishing Company, 1981.

[Bar83] M. Bartha. Linear deterministic attributed transformations. *Acta Cybernetica*, 6:125–147, 1983.

[BD77] R. M. Burstall and J. Darlington. A transformation system for developing recursive programs. *J. Assoc. Comput. Mach.*, 24:44–67, 1977.

[BEMV97] H. Braxmeier, D. Ernst, A. Moessle, and H. Vogler. The project NoName – a functional programming language with its development environment. Technical report, Technical University of Dresden, 1997. Nr. 97-08, April 1997.

[Bir86] R.S. Bird. An introduction to the theory of lists. In M. Broy (ed.), *Logic of Programming and Calculi of Discrete Design*, pages 3–42. Springer-Verlag, Berlin, 1986.

[Boo83] R.V. Book. Thue-systems and the Church-Rosser property: replacement systems, specification of formal languages and presentations of monoids. In L. Cummings (ed.), *Progress in combinatorics on words*, pages 1–38. Academic Press, New York, 1983.

[CF82] B. Courcelle and P. Franchi-Zannettacci. Attribute grammars and recursive program schemes I and II. *Theoret. Comput. Sci.*, 17:163–191 and 235–257, 1982.

[CM73] A. K. Chandra and Z. Manna. On the power of programming features. Report CS-333, AIM-185. Technical report, Stanford University, 1973.

[CM79] L. M. Chirica and D. F. Martin. An order-algebraic definition of Knuthian semantics. *Math. Systems Theory*, 13, 1979.

[Cou84] B. Courcelle. Attribute grammars: definitions, analysis of dependencies, proof methods. In B. Lorho (ed.), *Methods and tools for compiler construction*, pages 81–102. Cambridge University Press, Cambridge, UK, 1984.

[Cou90] B. Courcelle. Recursive applicative program schemes. In J. van Leeuwen (ed.), *Handbook of theoretical computer science, Vol B*, pages 459–492. Elsevier, Amsterdam, The Netherlands, 1990.

[DF96] G. Dányi and Z. Fülöp. Super linear deterministic top-down tree transducers. *Math. Systems Theory*, 29:507–534, 1996.

[DF98] G. Dányi and Z. Fülöp. Compositions with superlinear deterministic top-down tree transformations. *Theoret. Comput. Sci.*, 194:57–85, 1998.

[DJ90] P. Deransart and M. Jourdan (eds.). Lecture Notes in Computer Science, 461. *Attribute grammars and their applications.* Springer-Verlag, Berlin, 1990.

[DJL88] P. Deransart, M. Jourdan, and B. Lorho. *Attribute grammars, definitions and bibliography.* Lecture Notes in Computer Science, 323. Springer-Verlag, 1988.

[EF81] J. Engelfriet and G. File. The formal power of one-visit attribute grammars. *Acta Informatica*, 16:275–302, 1981.

[EF89] J. Engelfriet and G. File. Passes, sweeps, and visits in attribute grammars. *J. Assoc. Comput. Sci.*, 36:841–869, 1989.

[Eng74] J. Engelfriet. *Simple program schemes and formal languages.* Lecture Notes in Computer Science, 20. Springer-Verlag, Berlin, 1974.

[Eng75] J. Engelfriet. Bottom-up and top-down tree transformations – a comparison. *Math. Systems Theory*, 9:198–231, 1975.

[Eng76] J. Engelfriet. Surface tree languages and parallel derivation trees. *Theoret. Comput. Sci.*, 2:9–27, 1976.

[Eng77] J. Engelfriet. Top-down tree transducers with regular look-ahead. *Math. Systems Theory*, 10:289–303, 1977.

[Eng78] J. Engelfriet. On tree transducers for partial functions. *Inf. Process. Lett.*, 7:170–172, 1978.

[Eng80] J. Engelfriet. Some open questions and recent results on tree transducers and tree languages. In R.V. Book (ed.), *Formal language theory; perspectives and open problems*, pages 241–286. Academic Press, New York, 1980.

[Eng81] J. Engelfriet. Tree transducers and syntax-directed semantics. Technical Report Memorandum 363, Technische Hogeschool Twente, 1981. Also in: *Proceedings of 7th Colloquium on Trees in Algebra and Programming (CAAP 1992)*, Lille, France, March 4-6, 1992.

[Eng82a] J. Engelfriet. The copying power of one-state tree transducers. *J. Comput. Syst. Sci.*, 25:418–435, 1982.

[Eng82b] J. Engelfriet. Three hierarchies of transducers. *Math. Systems Theory*, 15:95–125, 1982.

[Eng84] J. Engelfriet. Attribute grammars: attribute evaluation methods. In B. Lorho (ed.), *Methods and tools for compiler construction*, pages 103–138. Cambridge University Press, Cambridge, UK, 1984.

[Eng86] J. Engelfriet. Context-free grammars with storage. Technical Report 86-11, University of Leiden, 1986.

[Ern95] D. Ernst. Implementierung eines syntaxgesteuerten Editors für die Sprache NoName. Master's thesis, University of Ulm, 1995.

[ERS80] J. Engelfriet, G. Rozenberg, and G. Slutzki. Tree transducers, L systems, and two-way machines. *J. Comput. Syst. Sci.*, 20:150–202, 1980.

[ES78] J. Engelfriet and E.M. Schmidt. IO and OI, Parts I and II. *J. Comput. Syst. Sci.*, 15:328–353, 1977, 16:67–99, 1978.

[EV85] J. Engelfriet and H. Vogler. Macro tree transducers. *J. Comput. Syst. Sci.*, 31:71–146, 1985.

[EV86] J. Engelfriet and H. Vogler. Pushdown machines for the macro tree transducer. *Theoret. Comput. Sci.*, 42:251–368, 1986.

[EV88] J. Engelfriet and H. Vogler. High level tree transducers and iterated pushdown tree transducers. *Acta Informatica*, 26:131–192, 1988.

[EV91] J. Engelfriet and H. Vogler. Modular tree transducers. *Theoret. Comput. Sci.*, 78:267–304, 1991.

[EV94] J. Engelfriet and H. Vogler. The translation power of top-down tree-to-graph transducers. *J. Comput. Syst. Sci.*, 49:258–305, 1994.

[EV96] J. Engelfriet and H. Vogler. The equivalence of bottom-up and top-down tree-to-graph transducers. Technical report, TUD/FI/96-17, 1996. Dresden University of Technology, to appear: J. Comput. Syst. Sci.

[FHVV93] Z. Fülöp, F. Herrmann, S. Vágvölgyi, and H. Vogler. Tree transducers with external functions. *Theoret. Comput. Sci.*, 108:185–236, 1993.

[Fil83] G. File. *Theory of attribute grammars.* PhD thesis, Twente University of Technology, The Netherlands, 1983.

[Fok92] M. Fokkinga. *Law and order in algorithmics.* PhD thesis, Twente University of Technology, The Netherlands, 1992.

[Fra82] P. Franchi-Zannettacci. *Attributs sémantiques et schémas de programmes.* PhD thesis, Université de Bordeaux I, 1982. Thèse d'Etat.

[Fül81] Z. Fülöp. On attributed tree transducers. *Acta Cybernetica*, 5:261–279, 1981.

[Fül91] Z. Fülöp. A complete description for a monoid of deterministic bottom-up tree transformation classes. *Theoret. Comput. Sci.*, 88:253–268, 1991.

[FV87a] Z. Fülöp and S. Vágvölgyi. An infinite hierarchy of tree transformations in the class NDR. *Acta Cybernetica*, 8:153–168, 1987.

[FV87b] Z. Fülöp and S. Vágvölgyi. Results on compositions of deterministic root-to-frontier tree transformations. *Acta Cybernetica*, 8:49–61, 1987.

[FV88a] Z. Fülöp and S. Vágvölgyi. A finite presentation for a monoid of tree transformation classes. In F. Gécseg and I. Peák (eds.), *Proc. of 2nd Conference on Automata, Languages and Programming Systems*, pages 115–124, Budapest, 1988.

[FV88b] Z. Fülöp and S. Vágvölgyi. On ranges of compositions of deterministic root-to-frontier tree transformations. *Acta Cybernetica*, 8:259–266, 1988.

[FV89a] Z. Fülöp and S. Vágvölgyi. Iterated deterministic top-down look-ahead. In J. Csirik, J. Demetrovics, and F. Gécseg (eds.), *Proceedings of FTC 89*, Lecture Notes in Computer Science, 380, pages 175–184, Springer-Verlag, Berlin, 1989.

[FV89b] Z. Fülöp and S. Vágvölgyi. Top-down tree transducers with deterministic top-down look-ahead. *Inf. Process. Lett.*, 33:3–5, 1989.

[FV89c] Z. Fülöp and S. Vágvölgyi. Variants of top-down tree transducers with look-ahead. *Math. Systems Theory*, 21:125–145, 1989.

[FV90] Z. Fülöp and S. Vágvölgyi. A complete rewriting system for a monoid of tree transformation classes. *Inf. and Comput.*, 86:195–212, 1990.

[FV91] Z. Fülöp and S. Vágvölgyi. A complete classification of deterministic root-to-forntier tree transformation classes. *Theoret. Comput. Sci.*, 81:1–15, 1991.

[FV92a] H. Faßbender and H. Vogler. An implementation of syntax-directed functional programming on nested-stack machines. *Formal Aspects of Computing*, 4:341–375, 1992.

[FV92b] Z. Fülöp and S. Vágvölgyi. Decidability of the inclusion in monoids generated by tree transformation classes. In M. Nivat and A. Podelski (eds.), *Tree Automata and Languages*. Elsevier, Amsterdam, The Netherlands, 1992.

[Gan80] H. Ganzinger. Transforming denotational semantics into practical attribute grammars. In N.D. Jones (ed.), *Semantics-directed compiler generation*, Lecture Notes in Computer Science, 94, pages 1–69. Springer-Verlag, Berlin, 1980.

[Gan83] H. Ganzinger. Increasing modularity and language-independency in automatically generated compilers. *Science of Computer Programming*, 3:223–278, 1983.

[GG84] H. Ganzinger and R. Giegerich. Attribute coupled grammars. In *Proceedings of the ACM SIGPLAN'84, Symposium on Compiler Construction, SIGPLAN Notices Vol.19, No.6*, pages 157–170, ACM Press, 1984.

[GGV86] H. Ganzinger, R. Giegerich, and M. Vach. MARVIN – a tool for applicative and modular compiler specifications, report 220. Technical Report, University of Dortmund, 1986.

[Gib91] J. Gibbons. *Algebras for tree algorithms*. PhD thesis, Oxford University, 1991.

[Gib93] J. Gibbons. Upwards and downwards accumulation on trees. In *Mathematics of program construction*, Lecture Notes in Computer Science, 669, pages 122–138. Springer-Verlag, Berlin, 1993. Revised version in: Proc. of the Massey Functional Programming Workshop, E. Ireland and N. Perry (eds.), 1992.

[Gie88] R. Giegerich. Composition and evaluation of attribute coupled grammars. *Acta Informatica*, 25:355–423, 1988.

[GM96] J. A. Goguen and G. Malcolm. *Algebraic semantics of imperative programs*. MIT Press, Cambridge, MA, 1996.

[Gor79] M. J. C. Gordon. *The denotational description of programming languages; an introduction*. Springer-Verlag, Berlin, 1979.

[Gre75] S.A. Greibach. *Theorey of program structures: schemes, semantics, verification*. Lecture Notes in Computer Science, 36. Springer-Verlag, Berlin, 1975.

[GS84] F. Gécseg and M. Steinby. *Tree Automata*. Akadémiai Kiadó, Budapest, 1984.

[GS97] F. Gécseg and M. Steinby. Tree languages. In G. Rozenberg and A. Salomaa (eds.), *Handbook of formal languages*, volume 3, pages 1–68. Springer-Verlag, Berlin, 1997.

[Gue83] I. Guessarian. Pushdown tree automata. *Math. Systems Theory*, 16:237–263, 1983.

[GV96] P. Gyenizse and S. Vágvölgyi. Compositions of deterministic bottom-up, top-down, and regular look-ahead tree transformations. *Theoret. Comput. Sci.*, 156:71–97, 1996.

[Hou94] F. Houdek. *KW-Auswerter für macro attributed tree transducers*. Studienarbeit, University of Ulm, 1994.

[Hue80] G. Huet. Confluent reductions: abstract properties and applications to term rewriting systems. *J. Assoc. Comput. Mach.*, 27:797–821, 1980.

[Hup78] U. Hupbach. Rekursive Funktionen in mehrsortigen Algebren. *Elektron. Informationsverarb. Kybernetik*, 15:491–506, 1978.

[Ind85] K. Indermark. Functional compiler description. Technical report, Aachen University of Technology, 1985. Schriften zur Informatik und angewandten Mathematik.

[Ind88] K. Indermark. Functional compiler description. *Banach Center Publications*, 21:257–275, 1988.

[Iro61] E.T. Irons. A syntax-directed compiler for ALGOL 60. *Comm. ACM*, 4:51–55, 1961.

[Jaz81] M. Jazayeri. Simpler construction for showing the intrinsically exponential complexity of the circularity problem for attribute grammars. *J. ACM*, 28:715–720, 1981.

[JOR75] M. Jazayeri, W. F. Ogden, and W. C. Rounds. The intrinsically exponentenial complexity of the circularity problem for attribute grammars. *Comm. ACM*, 18:697–706, 1975.

[Kas90] U. Kastens. *Übersetzerbau*. Oldenbourg-Verlag, München, 1990.

[Klo92] J. W. Klop. Term rewriting systems. In S. Abramski, D.M. Gabbay, and T.S.E. Maibaum (eds.), *Handbook of logic in computer science, Volume 2*, pages 1–116. Clarendon Press, Oxford, 1992.

[Knu68] D.E. Knuth. Semantics of context-free languages. *Math. Systems Theory*, 2:127–145, 1968.

[KSV89] M.F. Kuiper, S.D. Swierstra, and H.H. Vogt. Higher order attribute grammars. In *Proc. SIGPLAN '89*, pages 131–145. ACM Press, 1989. Symp. Language Design and Implementation.

[Küh95a] A. Kühnemann, 1995. Personal communication.

[Küh95b] A. Kühnemann. A pumping lemma for output languages of macro tree transducers. TUD/FI 95-08. Technical report, Dresden University of Technology, September 1995.

[Küh96] A. Kühnemann. A pumping lemma for output languages of macro tree transducers. In H. Kirchner (ed.), *21st International Colloquium on Trees in Algebra and Programming, CAAP, Linköping, Sweden, April 1996*. Lecture Notes in Computer Science, 1059, pages 44–58. Springer, Berlin, 1996.

[Küh97] A. Kühnemann. *Berechnungsstärken von Teilklassen primitiv-rekursiver Programmschemata*. PhD thesis, Dresden University of Technology, 1997.

[KV94a] A. Kühnemann and H. Vogler. A pumping lemma for output languages of attributed tree transducers. *Acta Cybernetica*, 11:261–305, 1994.

[KV94b] A. Kühnemann and H. Vogler. Synthesized and inherited functions – a new computational model for syntax-directed semantics. *Acta Informatica*, 31:431–477, 1994.

[KV97] A. Kühnemann and H. Vogler. *Attributgrammatiken*. Vieweg-Verlag, Braunschweig, 1997.

[LRS74] P. M. Lewis, D. J. Rosenkrantz, and R. E. Stearns. Attributed translations. *J. Comput. Syst. Sci.*, 9:279–307, 1974.

[Mad80] O.L. Madsen. On defining semantics by means of extended attribute grammars. In N. D. Jones (ed.), *Semantics directed compiler generation*. Lecture Notes in Computer Science, 94, pages 259–299. Springer-Verlag, Berlin, 1980.

[Man73] Z. Manna. Program schemes. In A. V. Aho (ed.), *Currents in the theory of computing*, pages 90–142. Prentice Hall, Englewood Cliffs, NJ, 1973.

[Man74] Z. Manna. *Mathematical theory of computation*. McGraw-Hill, New York, 1974.

[Man96] S. Maneth. On the generating power of deterministic tree transducers. Technical report TUD/FI96/19, Dresden Technical University, 1996.

[May81] B. H. Mayoh. Attribute grammars and mathematical semantics. *SIAM J. Computing 10 (1981), 5*, 1981.

[Mee86] L. Meertens. Algorithmics – towards programming as a mathematical activity. In J.W. de Bakker and J.C. van Vliet (eds.), *Proc. of the CWI Symposium on Mathematics and Computer Science*, pages 289–334. North-Holland, Amsterdam, 1986.

[Mos90] P.D. Mosses. Denotational semantics. In J. van Leeuwen (ed.), *Handbook of Theoretical Computer Science, Vol. B*, pages 575–631. Elsevier, Amsterdam, 1990.

[MS97] A. Mateescu and A. Salomaa. Formal languages: an introduction and a synopsis. In G. Rozenberg and A. Salomaa (eds.), *Handbook of Formal Languages, Vol. 1: Word, Language, Grammar*, volume 1, pages 1–39. Springer-Verlag, Berlin, 1997.

[Mül95] E. Müller. Entwicklung eines Modul-Informationssystems für den NoName-Editor. Master's thesis, University of Ulm, 1995.

[MV95a] A. Mößle and H. Vogler. Efficient call-by-value evaluation strategy of primitive recursive program schemes. In M. Takeichi and T. Ida (eds.), *Fuji International Workshop on Functional and Logic Programming, Susono Japan, July 17-19, 1995.* World Scientific, Singapore, 1995.

[MV95b] A. Mößle and H. Vogler. Efficient call-by-value evaluation strategy of primitive recursive program schemes. Technical report TUD/FI 95/19. Dresden University of Technology, December 1995.

[Nol95] T. Noll. *Klassen applikativer Programmschemata ind ihre Berechnungsstärke.* PhD thesis, Technical University of Aachen, 1995.

[Now96] D. Nowottka. Personal comunications. Dresden University of Technology, 1996.

[Paa95] J. Paakki. Attribute grammar paradigms – a high-level methodology in language implementation. *ACM Computing Surveys*, 27:196–255, 1995.

[Par90] H. Partsch. *Specification and Transformation of Programs – A formal approach to software development.* Springer-Verlag, Berlin, 1990.

[Pat72] M. S. Paterson. Decision problems in computational models. *Proc. of the ACM Conf. on Proving Assertions about Programs*, pages 74–82, ACM Press, 1972.

[PP86] H. Partsch and P. Pepper. Program transformations expressed by algebraic type manipulations. *Technique et Science Informatiques*, 3:197–212, 1986.

[Rou69] W. C. Rounds. Context-free grammars on trees. *Proc. ACM Symp. on Theory of Comput.*, pages 143-148, ACM Press, 1969.

[Rou70a] W. C. Rounds. Tree-oriented proofs of some theorems on context-free and indexed languages *Proc. ACM Symp. on Theory of Comput.*, pages 109-116, ACM Press, 1970.

[Rou70b] W.C. Rounds. Mappings and grammars on trees. *Math. Systems Theory*, 4:257–287, 1970.

[RS81] H. Riis and S. Skyum. k-visit attribute grammars. *Math. Systems Theory*, 15:17–28, 1981.

[Sch95] L. Schmitz. *Syntaxbasierte Programmierwerkzeuge.* Teubner, Stuttgart, 1995.

[SG85] K. M. Schimpf and J. H. Gallier. Tree pushdown automata. *J. Comput. Syst. Sci.*, 30:25–40, 1985.

[Son87] M. Sonnenschein. Graph translation schemes to generate compiler parts. *ACM Trans. on Progr. Languages*, 9:473–490, 1987.

[SS71] D. Scott and C. Strachey. Toward a mathematical semantics for computer languages. In J. Fox (ed.), *Proc. of the 21st Symposium on Computers and Automata*, pages 19–46. Wiley, New York, 1971.

[Sto77] J.E. Stoy. *Denotational semantics: The Scott-Stratchey approach to programming language theory.* MIT Press, Cambridge, MA, 1977.

[Str66] C. Strachey. Towards a formal semantics. In *IFIP TC2 Working Conference on Formal Language Description Languages for Computer Programming*, pages 198–220. North-Holland, Amsterdam, 1966.

[SV93] G. Slutzki and S. Vágvölgyi. A hierarchy of deterministic top-down tree transformations. *Math. Systems Theory*, 29:169–188, 1993.

[Ten91] R.D. Tennent. *Semantics of programming languages.* Prentice Hall, Englewood Cliffs, NJ, 1991.

[Tha70] J.W. Thatcher. Generalized^2 sequential machine maps. *J. Comput. Syst. Sci.*, 4:339–367, 1970.

[Vog86] H. Vogler. *Tree transducers and pushdown machines.* PhD thesis, Twente University of Technology, The Netherlands, 1986.

[Vog87] H. Vogler. Basic tree transducers. *J. Comput. Syst. Sci.*, 34:87–128, 1987.

[Vog91] H. Vogler. Functional description of the contextual analysis in block-structured programming languages: a case study of tree transducers. *Science of Computer Programming*, 16:251–275, 1991.

[Wat91] D.A. Watt. *Programming language syntax and semantics.* Prentice-Hall, Englewood Cliffs, NJ, 1991.

[WG84] W. M. Waite and G. Goos. *Compiler Construction.* Springer-Verlag, Berlin, 1984.

[Win93] G. Winskel. *The formal semantics of programming languages.* MIT Press, Cambridge, MA, 1993.

[WM97] R. Wilhelm and D. Maurer. *Übersetzerbau – Theorie, Konstruktion, Generierung.* Springer-Verlag, Berlin, 2nd edition 1997.

Index

Terms

Formulas

Symbols

Monographs in Theoretical Computer Science · An EATCS Series

C. Calude
Information and Randomness
An Algorithmic Perspective

K. Jensen
Coloured Petri Nets
***Basic Concepts,* Analysis Methods**
and Practical Use, Vol. 1
2nd ed.

K. Jensen
Coloured Petri Nets
Basic Concepts, *Analysis Methods*
and Practical Use, Vol. 2

K. Jensen
Coloured Petri Nets
Basic Concepts, Analysis Methods
and *Practical Use,* Vol. 3

A. Nait Abdallah
The Logic of Partial Information

Z. Fülöp, H. Vogler
Syntax-Directed Semantics
Formal Models
Based on Tree Transducers

Texts in Theoretical Computer Science · An EATCS Series

J. L. Balcázar, J. Díaz, J. Gabarró
Structural Complexity I
2nd ed. (see also overleaf, Vol. 22)

M. Garzon
Models of Massive Parallelism
Analysis of Cellular Automata
and Neural Networks

J. Hromkovič
Communication Complexity
and Parallel Computing

A. Leitsch
The Resolution Calculus

G. Păun, G. Rozenberg, A. Salomaa
DNA Computing
New Computing Paradigms

A. Salomaa
Public-Key Cryptography
2nd ed.

K. Sikkel
Parsing Schemata
A Framework for Specification
and Analysis of Parsing Algorithms

Former volumes appeared as EATCS Monographs on Theoretical Computer Science

Vol. 5: W. Kuich, A. Salomaa
Semirings, Automata, Languages

Vol. 6: H. Ehrig, B. Mahr
Fundamentals of Algebraic Specification 1
Equations and Initial Semantics

Vol. 7: F. Gécseg
Products of Automata

Vol. 8: F. Kröger
Temporal Logic of Programs

Vol. 9: K. Weihrauch
Computability

Vol. 10: H. Edelsbrunner
Algorithms in Combinatorial Geometry

Vol. 12: J. Berstel, C. Reutenauer
Rational Series and Their Languages

Vol. 13: E. Best, C. Fernández C.
Nonsequential Processes
A Petri Net View

Vol. 14: M. Jantzen
Confluent String Rewriting

Vol. 15: S. Sippu, E. Soisalon-Soininen
Parsing Theory
Volume I: Languages and Parsing

Vol. 16: P. Padawitz
Computing in Horn Clause Theories

Vol. 17: J. Paredaens, P. DeBra, M. Gyssens, D. Van Gucht
The Structure of the Relational Database Model

Vol. 18: J. Dassow, G. Páun
Regulated Rewriting in Formal Language Theory

Vol. 19: M. Tofte
Compiler Generators
What they can do, what they might do, and what they will probably never do

Vol. 20: S. Sippu, E. Soisalon-Soininen
Parsing Theory
Volume II: LR(k) and LL(k) Parsing

Vol. 21: H. Ehrig, B. Mahr
Fundamentals of Algebraic Specification 2
Module Specifications and Constraints

Vol. 22: J. L. Balcázar, J. Díaz, J. Gabarró
Structural Complexity II

Vol. 24: T. Gergely, L. Úry
First-Order Programming Theories

R. Janicki, P. E. Lauer
Specification and Analysis of Concurrent Systems
The COSY Approach

O. Watanabe (Ed.)
Kolmogorov Complexity and Computational Complexity

G. Schmidt, Th. Ströhlein
Relations and Graphs
Discrete Mathematics for Computer Scientists

S. L. Bloom, Z. Ésik
Iteration Theories
The Equational Logic of Iterative Processes

Printing: Weihert-Druck GmbH, Darmstadt
Binding: Buchbinderei Schäffer, Grünstadt